SCHAUM'S OUTLINE OF

THEORY AND PROBLEMS

of

PARTIAL DIFFERENTIAL EQUATIONS

•

PAUL DuCHATEAU, Ph.D.
DAVID W. ZACHMANN, Ph.D.
Professors of Mathematics
Colorado State University

•

SCHAUM'S OUTLINE SERIES
McGRAW-HILL PUBLISHING COMPANY

New York St. Louis San Francisco Auckland Bogotá Caracas Hamburg Lisbon London Madrid Mexico Montreal New Delhi Paris San Juan São Paulo Singapore Sydney Tokyo Toronto

PAUL DuCHATEAU is currently Professor of Mathematics at Colorado State University in Fort Collins, Colorado. He received his B.Sc. in engineering science and his Ph.D. in mathematics at Purdue University in 1962 and 1970 respectively. In addition to teaching, he has worked as an applied mathematician for General Motors and United Aircraft corporations and has held visiting positions at Argonne National Laboratory. He has received government funding for research in applied mathematics and has published extensively in the area of partial differential equations.

DAVID ZACHMANN is Professor of Mathematics at Colorado State University. He received his Ph.D. in applied mathematics from the University of Arizona in 1970 and his B.S. in mathematics from Colorado State University in 1965. During 1983 he was a visiting senior research scientist at the CSIRO Environmental Mechanics Division in Canberra, Australia. His research in applied mathematics has been supported by various government agencies. In addition to his teaching and research activities, he frequently serves as a consultant to industry and government.

Schaum's Outline of Theory and Problems of
PARTIAL DIFFERENTIAL EQUATIONS

4 5 6 7 8 9 10 11 12 13 14 15 16 17 18 19 20 SHP SHP 8 9

ISBN 0-07-017897-6

Sponsoring Editor, David Beckwith
Editing Supervisor, Marthe Grice
Production Manager, Nick Monti

Library of Congress Cataloging-in-Publication Data

DuChateau, Paul.
Schaum's outline of partial differential equations.

Includes index.
1. Differential equations, Partial. I. Zachmann, D. W. II. Title. III. Title: Partial differential equations.
QA374.D8 1986 515.3'53 85-24156
ISBN 0-07-017897-6

Preface

The importance of partial differential equations among the topics of applied mathematics has been recognized for many years. However, the increasing complexity of today's technology is demanding of the mathematician, the engineer, and the scientist an understanding of the subject previously attained only by specialists. This book is intended to serve as a supplemental or primary text for a course aimed at providing this understanding. It has been organized so as to provide a helpful reference for the practicing professional, as well.

After the introductory Chapter 1, the book is divided into three parts. Part I, consisting of Chapters 2 through 5, is devoted primarily to qualitative aspects of the subject. Chapter 2 discusses the classification of problems, while Chapters 3 and 4 characterize the behavior of solutions to elliptic boundary value problems and evolution equations, respectively. Chapter 5 focuses on hyperbolic systems of equations of order one.

Part II comprises Chapters 6 through 8, which present the principal techniques for constructing exact solutions to linear problems in partial differential equations. Chapter 6 contains the essential ideas of eigenfunction expansions and integral transforms, which are then applied to partial differential equations in Chapter 7. Chapter 8 provides a practical treatment of the important topic of Green's functions and fundamental solutions.

Part III, Chapters 9 through 14, deals with the construction of approximate solutions. Chapters 9, 10, and 11 focus on finite-difference methods and, for hyperbolic problems, the numerical method of characteristics. Some of these methods are implemented in FORTRAN 77 programs. Chapters 12, 13, and 14 are devoted to approximation methods based on variational principles, Chapter 14 constituting a very elementary introduction to the finite element method.

In every chapter, the solved and supplementary problems have the vital function of applying, reinforcing, and sometimes expanding the theoretical concepts.

It is the authors' good fortune to have long been associated with a large, active group of users of partial differential equations, and the development of this Outline has been considerably influenced by this association. Our aim has been to create a book that would provide answers to all the questions—or, at least, those most frequently asked—of our students and colleagues. As a result, the level of the material included varies from rather elementary and practical to fairly advanced and theoretical. The novel feature is that it is all collected in a single source, from which, we believe, the student and the technician alike can benefit.

We would like to express our gratitude to the McGraw-Hill staff and the Colorado State University Department of Mathematics for their cooperation and support during the preparation of this book. In particular, we thank David Beckwith of McGraw-Hill for his many helpful suggestions.

PAUL DUCHATEAU
DAVID W. ZACHMANN

Contents

Chapter 1

Introduction

1.1 NOTATION AND TERMINOLOGY

Let u denote a function of several independent variables; say, $u = u(x, y, z, t)$. At (x, y, z, t), the *partial derivative of u with respect to x* is defined by

$$\frac{\partial u}{\partial x} = \lim_{h \to 0} \frac{u(x+h, y, z, t) - u(x, y, z, t)}{h}$$

provided the limit exists. We will use the following subscript notation:

$$\frac{\partial u}{\partial x} \equiv u_x \qquad \frac{\partial u}{\partial y} \equiv u_y \qquad \frac{\partial^2 u}{\partial x^2} \equiv u_{xx} \qquad \frac{\partial^2 u}{\partial y^2} \equiv u_{yy} \qquad \frac{\partial^2 u}{\partial x \partial y} \equiv u_{xy} \qquad \frac{\partial^3 u}{\partial x^3} \equiv u_{xxx} \qquad \cdots$$

If all partial derivatives of u up through order m are continuous in some region Ω, we say *u is in the class* $C^m(\Omega)$, or *u is* C^m *in* Ω.

A *partial differential equation* (abbreviated *PDE*) is an equation involving one or more partial derivatives of an unknown function of several variables. The *order* of a PDE is the order of the highest-order derivative that appears in the equation.

The partial differential equation $F(x, y, z, t; u, u_x, u_y, u_z, u_t, u_{xx}, u_{xy}, \ldots) = 0$ is said to be *linear* if the function F is algebraically linear in each of the variables $u, u_x, u_y, \ldots$, and if the coefficients of u and its derivatives are functions only of the independent variables. An equation that is not linear is said to be *nonlinear*; a nonlinear equation is *quasilinear* if it is linear in the highest-order derivatives. Some of the qualitative theory of linear PDEs carries over to quasilinear equations.

The spatial variables in a PDE are usually restricted to some open region Ω with boundary S; the union of Ω and S is called the *closure* of Ω and is denoted $\bar{\Omega}$. If present, the time variable is considered to run over some interval, $t_1 < t < t_2$. A function $u = u(x, y, z, t)$ is a *solution* for a given mth-order PDE if, for (x, y, z) in Ω and $t_1 < t < t_2$, u is C^m and satisfies the PDE.

In problems of mathematical physics, the region Ω is often some subset of Euclidean n-space, $\mathbf{R}^n$. In this case a typical point in Ω is denoted by $\mathbf{x} = (x_1, x_2, \ldots, x_n)$, and the integral of u over Ω is denoted by

$$\int \cdots \int_{\Omega} u(x_1, x_2, \ldots, x_n)\, dx_1\, dx_2 \cdots dx_n = \int_{\Omega} u\, d\Omega$$

1.2 VECTOR CALCULUS AND INTEGRAL IDENTITIES

If $F = F(x, y, z)$ is a C^1 function defined on a region Ω of $\mathbf{R}^3$, the *gradient* of F is defined by

$$\text{grad } F \equiv \nabla F = \frac{\partial F}{\partial x}\mathbf{i} + \frac{\partial F}{\partial y}\mathbf{j} + \frac{\partial F}{\partial z}\mathbf{k} \tag{1.1}$$

If $\mathbf{n}$ denotes a unit vector in $\mathbf{R}^3$, the *directional derivative of F* in the direction $\mathbf{n}$ is given by

$$\frac{\partial F}{\partial n} = \nabla F \cdot \mathbf{n} \tag{1.2}$$

Suppose $\mathbf{w} = \mathbf{w}(x, y, z)$ is a C^1 *vector field* on Ω, which means that

$$\mathbf{w} = w_1(x, y, z)\mathbf{i} + w_2(x, y, z)\mathbf{j} + w_3(x, y, z)\mathbf{k}$$

for continuously differentiable scalar functions w_1, w_2, w_3. The *divergence* of $\mathbf{w}$ is defined to be

$$\operatorname{div}\mathbf{w} \equiv \nabla\cdot\mathbf{w} = \frac{\partial w_1}{\partial x} + \frac{\partial w_2}{\partial y} + \frac{\partial w_3}{\partial z} \tag{1.3}$$

In particular, for $\mathbf{w} = \operatorname{grad} F$, we have

$$\operatorname{div}\operatorname{grad} F \equiv \nabla\cdot\nabla F \equiv \nabla^2 F = \frac{\partial^2 F}{\partial x^2} + \frac{\partial^2 F}{\partial y^2} + \frac{\partial^2 F}{\partial z^2} = F_{xx} + F_{yy} + F_{zz} \tag{1.4}$$

The expression $\nabla^2 F$ is called the *Laplacian* of F.

Theorem 1.1 (*Divergence Theorem*): Let Ω be a bounded region with piecewise smooth boundary surface S. Suppose that any line intersects S in a finite number of points or has a whole interval in common with S. Let $\mathbf{n} = \mathbf{n}(\mathbf{x})$ be the unit outward normal vector to S and let $\mathbf{w}$ be a vector field that is C^1 in Ω and C^0 on $\bar{\Omega}$. Then,

$$\int_\Omega \nabla\cdot\mathbf{w}\,d\Omega = \int_S \mathbf{w}\cdot\mathbf{n}\,dS \tag{1.5}$$

If u and v are scalar functions that are C^2 in Ω and C^1 on $\bar{\Omega}$, then the divergence theorem and the differential identity

$$\nabla\cdot(u\nabla v) = \nabla u\cdot\nabla v + u\,\nabla^2 v \tag{1.6}$$

lead to *Green's first and second integral identities*:

$$\int_\Omega u\,\nabla^2 v\,d\Omega = \int_S u\frac{\partial v}{\partial n}\,dS - \int_\Omega \nabla u\cdot\nabla v\,d\Omega \tag{1.7}$$

$$\int_\Omega (u\,\nabla^2 v - v\,\nabla^2 u)\,d\Omega = \int_S \left(u\frac{\partial v}{\partial n} - v\frac{\partial u}{\partial n}\right) dS \tag{1.8}$$

1.3 AUXILIARY CONDITIONS; WELL-POSED PROBLEMS

The PDEs that model physical systems usually have infinitely many solutions. To select the single function that represents the solution to a physical problem, one must impose certain auxiliary conditions that further characterize the system being modeled. These fall into two categories.

Boundary conditions. These are conditions that must be satisfied at points on the boundary S of the spatial region Ω in which the PDE holds. Special names have been given to three forms of boundary conditions:

Dirichlet condition $u = g$

Neumann (or **flux**) **condition** $\dfrac{\partial u}{\partial n} = g$

Mixed (or **Robin** or **radiation**) **condition** $\alpha u + \beta\dfrac{\partial u}{\partial n} = g$

in which g, α, and β are functions prescribed on S.

Initial conditions. These are conditions that must be satisfied throughout Ω at the instant when consideration of the physical system begins. A typical initial condition prescribes some combination of u and its time derivatives.

The prescribed initial- and boundary-condition functions, together with the coefficient functions and any inhomogeneous term in the PDE, are said to comprise the *data* in the problem modeled by the PDE. The solution is said to *depend continuously on the data* if small changes in the data produce correspondingly small changes in the solution. The problem itself is said to be *well-posed* if (i) a solution to the problem exists, (ii) the solution is unique, (iii) the solution depends continuously on the data. If any of these conditions is not satisfied, then the problem is said to be *ill-posed.*

The auxiliary conditions that, together with a PDE, comprise a well-posed problem must not be too many, or the problem will have no solution. They must not be too few, or the solution will not be unique. Finally, they must be of the correct type, or the solution will not depend continuously on the data. The well-posedness of some common *boundary value problems* (no initial conditions) and *initial–boundary value problems* is discussed in Chapters 3 and 4.

Chapter 2

Classification and Characteristics

2.1 TYPES OF SECOND-ORDER EQUATIONS

In the linear PDE of order two in two variables,

$$au_{xx} + 2bu_{xy} + cu_{yy} + du_x + eu_y + fu = g \tag{2.1}$$

if u_{xx} is formally replaced by α^2, u_{xy} by $\alpha\beta$, u_{yy} by β^2, u_x by α, and u_y by β, then associated with (*2.1*) is a polynomial of degree two in α and β:

$$P(\alpha, \beta) \equiv a\alpha^2 + 2b\alpha\beta + c\beta^2 + d\alpha + e\beta + f$$

The mathematical properties of the solutions of (*2.1*) are largely determined by the algebraic properties of the polynomial $P(\alpha, \beta)$. $P(\alpha, \beta)$—and along with it, the PDE (*2.1*)—is classified as *hyperbolic*, *parabolic*, or *elliptic* according as its discriminant, $b^2 - ac$, is *positive*, *zero*, or *negative*. Note that the type of (*2.1*) is determined solely by its *principal part* (the terms involving the highest-order derivatives of u) and that the type will generally change with position in the xy-plane unless a, b, and c are constants.

EXAMPLE 2.1 (*a*) The PDE $3u_{xx} + 2u_{xy} + 5u_{yy} + xu_y = 0$ is elliptic, since

$$b^2 - ac = 1^2 - 3(5) = -14 < 0$$

(*b*) The *Tricomi equation* for transonic flow, $u_{xx} + yu_{yy} = 0$, has

$$b^2 - ac = 0^2 - (1)y = -y$$

Thus, the equation is elliptic for $y > 0$, parabolic for $y = 0$, and hyperbolic for $y < 0$.

The general linear PDE of order two in n variables has the form

$$\sum_{i,j=1}^{n} a_{ij} u_{x_i x_j} + \sum_{i=1}^{n} b_i u_{x_i} + cu = d \tag{2.2}$$

If $u_{x_i x_j} = u_{x_j x_i}$, then the principal part of (*2.2*) can always be arranged so that $a_{ij} = a_{ji}$; thus, the $n \times n$ matrix $\mathbf{A} = [a_{ij}]$ can be assumed symmetric. In linear algebra it is shown that every real, symmetric $n \times n$ matrix has n real eigenvalues. These eigenvalues are the (possibly repeated) zeros of an nth-degree polynomial in λ, $\det(\mathbf{A} - \lambda \mathbf{I})$, where $\mathbf{I}$ is the $n \times n$ identity matrix. Let P denote the number of positive eigenvalues, and Z the number of zero eigenvalues (i.e., the multiplicity of the eigenvalue zero), of the matrix $\mathbf{A}$. Then (*2.2*) is:

hyperbolic	if	$Z = 0$ and $P = 1$ *or* $Z = 0$ and $P = n - 1$
parabolic	if	$Z > 0$ (equivalently, if $\det \mathbf{A} = 0$)
elliptic	if	$Z = 0$ and $P = n$ *or* $Z = 0$ and $P = 0$
ultrahyperbolic	if	$Z = 0$ and $1 < P < n - 1$

If any of the a_{ij} is nonconstant, the type of (*2.2*) can vary with position.

EXAMPLE 2.2 For the PDE $3u_{x_1x_1} + u_{x_2x_2} + 4u_{x_2x_3} + 4u_{x_3x_3} = 0$,

$$\mathbf{A} = \begin{bmatrix} 3 & 0 & 0 \\ 0 & 1 & 2 \\ 0 & 2 & 4 \end{bmatrix} \qquad \text{and} \qquad \det \begin{bmatrix} 3-\lambda & 0 & 0 \\ 0 & 1-\lambda & 2 \\ 0 & 2 & 4-\lambda \end{bmatrix} = (3-\lambda)(\lambda)(\lambda-5)$$

Because $\lambda = 0$ is an eigenvalue, the PDE is parabolic (throughout $x_1x_2x_3$-space).

2.2 CHARACTERISTICS

The following, seemingly unrelated, questions both naturally lead to a consideration of special curves associated with (*2.1*), called *characteristic curves* or, simply, *characteristics.* (1) How can a coordinate transformation be used to simplify the principal part of (*2.1*)? (2) Along what curves is a knowledge of u, u_x, and u_y, together with (*2.1*), insufficient uniquely to determine u_{xx}, u_{xy}, and u_{yy}?

To address the first question, suppose a locally invertible change of independent variables,

$$\xi = \phi(x, y) \qquad \eta = \psi(x, y) \tag{2.3}$$

$(\phi_x\psi_y - \phi_y\psi_x \neq 0)$, is used to transform the principal part of (*2.1*) from

$$au_{xx} + 2bu_{xy} + cu_{yy} \qquad \text{to} \qquad Au_{\xi\xi} + 2Bu_{\xi\eta} + Cu_{\eta\eta} + \textit{lower-order terms}$$

which implies that the principal part of the transformed equation is

$$Au_{\xi\xi} + 2Bu_{\xi\eta} + Cu_{\eta\eta}$$

As found in Problem 2.3,

$$A = a\phi_x^2 + 2b\phi_x\phi_y + c\phi_y^2 \qquad C = a\psi_x^2 + 2b\psi_x\psi_y + c\psi_y^2 \qquad B = a\phi_x\psi_x + b(\phi_x\psi_y + \phi_y\psi_x) + c\phi_y\psi_y$$

Since the transformed and original discriminants are related by

$$B^2 - AC = (b^2 - ac)(\phi_x\psi_y - \phi_y\psi_x)^2$$

the type of (*2.1*) is invariant under an invertible change of independent variables. The principal part of the transformed equation will take a particularly simple form if $A = C = 0$, which will be the case if ϕ and ψ are both solutions of

$$az_x^2 + 2bz_xz_y + cz_y^2 = 0 \tag{2.4}$$

(*2.4*) is called the *characteristic equation* of (*2.1*); the level curves, $z(x, y) = \text{const.}$, of a solution of (*2.4*) are called *characteristic curves* of (*2.1*).

Turning to question (2), suppose that u, u_x, and u_y are known along some curve Γ. Then, as shown in Problem 2.10, u_{xx}, u_{xy}, and u_{yy} are uniquely determined along Γ unless

$$a\,dy^2 - 2b\,dx \cdot dy + c\,dx^2 = 0 \tag{2.5}$$

holds along (i.e., is the ordinary differential equation for) Γ.

Theorem 2.1: $z(x, y) = \text{const.}$ is a characteristic of (*2.1*) if and only if $z(x, y) = \text{const.}$ is a solution of (*2.5*).

For the proof, see Problem 2.4.

Related to the indeterminacy of the second derivatives along a characteristic is the fact that *physically significant discontinuities in the solution of* (*2.1*) *can propagate only along characteristics.*

The number of real solutions of (*2.4*) or (*2.5*) is dictated by the sign of the discriminant,

$$b^2(x, y) - a(x, y)c(x, y)$$

Thus, when (*2.1*) is hyperbolic, parabolic, or elliptic, there are, respectively, two, one, or zero characteristics passing through (x, y). In the hyperbolic case, the two families of characteristics define natural coordinates (ξ, η) in which to study (*2.1*). The absence of characteristics for elliptic equations implies that there are no curves along which discontinuities in a solution can propagate; solutions of elliptic equations are generally smooth.

EXAMPLE 2.3

(*a*) By Theorem 2.1, the characteristics of the (hyperbolic) *one-dimensional wave equation,* $a^2u_{xx} - u_{tt} = 0$, are defined by $a^2\,dt^2 - dx^2 = 0$. Thus, the characteristics are the lines

$$x + at \equiv \xi = \text{const.} \qquad x - at \equiv \eta = \text{const.}$$

(*b*) The characteristics of the (parabolic) *one-dimensional heat equation*,

$$\kappa u_{xx} - u_t = 0$$

are defined by $\kappa\, dt^2 = 0$. Thus, the characteristics are the lines $t \equiv \eta = \text{const.}$

(*c*) Characteristics of the (elliptic) *two-dimensional Laplace's equation*,

$$u_{xx} + u_{yy} = 0$$

must satisfy $dy^2 + dx^2 = 0$, which has no nonzero real solution. Thus, Laplace's equation has no real characteristics.

For the n-dimensional PDE (*2.2*), the *characteristic surfaces* are the level surfaces,

$$z(x_1, x_2, \ldots, x_n) = \text{const.}$$

of the solutions of the *characteristic equation*

$$\sum_{i,j=1}^{n} a_{ij} z_{x_i} z_{x_j} = 0$$

For $n > 2$, this characteristic equation cannot generally be reduced to an ordinary differential equation as in Theorem 2.1; so, the characteristics are often difficult to determine. As in the case $n = 2$, the characteristics of (*2.2*) are the surfaces along which discontinuities in derivatives of the solution propagate.

EXAMPLE 2.4 The characteristics of the hyperbolic equation

$$u_{x_1x_1} - u_{x_2x_2} - u_{x_3x_3} = 0 \tag{1}$$

(a two-dimensional wave equation with x_1 taking the role of time) are the level surfaces of the solutions of

$$z_{x_1}^2 - z_{x_2}^2 - z_{x_3}^2 = 0 \tag{2}$$

By direct substitution it may be verified that

$$z = F(x_1 + x_2 \sin\alpha + x_3 \cos\alpha) \tag{3}$$

with F an arbitrary C^1 function and α an arbitrary parameter, solves (*2*). This solution is constant when

$$x_1 + x_2 \sin\alpha + x_3 \cos\alpha = \text{const.}$$

which may be rewritten as

$$(x - \bar{x}_1) + (x_2 - \bar{x}_2)\sin\alpha + (x_3 - \bar{x}_3)\cos\alpha = 0 \tag{4}$$

where $(\bar{x}_1, \bar{x}_2, \bar{x}_3)$ is an arbitrary point in $x_1x_2x_3$-space. Equation (*4*) represents a one-parameter family of planes, each plane containing the point $(\bar{x}_1, \bar{x}_2, \bar{x}_3)$ and making a 45° angle with the positive x_1-axis. As is geometrically obvious, the family has as its envelope the right circular cone

$$(x_1 - \bar{x}_1)^2 - (x_2 - \bar{x}_2)^2 - (x_3 - \bar{x}_3)^2 = 0 \tag{5}$$

On the cones (*5*), *all* solutions (*3*) are constant; hence, these cones are the characteristic surfaces of (*1*).

2.3 CANONICAL FORMS

We have already seen, in Section 2.2, how a hyperbolic second-order PDE may be simplified by choosing the characteristics as the new coordinate curves. In general, if a, b, and c in (*2.1*) are sufficiently smooth functions of x and y, there will always exist a locally one-one coordinate transformation, $\xi = \phi(x, y)$, $\eta = \psi(x, y)$, which transforms the principal part to the *canonical form*

hyperbolic PDE	$u_{\xi\eta}$ or $u_{\xi\xi} - u_{\eta\eta}$
parabolic PDE	$u_{\xi\xi}$
elliptic PDE	$u_{\xi\xi} + u_{\eta\eta}$

The canonical forms $u_{\xi\xi} - u_{\eta\eta}$, $u_{\xi\xi}$, and $u_{\xi\xi} + u_{\eta\eta}$ are the principal parts of the wave, heat, and Laplace equations, which serve as prototypes of hyperbolic, parabolic, and elliptic equations, respectively. Methods for choosing ϕ and ψ to reduce (*2.1*) to canonical form are illustrated in Problems 2.6–2.9.

If (*2.1*), or more generally (*2.2*), has constant coefficients in its principal part, reduction to a canonical form can be accomplished using a *linear* change of independent variables. Specifically, there will exist an invertible linear coordinate transformation,

$$\xi_r = \sum_{i=1}^{n} b_{ir} x_i \qquad (r = 1, 2, \ldots, n)$$

that takes (*2.2*) into an equation with principal part

$$\sum_{i=1}^{n} \lambda_i u_{\xi_i \xi_i}$$

where λ_i $(i = 1, 2, \ldots, n)$ are the eigenvalues of the symmetric matrix $\mathbf{A}$. (If λ is an eigenvalue of multiplicity $q > 1$, then q of the ξ-variables will correspond to λ.) A rescaling of the independent variables,

$$\eta_i = \begin{cases} \xi_i / \sqrt{|\lambda_i|} & \lambda_i \neq 0 \\ \xi_i & \lambda_i = 0 \end{cases}$$

then yields one of the canonical forms

hyperbolic PDE $\qquad u_{\eta_1\eta_1} - \sum_{i=2}^{n} u_{\eta_i\eta_i}$

parabolic PDE $\qquad \sum_{i=1}^{n-m} \pm u_{\eta_i\eta_i} \qquad (m = Z > 0)$

elliptic PDE $\qquad \sum_{i=1}^{n} u_{\eta_i\eta_i}$

ultrahyperbolic PDE $\qquad \sum_{i=1}^{m} u_{\eta_i\eta_i} - \sum_{i=m+1}^{n} u_{\eta_i\eta_i} \qquad (1 < m = P < n-1)$

If (*2.2*) has all coefficients constant and has been reduced to one of the above canonical forms, a further simplification is always possible in the elliptic or hyperbolic case (see Problem 2.14) and is sometimes possible in the parabolic case (see Problem 2.15).

2.4 DIMENSIONAL ANALYSIS

The reduction of a PDE to canonical form does not change the order of the equation or the number of independent variables. However, by seeking a solution of a particular form it is often possible to reduce the number of independent variables in a problem.

EXAMPLE 2.5 (*a*) If we look for oscillatory solutions to the wave equation,

$$u_{xx} + u_{yy} - u_{tt} = 0$$

of the form $u(x, y, t) = v(x, y)e^{ikt}$ $(i = \sqrt{-1})$, then v satisfies the *Helmholtz equation*,

$$v_{xx} + v_{yy} + k^2 v = 0$$

(*b*) Traveling-wave solutions to $u_{xx} - u_{tt} + u = 0$, in the form $u(x, t) = v(x - at)$ $(a = \text{const.})$, satisfy the ordinary differential equation $(1 - a^2)v'' + v = 0$. (*c*) Radially symmetric solutions of Laplace's equation, $u_{xx} + u_{yy} = 0$, of the form

$$u(x, y) = v(r) \qquad \text{where} \qquad r = (x^2 + y^2)^{1/2}$$

satisfy $v'' + r^{-1}v' = 0$.

Suppose that a physical problem is modeled by a PDE that involves dependent variable u; independent variables $x_1, x_2, \ldots, x_n$; and parameters $p_1, p_2, \ldots, p_m$. The general expression for the solution of the PDE is

$$F(u, x_1, x_2, \ldots, x_n, p_1, p_2, \ldots, p_m) = 0 \tag{2.6}$$

which can usually be "solved" to give $u = f(x_1, \ldots, x_n, p_1, \ldots, p_m)$. Consider a fundamental system of *physical dimensions*, each with its corresponding *base unit*; specifically, consider the International System (SI), as indicated in Table 2-1.

Table 2-1

Fundamental Dimension	Base Unit
Length (L)	meter, m
Mass (M)	kilogram, kg
Time (T)	second, s
Electric current (A)	ampere, A
Thermodynamic temperature (Θ)	kelvin, K
Amount of substance (X)	mole, mol
Luminous intensity (I)	candela, cd

Each quantity in (*2.6*) is either *dimensionless* (i.e., a pure number) or has as its physical dimension a product of powers of the fundamental dimensions of Table 2-1.

EXAMPLE 2.6 Let $F(u, x, t, \rho c, k) = 0$ be the general solution of the one-dimensional heat equation

$$\rho c u_t - k u_{xx} = 0 \tag{1}$$

The dependent variable is the temperature u, while the independent variables and physical parameters are $x_1 \equiv x$, $x_2 \equiv t$, $p_1 \equiv \rho c$ (density times specific heat capacity), and $p_2 \equiv k$ (thermal conductivity). The physical dimensions of these quantities are:

$$\{u\} = \Theta \qquad \{x_1\} = L \qquad \{x_2\} = T$$
$$\{p_1\} = ML^{-1}T^{-2}\Theta^{-1} \qquad \{p_2\} = MLT^{-3}\Theta^{-1}$$

There are in all $N = 5$ dimensional quantities, which are specified in terms of $R = 4$ fundamental dimensions (the three mechanical dimensions L, T, M and the thermal dimension Θ). It happens that only integral powers of the fundamental dimensions enter.

If we define $\kappa \equiv k/\rho c$ (thermal diffusivity) and rewrite (*1*) as

$$u_t - \kappa u_{xx} = 0 \tag{2}$$

there are present in (*2*) one fewer physical parameter and one fewer fundamental dimension ($\{\kappa\} = L^2T^{-1}$); hence, $N - R$ is unchanged. This invariance reflects the mass independence of the thermal process, and should not be expected in general.

When, as in Example 2.6, a PDE involves fewer fundamental dimensions than dimensional quantities, it must admit a simplified, *similarity solution*, in accordance with

Theorem 2.2 (*Buckingham Pi Theorem*): If (i) the function F in (*2.6*) is continuously differentiable with respect to each argument; (ii) given $N - 1$ of the $N = 1 + n + m$ quantities u, x_i, p_j, the equation (*2.6*) can be uniquely solved for the remaining quantity; and (iii) u, x_i, p_j collectively involve R fundamental units ($0 < R < N$); then (*2.6*) is equivalent to

$$G(\pi_1, \pi_2, \ldots, \pi_{N-R}) = 0$$

where the π_α are dimensionless and

$$\pi_\alpha = u^{\gamma_{\alpha 0}} x_1^{\gamma_{\alpha 1}} x_2^{\gamma_{\alpha 2}} \cdots x_n^{\gamma_{\alpha n}} p_1^{\gamma_{\alpha, n+1}} p_2^{\gamma_{\alpha, n+2}} \cdots p_m^{\gamma_{\alpha, N-1}}$$

for some real numbers $\gamma_{\alpha\beta}$ ($\alpha = 1, 2, \ldots, N-R$, $\beta = 0, 1, \ldots, N-1$) such that $[\gamma_{\alpha\beta}]$ is of rank $N-R$.

(For the case $R = 0$, Theorem 2.2 holds trivially, with $G = F$.)

Solved Problems

2.1 Classify according to type:

$$(a)\quad u_{xx} + 2yu_{xy} + xu_{yy} - u_x + u = 0$$
$$(b)\quad 2xyu_{xy} + xu_y + yu_x = 0$$
$$(c)\quad u_{xx} + u_{xy} + 5u_{yx} + u_{yy} + 2u_{yz} + u_{zz} = 0$$

(*a*) In the notation of (*2.1*), $a = 1$, $b = y$, and $c = x$. Since $b^2 - ac = y^2 - x$, the equation is hyperbolic in the region $y^2 > x$, parabolic on the curve $y^2 = x$, and elliptic in the region $y^2 < x$.

(*b*) Here, $a = 0$, $b = xy$, and $c = 0$. Since $b^2 - ac = x^2y^2$, which is positive except on the coordinate axes, the equation is hyperbolic for all x and y except $x = 0$ or $y = 0$. Along the coordinate axes the equation degenerates to first-order and the second-order categories do not apply.

(*c*) Rewrite the equation in symmetrical form:

$$u_{x_1x_1} + 3u_{x_1x_2} + 3u_{x_2x_1} + u_{x_2x_2} + u_{x_2x_3} + u_{x_3x_2} + u_{x_3x_3} = 0 \tag{1}$$

where $x_1 \equiv x$, $x_2 \equiv y$, $x_3 \equiv z$. The matrix corresponding to the principal part of (*1*) is

$$\mathbf{A} = \begin{bmatrix} 1 & 3 & 0 \\ 3 & 1 & 1 \\ 0 & 1 & 1 \end{bmatrix}$$

Since $\det(\mathbf{A} - \lambda\mathbf{I}) = (1-\lambda)^3 - 10(1-\lambda)$, the eigenvalues of $\mathbf{A}$ are 1 and $1 \pm \sqrt{10}$. Thus, $Z = 0$ and $P = 3 - 1$, making the PDE hyperbolic (everywhere).

2.2 Use the transformation (*2.3*) to express all the x- and y-derivatives in (*2.1*) in terms of ξ and η.

By the chain rule,

$$\frac{\partial u}{\partial x} = \frac{\partial u}{\partial \xi}\frac{\partial \xi}{\partial x} + \frac{\partial u}{\partial \eta}\frac{\partial \eta}{\partial x} \qquad \text{or} \qquad u_x = u_\xi \phi_x + u_\eta \psi_x$$

and

$$\frac{\partial u}{\partial y} = \frac{\partial u}{\partial \xi}\frac{\partial \xi}{\partial y} + \frac{\partial u}{\partial \eta}\frac{\partial \eta}{\partial y} \qquad \text{or} \qquad u_y = u_\xi \phi_y + u_\eta \psi_y$$

By the product rule,

$$u_{xx} = u_\xi \phi_{xx} + (u_\xi)_x \phi_x + u_\eta \psi_{xx} + (u_\eta)_x \psi_x$$

which, after using the chain rule to find $(u_\xi)_x$ and $(u_\eta)_x$, yields

$$\begin{aligned} u_{xx} &= u_\xi \phi_{xx} + (u_{\xi\xi}\phi_x + u_{\xi\eta}\psi_x)\phi_x + u_\eta \psi_{xx} + (u_{\eta\xi}\phi_x + u_{\eta\eta}\psi_x)\psi_x \\ &= u_{\xi\xi}\phi_x^2 + 2u_{\xi\eta}\phi_x\psi_x + u_{\eta\eta}\psi_x^2 + u_\xi\phi_{xx} + u_\eta\psi_{xx} \end{aligned}$$

Similarly,

$$\begin{aligned} u_{yy} &= u_\xi\phi_{yy} + (u_\xi)_y\phi_y + u_\eta\psi_{yy} + (u_\eta)_y\psi_y \\ &= u_\xi\phi_{yy} + (u_{\xi\xi}\phi_y + u_{\xi\eta}\psi_y)\phi_y + u_\eta\psi_{yy} + (u_{\eta\xi}\phi_y + u_{\eta\eta}\psi_y)\psi_y \\ &= u_{\xi\xi}\phi_y^2 + 2u_{\xi\eta}\phi_y\psi_y + u_{\eta\eta}\psi_y^2 + u_\xi\phi_{yy} + u_\eta\psi_{yy} \end{aligned}$$

Finally,

$$\begin{aligned} u_{xy} &= u_\xi\phi_{xy} + (u_\xi)_y\phi_x + u_\eta\psi_{xy} + (u_\eta)_y\psi_x \\ &= u_\xi\phi_{xy} + (u_{\xi\xi}\phi_y + u_{\xi\eta}\psi_y)\phi_x + u_\eta\psi_{xy} + (u_{\eta\xi}\phi_y + u_{\eta\eta}\psi_y)\psi_x \\ &= u_{\xi\xi}\phi_x\phi_y + u_{\xi\eta}(\phi_x\psi_y + \phi_y\psi_x) + u_{\eta\eta}\psi_x\psi_y + u_\xi\phi_{xy} + u_\eta\psi_{xy} \end{aligned}$$

2.3 Use the results of Problem 2.2 to find the principal part of (*2.1*) when that equation is written in terms of ξ and η.

Since $u_{\xi\xi}$, $u_{\xi\eta}$, and $u_{\eta\eta}$ occur only in the transformations of u_{xx}, u_{xy}, and u_{yy}, it suffices to transform only the principal part of (*2.1*):

$$\begin{aligned} au_{xx} + 2bu_{xy} + cu_{yy} &= (a\phi_x^2 + 2b\phi_x\phi_y + c\phi_y^2)u_{\xi\xi} \\ &\quad + 2[a\phi_x\psi_x + b(\phi_x\psi_y + \phi_y\psi_x) + c\phi_y\psi_y]u_{\xi\eta} \\ &\quad + (a\psi_x^2 + 2b\psi_x\psi_y + c\psi_y^2)u_{\eta\eta} \\ &\quad + R \\ &\equiv Au_{\xi\xi} + 2Bu_{\xi\eta} + Cu_{\eta\eta} + R \end{aligned}$$

where $R \equiv (a\phi_{xx} + 2b\phi_{xy} + c\phi_{yy})u_\xi + (a\psi_{xx} + 2b\psi_{xy} + c\psi_{yy})u_\eta$ is first-order in u. Note that $R = 0$ if both ϕ and ψ are linear functions of x and y.

2.4 Prove Theorem 2.1.

First assume that $z(x, y)$ satisfies (*2.4*) and that $z_y(x, y) \neq 0$, so that the relation

$$z(x, y) = \gamma = \text{const.}$$

defines at least one single-valued function $y = f(x, \gamma)$. Then, for $y = f(x, \gamma)$,

$$\frac{dy}{dx} = -\frac{z_x(x, y)}{z_y(x, y)}$$

Dividing each term of (*2.4*) by z_y^2 yields

$$a\left(\frac{z_x}{z_y}\right)^2 + 2b\frac{z_x}{z_y} + c = 0$$

which on $y = f(x, \gamma)$ is equivalent to

$$a\left(\frac{dy}{dx}\right)^2 - 2b\frac{dy}{dx} + c = 0$$

This shows that $y = f(x, \gamma)$ is a particular solution of (*2.5*), and so $z(x, y) = \gamma$ is an implicit solution of (*2.5*). If $z_y(x, y) = 0$ but (*2.4*) is not identically satisfied, then $z_x(x, y) \neq 0$ and the above argument can be repeated with the roles of x and y interchanged.

To complete the proof, let $z(x, y) = \text{const.}$ be a general integral of (*2.5*). To show that $z(x, y)$ satisfies (*2.4*) at an arbitrary point (x_0, y_0), let $\gamma_0 = z(x_0, y_0)$ and consider the curve $y = f(x, \gamma_0)$. Along this curve,

$$0 = a\left(\frac{dy}{dx}\right)^2 - 2b\frac{dy}{dx} + c = a\left(\frac{z_x}{z_y}\right)^2 - 2b\left(-\frac{z_x}{z_y}\right) + c$$

from which it follows, upon setting $x = x_0$, that (*2.4*) holds at (x_0, y_0).

2.5 Classify according to type and determine the characteristics of:

$$(a)\ 2u_{xx} - 4u_{xy} - 6u_{yy} + u_x = 0 \qquad (c)\ u_{xx} - x^2yu_{yy} = 0 \quad (y > 0)$$
$$(b)\ 4u_{xx} + 12u_{xy} + 9u_{yy} - 2u_x + u = 0 \qquad (d)\ e^{2x}u_{xx} + 2e^{x+y}u_{xy} + e^{2y}u_{yy} = 0$$

(*a*) In the notation of (*2.1*) $a = 2$, $b = -2$, $c = -6$; so $b^2 - ac = 16$ and the equation is hyperbolic. From Theorem 2.1, the characteristics are determined by

$$\frac{dy}{dx} = \frac{b \pm \sqrt{b^2 - ac}}{a} = -1 \pm 2$$

Thus, the lines $x - y = \text{const.}$ and $3x + y = \text{const.}$ are the characteristics of the equation.

(*b*) In this case, $a = 4$, $b = 6$, $c = 9$; so $b^2 - ac = 0$ and the equation is parabolic. From Theorem 2.1 it follows that there is a single family of characteristics, given by

$$\frac{dy}{dx} = \frac{b}{a} = \frac{3}{2} \qquad \text{or} \qquad 2y - 3x = \text{const.}$$

(*c*) In the region $y > 0$, $b^2 - ac = x^2y$ is positive, so that the equation is of hyperbolic type. The characteristics are determined by

$$\frac{dy}{dx} = \pm x\sqrt{y} \qquad \text{or} \qquad \frac{dy}{\sqrt{y}} \mp x\,dx = 0$$

from which it follows that the characteristics are $x^2 \pm 4\sqrt{y} = \text{const.}$

(*d*) $b^2 - ac = (e^{x+y})^2 - e^{2x}e^{2y} = 0$, and the equation is parabolic. Theorem 2.1 implies that the characteristics are given by

$$\frac{dy}{dx} = \frac{e^{x+y}}{e^{2x}} \qquad \text{or} \qquad e^{-x}\,dx - e^{-y}\,dy = 0$$

from which $e^{-x} - e^{-y} = \text{const.}$

2.6 Transform the hyperbolic equations

$$(a)\ 2u_{xx} - 4u_{xy} - 6u_{yy} + u_x = 0 \qquad (b)\ u_{xx} - x^2yu_{yy} = 0 \quad (y > 0)$$

to a canonical form with principal part $u_{\xi\eta}$.

(*a*) In the notation of Section 2.2, if $\xi = \phi(x, y) = \text{const.}$ and $\eta = \psi(x, y) = \text{const.}$ are independent families of characteristics, then $A = C = 0$. In Problem 2.5(*a*), the characteristics of the given equation were shown to be $x - y = \text{const.}$ and $3x + y = \text{const.}$ Therefore, we take

$$\xi = \phi(x, y) = x - y \qquad \eta = \psi(x, y) = 3x + y$$

Transforming the equation with the aid of Problem 2.3,

$$2u_{xx} - 4u_{xy} - 6u_{yy} + u_x = 16u_{\xi\eta} + u_\xi + 3u_\eta$$

The desired canonical form is therefore

$$u_{\xi\eta} + \frac{1}{16}u_\xi + \frac{3}{16}u_\eta = 0$$

(*b*) In Problem 2.5(*c*) the characteristics were found to be $x^2 \pm 4\sqrt{y} = \text{const.}$; therefore, we take

$$\xi = \phi(x, y) = x^2 + 4\sqrt{y} \qquad \eta = \psi(x, y) = x^2 - 4\sqrt{y}$$

With $\phi_x = 2x$, $\phi_y = 2y^{-1/2}$, $\psi_x = 2x$, $\psi_y = -2y^{-1/2}$, $\phi_{xx} = 2 = \psi_{xx}$, $\phi_{yy} = -y^{-3/2} = -\psi_{yy}$, and $\phi_{xy} = 0 = \psi_{xy}$, Problem 2.3 gives

$$\begin{aligned} u_{xx} - x^2yu_{yy} &= 16x^2u_{\xi\eta} + (2 + x^2y^{-1/2})u_\xi + (2 - x^2y^{-1/2})u_\eta \\ &= 8(\xi + \eta)u_{\xi\eta} + \frac{6\xi + 2\eta}{\xi - \eta}u_\xi - \frac{2\xi + 6\eta}{\xi - \eta}u_\eta \end{aligned}$$

where the last equality follows from $x^2 = (\xi + \eta)/2$ and $y^{1/2} = (\xi - \eta)/8$. The desired canonical form is then

$$u_{\xi\eta} + \frac{3\xi + \eta}{4(\xi^2 - \eta^2)} u_\xi - \frac{\xi + 3\eta}{4(\xi^2 - \eta^2)} u_\eta = 0 \qquad (\xi > \eta)$$

2.7 Transform the parabolic equations

$$(a) \quad 4u_{xx} + 12u_{xy} + 9u_{yy} - 2u_x + u = 0 \qquad (b) \quad e^{2x}u_{xx} + 2e^{x+y}u_{xy} + e^{2y}u_{yy} = 0$$

to a canonical form with principal part $u_{\xi\xi}$.

(*a*) In the notation of Section 2.2, $C = 0$ if $\eta = \psi(x, y) = \text{const.}$ is a characteristic of the equation. Since $B^2 - AC = 0$, this assignment of η will also make $B = 0$. From Problem 2.5(*b*),

$$\eta = \psi(x, y) = 3x - 2y$$

Any $\phi(x, y)$ satisfying $\phi_x\psi_y - \phi_y\psi_x \neq 0$ can be chosen as the second new variable; a convenient choice is the linear function

$$\xi = \phi(x, y) = y$$

From Problem 2.3,

$$4u_{xx} + 12u_{xy} + 9u_{yy} - 2u_x + u = 9u_{\xi\xi} - 3u_\eta + u$$

whence the canonical form

$$u_{\xi\xi} - \frac{1}{3}u_\eta + \frac{1}{9}u = 0$$

(*b*) In Problem 2.5(*d*) the characteristics were shown to be $e^{-x} - e^{-y} = \text{const.}$, so $C = B = 0$ if we set

$$\eta = \psi(x, y) = e^{-x} - e^{-y}$$

A convenient choice for the other new variable is

$$\xi = \phi(x, y) = x$$

From Problem 2.3,

$$e^{2x}u_{xx} + 2e^{x+y}u_{xy} + e^{2y}u_{yy} = e^{2x}u_{\xi\xi} + 2u_\eta = e^{2\xi}u_{\xi\xi} + 2u_\eta$$

whence the canonical form $u_{\xi\xi} + 2e^{-2\xi}u_\eta = 0$.

2.8 If (*2.1*) is elliptic, show how to select ϕ and ψ in (*2.3*) so that the principal part of the transformed equation will be $u_{\xi\xi} + u_{\eta\eta}$.

When $b^2 - ac < 0$, (*2.5*) has complex conjugate solutions for dy/dx, one of which is ($i = \sqrt{-1}$)

$$\frac{dy}{dx} = \frac{b + i\sqrt{|b^2 - ac|}}{a} \qquad (1)$$

The ordinary differential equation (*1*) will have a solution of the form

$$z(x, y) = \phi(x, y) + i\psi(x, y) = \text{const.}$$

for real functions ϕ and ψ. Then, by Theorem 2.1,

$$\begin{aligned}
0 &= az_x^2 + 2bz_xz_y + cz_y^2 \\
&= a(\phi_x + i\psi_x)^2 + 2b(\phi_x + i\psi_x)(\phi_y + i\psi_y) + c(\phi_y + i\psi_y)^2 \\
&= [(a\phi_x^2 + 2b\phi_x\phi_y + c\phi_y^2) - (a\psi_x^2 + 2b\psi_x\psi_y + c\psi_y^2)] + 2i[a\phi_x\psi_x + b(\phi_x\psi_y + \phi_y\psi_x) + c\phi_y\psi_y] \\
&\equiv [A - C] + 2i[B]
\end{aligned}$$

which holds only if $A = C$ and $B = 0$. Thus, if we set $\xi = \phi(x, y)$ and $\eta = \psi(x, y)$, the transformed PDE will have principal part

$$A(u_{\xi\xi} + u_{\eta\eta})$$

and division by A will yield the required canonical form.

For the above analysis to be strictly valid, one must require the coefficients a, b, c to be analytic functions (see Section 3.1).

2.9 Using Problem 2.8, transform the elliptic equations

$$(a)\quad u_{xx} + 2u_{xy} + 17u_{yy} = 0 \qquad (b)\quad x^2u_{xx} + y^2u_{yy} = 0 \quad (x>0,\ y>0)$$

to canonical form with principal part $u_{\xi\xi} + u_{\eta\eta}$.

(*a*) Here, $a = 1$, $b = 1$, and $c = 17$; (*1*) of Problem 2.8 becomes

$$\frac{dy}{dx} = 1 + i4$$

which has the solution $z = (x - y) + i4x = \text{const}$. Thus, setting

$$\xi = \phi(x, y) = x - y \qquad \eta = \psi(x, y) = 4x$$

we obtain, as in Problem 2.3,

$$u_{xx} + 2u_{xy} + 17u_{yy} = 16u_{\xi\xi} + 16u_{\eta\eta}$$

whence the canonical form $u_{\xi\xi} + u_{\eta\eta} = 0$ (Laplace's equation).

(*b*) In this case, (*1*) of Problem 2.8 becomes

$$\frac{dy}{dx} = i\frac{y}{x}$$

with solution $z = \log x + i \log y = \text{const}$. Setting

$$\xi = \phi(x) = \log x \qquad \eta = \psi(y) = \log y$$

we calculate, following Problem 2.3,

$$x^2u_{xx} + y^2u_{yy} = u_{\xi\xi} + u_{\eta\eta} - u_\xi - u_\eta = 0$$

as the required canonical form.

2.10 Show that a characteristic curve of (*2.1*) is an exceptional curve in the sense that the values of u, u_x, and u_y along the curve, together with the PDE, do not uniquely determine the values of u_{xx}, u_{xy}, and u_{yy} along the curve.

Let Γ be a smooth curve in the xy-plane, given parametrically by $x = x(s)$, $y = y(s)$, $s_1 < s < s_2$. Suppose u, u_x, and u_y are specified along Γ as $u = F(s)$, $u_x = G(s)$, and $u_y = H(s)$. Then,

$$\frac{du_x}{ds} = u_{xx}x'(s) + u_{xy}y'(s) = G'(s)$$

$$\frac{du_y}{ds} = u_{xy}x'(s) + u_{yy}y'(s) = H'(s)$$

These two equations and (*2.1*) comprise three linear equations, which may be solved *uniquely* for the three unknowns u_{xx}, u_{xy}, and u_{yy} along Γ, unless the coefficient matrix

$$\begin{bmatrix} a & 2b & c \\ x'(s) & y'(s) & 0 \\ 0 & x'(s) & y'(s) \end{bmatrix}$$

has determinant zero; that is, unless

$$a\left(\frac{dy}{ds}\right)^2 - 2b\frac{dx}{ds}\frac{dy}{ds} + c\left(\frac{dx}{ds}\right)^2 = 0$$

This last equation is equivalent to (*2.5*), which defines the characteristics of (*2.1*).

2.11 If the variables $\xi_1, \xi_2, \ldots, \xi_n$ and $x_1, x_2, \ldots, x_n$ are related by the linear transformation

$$\boldsymbol{\xi} = \mathbf{B}^T\mathbf{x} \qquad \text{or} \qquad \xi_r = \sum_{i=1}^{n} b_{ir}x_i \qquad (r = 1, 2, \ldots, n)$$

change (*2.2*) to the ξ-variables.

According to the chain rule,

$$u_{x_i} = \frac{\partial u}{\partial x_i} = \sum_{r=1}^{n} \frac{\partial u}{\partial \xi_r}\frac{\partial \xi_r}{\partial x_i} = \sum_{r=1}^{n} b_{ir}\frac{\partial u}{\partial \xi_r}$$

$$u_{x_i x_j} = \sum_{s=1}^{n} (u_{x_i})_{\xi_s}\frac{\partial \xi_s}{\partial x_j} = \sum_{s=1}^{n} b_{js}(u_{x_i})_{\xi_s} = \sum_{r,\,s=1}^{n} b_{ir}b_{js}\frac{\partial^2 u}{\partial \xi_r \partial \xi_s}$$

Thus, in terms of $\xi_1, \xi_2, \ldots, \xi_n$, (*2.2*) is

$$\sum_{r,\,s=1}^{n} \underbrace{\left[\sum_{i,\,j=1}^{n} b_{ir}a_{ij}b_{js}\right]}_{c_{rs}} u_{\xi_r\xi_s} + \sum_{r=1}^{n}\left[\sum_{i=1}^{n} b_{ir}b_i\right] u_{\xi_r} + cu = d$$

2.12 If (*2.2*) has constant coefficients a_{ij}, show that it is possible to choose matrix **B** in Problem 2.11 such that no mixed partials with respect to the ξ-variables occur in the transformed equation.

From Problem 2.11 it is seen that the matrix **C** defining the principal part of (*2.2*) in the ξ-variables is given by

$$\mathbf{C} = \mathbf{B}^T\mathbf{A}\mathbf{B}$$

According to the following result from linear algebra, **B** can be chosen to make **C** a diagonal matrix ($c_{rs} = 0$ for $r \neq s$), thereby removing all mixed partials from the transformed PDE.

Theorem 2.3: Let **A** be a real, symmetric matrix. Then there exists an orthogonal matrix **B** such that $\mathbf{C} = \mathbf{B}^T\mathbf{A}\mathbf{B}$ is diagonal. Moreover, the columns of **B** are the normalized eigenvectors of **A** and the diagonal entries of **C** are the corresponding eigenvalues of **A**.

(**B** is *orthogonal* if $\mathbf{B}^T = \mathbf{B}^{-1}$. It can be shown that to an m-fold eigenvalue of **A** there correspond precisely m linearly independent eigenvectors.)

2.13 Find an orthogonal change of coordinates that eliminates the mixed partial derivatives from

$$2u_{x_1x_1} + 2u_{x_2x_2} - 15u_{x_3x_3} + 8u_{x_1x_1} - 12u_{x_2x_3} - 12u_{x_1x_3} = 0$$

Then rescale to put the equation in canonical form.

The matrix corresponding to the principal part of this equation is

$$\mathbf{A} = \begin{bmatrix} 2 & 4 & -6 \\ 4 & 2 & -6 \\ -6 & -6 & -15 \end{bmatrix}$$

From

$$\det(\mathbf{A} - \lambda\mathbf{I}) = \lambda^3 + 11\lambda^2 - 144\lambda - 324 = (\lambda + 2)(\lambda + 18)(\lambda - 9)$$

it follows that the eigenvalues of **A** are $\lambda_1 = -2$, $\lambda_2 = -18$, $\lambda_3 = 9$. According to Theorem 2.3, the rth column-vector of the diagonalizing matrix **B**,

$$\mathbf{b}_r = (b_{1r}, b_{2r}, b_{3r})^T$$

satisfies $(\mathbf{A} - \lambda_r \mathbf{I})\mathbf{b}_r = \mathbf{0}$, or

$$\begin{aligned} (2-\lambda_r)b_{1r} + \quad 4b_{2r} - \quad 6b_{3r} &= 0 \\ 4b_{1r} + (2-\lambda_r)b_{2r} - \quad 6b_{3r} &= 0 \\ -6b_{1r} - \quad 6b_{2r} + (-15-\lambda_r)b_{3r} &= 0 \end{aligned} \tag{1}$$

together with the normalizing condition

$$\mathbf{b}_r \cdot \mathbf{b}_r = b_{1r}^2 + b_{2r}^2 + b_{3r}^2 = 1 \tag{2}$$

For $r = 1$, $\lambda_1 = -2$, (*1*) implies $b_{11} = -b_{21}$ and $b_{31} = 0$; then (*2*) is satisfied if

$$b_{11} = -b_{21} = \frac{1}{\sqrt{2}} \qquad b_{31} = 0$$

For $r = 2$, $\lambda_2 = -18$, and $4b_{12} = 4b_{22} = b_{32}$; normalizing,

$$b_{12} = b_{22} = \frac{1}{3\sqrt{2}} \qquad b_{32} = \frac{4}{3\sqrt{2}}$$

For $r = 3$, $\lambda_3 = 9$, and $b_{13} = b_{23} = -2b_{33}$; normalizing,

$$b_{13} = b_{23} = \frac{2}{3} \qquad b_{33} = -\frac{1}{3}$$

With **B** determined, the change of variables $\boldsymbol{\xi} = \mathbf{B}^T\mathbf{x}$ transforms the given equation to

$$-2u_{\xi_1\xi_1} - 18u_{\xi_2\xi_2} + 9u_{\xi_3\xi_3} = 0$$

Finally, by defining $\eta_1 \equiv \xi_1/\sqrt{2}$, $\eta_2 \equiv \xi_2/(3\sqrt{2})$ and $\eta_3 \equiv \xi_3/3$, and multiplying the equation by -1, we obtain

$$u_{\eta_1\eta_1} - u_{\eta_2\eta_2} - u_{\eta_3\eta_3} = 0$$

which is the canonical form for a hyperbolic PDE in three variables.

2.14 Find a change of dependent variable which eliminates the lower-order derivatives from

$$u_{x_1x_1} + u_{x_2x_2} - u_{x_3x_3} + 6u_{x_1} - 14u_{x_2} + 8u_{x_3} = 0$$

If $u(x_1, x_2, x_3) = v(x_1, x_2, x_3)\exp(\sum_{i=1}^{3} c_i x_i)$, then

$$u_{x_i} = (v_{x_i} + c_i v)\exp\left(\sum_{i=1}^{3} c_i x_i\right)$$

$$u_{x_ix_j} = (v_{x_ix_j} + 2c_i v_{x_i} + c_i^2 v)\exp\left(\sum_{i=1}^{3} c_i x_i\right)$$

and the PDE for v is

$$v_{x_1x_1} + v_{x_2x_2} - v_{x_3x_3} + (6+2c_1)v_{x_1} + (-14+2c_2)v_{x_2} + (8+2c_3)v_{x_3} + (c_1^2 + c_2^2 - c_3^2)v = 0$$

Now choose $c_1 = -3$, $c_2 = 7$, and $c_3 = -4$, to produce

$$v_{x_1x_1} + v_{x_2x_2} - v_{x_3x_3} + 42v = 0$$

2.15 Find a change of dependent variable that reduces the parabolic PDE

$$u_{xx} + 4u_x - 2u_t + 8u = 0$$

to the one-dimensional heat equation.

If we set $u(x, t) = v(x, t)\exp(c_0 t + c_1 x)$, then the PDE for v is

$$v_{xx} + (4 + 2c_1)v_x - 2v_t + (8 + c_1^2 + c_0)v = 0$$

Setting $c_1 = -2$ and $c_0 = -12$ we have $v_{xx} - 2v_t = 0$, a homogeneous heat equation with thermal diffusivity $\kappa = 1/2$.

2.16 Refer to Example 2.6. Apply Theorem 2.2 to the initial–boundary value problem

$$\begin{aligned} u_t - \kappa u_{xx} &= 0 && \text{for } x > 0,\ t > 0 \\ u(0, t) &= 0 && \text{for } t > 0 \\ u(x, 0) &= u_0 && \text{for } x > 0 \end{aligned}$$

obtaining two dimensionless groups, and then find a differential equation relating these groups.

The initial condition has added one parameter to (*2*) of Example 2.6, without increasing the number of fundamental dimensions involved ($\{u_0\} = \Theta$). Thus, $N = 5$ and $R = 3$ in the Buckingham Pi Theorem, which guarantees a solution $G(\pi_1, \pi_2) = 0$, with

$$\begin{aligned} 1 = \{\pi_\alpha\} &= \{u\}^{\gamma_{\alpha 0}}\{x\}^{\gamma_{\alpha 1}}\{t\}^{\gamma_{\alpha 2}}\{\kappa\}^{\gamma_{\alpha 3}}\{u_0\}^{\gamma_{\alpha 4}} \\ &= \Theta^{\gamma_{\alpha 0}} L^{\gamma_{\alpha 1}} T^{\gamma_{\alpha 2}} (L^2 T^{-1})^{\gamma_{\alpha 3}} \Theta^{\gamma_{\alpha 4}} \\ &= L^{\gamma_{\alpha 1} + 2\gamma_{\alpha 3}} T^{\gamma_{\alpha 2} - \gamma_{\alpha 3}} \Theta^{\gamma_{\alpha 0} + \gamma_{\alpha 4}} \end{aligned}$$

To make the exponents vanish, choose, for $\alpha = 1$, $\gamma_{10} = \gamma_{14} = 0$ and $\gamma_{11} = 1$; then $\gamma_{12} = \gamma_{13} = -1/2$. The dimensionless group

$$\pi_1 = \frac{x}{\sqrt{\kappa t}}$$

is known as the *Boltzmann variable* or *similarity variable* for the one-dimensional heat equation. For $\alpha = 2$, choose $\gamma_{20} = 1$, $\gamma_{24} = -1$, and $\gamma_{21} = \gamma_{22} = \gamma_{23} = 0$, to obtain

$$\pi_2 = \frac{u}{u_0}$$

Assuming that $G_{\pi_2}(\pi_1, \pi_2) \neq 0$, we can rewrite our solution as $\pi_2 = g(\pi_1)$ or $u = u_0 g(\pi_1)$. It then follows from the chain rule that

$$\frac{\partial u}{\partial t} = u_0 g'(\pi_1)\frac{\partial \pi_1}{\partial t} = -\frac{1}{2}u_0 g'(\pi_1)\frac{\pi_1}{t}$$

$$\frac{\partial u}{\partial x} = u_0 g'(\pi_1)\frac{\partial \pi_1}{\partial x} = u_0 g'(\pi_1)\frac{1}{\sqrt{\kappa t}}$$

$$\frac{\partial^2 u}{\partial x^2} = u_0 g''(\pi_1)\left(\frac{\partial \pi_1}{\partial x}\right)^2 = u_0 g''(\pi_1)\frac{1}{\kappa t}$$

The PDE $u_t - \kappa u_{xx} = 0$ now implies the following ordinary differential equation for π_2 as a function of π_1:

$$g''(\pi_1) + \frac{\pi_1}{2}g'(\pi_1) = 0 \tag{1}$$

2.17 Derive the similarity solution

$$u(x, t) = u_0 \operatorname{erf}\left(\frac{x}{2\sqrt{\kappa t}}\right)$$

for the initial–boundary value problem of Problem 2.16. The *error function*, erf z, is defined by

$$\operatorname{erf} z \equiv \frac{2}{\sqrt{\pi}} \int_0^z e^{-s^2}\, ds$$

Integrate (*1*) of Problem 2.16 once with respect to π_1 to find

$$\log|g'(\pi_1)| + \frac{\pi_1^2}{4} = \text{const.} \qquad \text{or} \qquad g'(\pi_1) = (\text{const.})\, e^{-\pi_1^2/4}$$

Integrate again and fix $g(0) = 0$ to make $u(0, t) = 0$:

$$g(\pi_1) = (\text{const.}) \int_0^{\pi_1} e^{-r^2/4}\, dr = (\text{const.}) \int_0^{\pi_1/2} e^{-s^2}\, ds = (\text{const.}) \operatorname{erf}(\pi_1/2)$$

Since $\lim_{z\to\infty} \operatorname{erf} z = 1$, the last constant should be set to unity, ensuring

$$\lim_{t\to 0^+} u(x, t) = u_0 \lim_{\pi_1\to+\infty} g(\pi_1) = u_0 \qquad \text{for } x > 0$$

This gives the required similarity solution.

2.18 Introduce dimensionless dependent and independent variables that transform the heat equation $u_t - \kappa u_{xx} = 0$ to the dimensionless form $v_\tau - v_{\xi\xi} = 0$.

The dimensions of x, t, u, and κ are:

$$\{x\} = L \qquad \{t\} = T \qquad \{u\} = \Theta \qquad \{\kappa\} = L^2T^{-1}$$

Choose dimensionless variables $\xi = x/x_0$, $\tau = t/t_0$, $v = u/u_0$; the PDE becomes

$$\frac{v_0}{t_0} v_\tau - \kappa \frac{u_0}{x_0^2} v_{\xi\xi} = 0 \qquad \text{or} \qquad v_\tau - \frac{\kappa}{x_0^2 t_0^{-1}} v_{\xi\xi} = 0$$

The coefficient of $v_{\xi\xi}$ is seen to be dimensionless. Therefore, by proper choice of x_0 and t_0, it can be made equal to unity, yielding the desired dimensionless equation. Note that this equation involves no fewer variables than did the original equation: neither ξ nor τ is the dimensionless Boltzmann variable of Problem 2.16.

Supplementary Problems

2.19 Describe the regions where the equation is hyperbolic (h.), parabolic (p.), and elliptic (e.).

(*a*) $u_{xx} - u_{xy} - 2u_{yy} = 0$

(*b*) $u_{xx} + 2u_{xy} + u_{yy} = 0$

(*c*) $2u_{xx} + 4u_{xy} + 3u_{yy} - 5u = 0$

(*d*) $u_{xx} + 2xu_{xy} + u_{yy} + (\cos xy)u_x = u$

(*e*) $yu_{xx} - 2u_{xy} + e^x u_{yy} + u = 3$

(*f*) $e^{xy}u_{xx} + (\sinh x)u_{yy} + u = 0$

(*g*) $xu_{xx} + 2xyu_{xy} - yu_{yy} = 0$

(*h*) $xu_{xx} + 2xyu_{xy} + yu_{yy} = 0$

2.20 Show that for (*2.2*) to be of ultrahyperbolic type, there must be at least four independent variables.

2.21 Let $p = p(x, y)$ be positive and continuously differentiable. Write out the principal part and classify the equation:

(*a*) $\nabla\cdot(p\nabla u) + qu = f$ (*b*) $u_t - \nabla\cdot(p\nabla u) + qu = f$ (*c*) $u_{tt} - \nabla\cdot(p\nabla u) + qu + ru = f$

2.22 Find the characteristic curves for the given equation.

(*a*) $u_{xx} - u_y + u = 0$
(*b*) $3u_{xx} + 8u_{xy} + 4u_{yy} = 0$
(*c*) $u_{xx} - u_{yy} + u_x + u_y = 0$
(*d*) $u_{xx} + yu_{yy} = 0$
(*e*) $u_{xx} - y^2u_{yy} = 0$
(*f*) $y^2u_{xx} - 2xyu_{xy} + x^2u_{yy} + yu_x + xu_y = 0$
(*g*) $y^3u_{xx} + u_{yy} = 0$
(*h*) $yu_{xx} + u_{yy} = 0$

2.23 Show that $5u_{xx} + 4u_{xy} + 4u_{yy} = 0$ is elliptic and use a transformation of independent variables to put it in canonical form.

2.24 Show that

$$3u_{x_1x_1} - 2u_{x_1x_2} + 2u_{x_2x_2} - 2u_{x_2x_3} + 3u_{x_3x_3} + 12u_{x_2} - 8u_{x_3} = 0$$

is elliptic and use a linear change of coordinates to transform its principal part to

$$u_{\xi_1\xi_1} + 3u_{\xi_2\xi_2} + 4u_{\xi_3\xi_3}$$

2.25 By rescaling the ξ-variables in Problem 2.24, transform the principal part to $\nabla^2 u$.

2.26 (*a*) Determine the type of the equation

$$3u_{x_1x_1} + 2u_{x_2x_2} + 3u_{x_3x_3} + 2u_{x_1x_3} = 0$$

and (*b*) use Theorem 2.3 to reduce it to canonical form.

2.27 Verify that the given equation is hyperbolic and then find a change of coordinates that reduces it to canonical form.

(*a*) $u_{xx} + 2u_{xy} - 8u_{yy} + u_x + 5 = 0$
(*b*) $u_{xx} + 2(x+1)u_{xy} + 2xu_{yy} = 0$
(*c*) $2u_{xx} + 4u_{xy} - u_{yy} = 0$
(*d*) $e^yu_{xx} + 2e^xu_{xy} - e^{2x-y}u_{yy} = 0$
(*e*) $(1+x^2)^2u_{xx} - (1+y^2)^2u_{yy} = 0$

2.28 Classify the given equation and then find a change of coordinates that puts it in canonical form.

(*a*) $u_{xx} + (1+x^2)^2u_{yy} = 0$
(*b*) $4u_{xx} - 4u_{xy} + 5u_{yy} = 0$
(*c*) $u_{xx} - 2u_{xy} + u_{yy} = 0$
(*d*) $u_{xx} - u_{yy} + u_x + u_y + 2x + y = 5$
(*e*) $x^2u_{xx} + 2xyu_{xy} + y^2u_{yy} = 4y^2$
(*f*) $x^2u_{xx} - y^2u_{yy} = xy$
(*g*) $(x^2u_x)_x - y^2u_{yy} = 0$
(*h*) $y^2u_{xx} - 2yu_{xy} + u_{yy} - u_y - 8y = 0$
(*i*) $u_{xx} + xyu_{yy} = 0$
(*j*) $yu_{xx} - xu_{yy} + u_x + yu_y = 0$
(*k*) $e^{2y}u_{xx} + 2e^{x+y}u_{xy} + e^{2x}u_{yy} = 0$
(*l*) $u_{xx} + (1+y)^2u_{yy} = 0$
(*m*) $xu_{xx} + 2\sqrt{xy}\, u_{xy} + yu_{yy} - u_y = 0 \quad (x > 0,\ y > 0)$
(*n*) $(\sin^2 x)u_{xx} + 2(\cos x)u_{xy} - u_{yy} = 0$
(*o*) $e^{2y}u_{xx} - x^2u_{yy} - u_x = 0$
(*p*) $(1+x^2)^2u_{xx} - 2(1+x^2)(1+y^2)u_{xy} + (1+y^2)^2u_{yy} = 0$

2.29 Use the results of Problem 2.3 to show that

$$B^2 - AC = (b^2 - ac)(\phi_x\psi_y - \phi_y\psi_x)^2$$

2.30 Show that if (*2.1*) is hyperbolic and in (*2.3*) ϕ and ψ are chosen to make A and C, the coefficients of $u_{\xi\xi}$ and $u_{\eta\eta}$, zero, then $2B$, the coefficient of $u_{\xi\eta}$, is not zero.

2.31 Show that $\quad ax^2u_{xx} + 2bxyu_{xy} + cy^2u_{yy} + dxu_x + eyu_y + fu = g \quad$ ($a, \ldots, f$ constants)

is transformed into a constant-coefficient equation under $\xi = \log x$, $\eta = \log y$.

2.32 Show that the two canonical forms for the wave equation, $u_{\xi\eta}$ and $u_{\alpha\alpha} - u_{\beta\beta}$, are related by a 45° rotation of coordinates.

2.33 Use a change of dependent variable to reduce $u_{xx} - u_t + 4u_x + 6u = 0$ to the heat equation, $v_{xx} - v_t = 0$.

Chapter 3

Qualitative Behavior of Solutions to Elliptic Equations

3.1 HARMONIC FUNCTIONS

Because the canonical example of an elliptic PDE is Laplace's equation, $\nabla^2 u = 0$, we begin with the following

Definition: A function $u = u(\mathbf{x})$ is *harmonic* in an open region, Ω, if u is twice continuously differentiable in Ω and satisfies Laplace's equation in Ω. u is harmonic in $\bar{\Omega}$, the closure of Ω, if u is harmonic in Ω and continuous in $\bar{\Omega}$.

EXAMPLE 3.1

(*a*) $u(x, y) = x^2 - y^2$ is harmonic in any region Ω of the xy-plane.

(*b*) $u(x, y, z) = (x^2 + y^2 + z^2)^{-1/2}$ is harmonic in any three-dimensional region which does not contain the origin. If Ω denotes the ball of radius one centered at $(1, 0, 0)$, then u is harmonic in Ω but not in $\bar{\Omega}$.

Let $\mathbf{x}_0$ be a point in Ω and let $B_R(\mathbf{x}_0)$ denote the open ball having center $\mathbf{x}_0$ and radius R. Let $\Sigma_R(\mathbf{x}_0)$ denote the boundary of $B_R(\mathbf{x}_0)$ and let $A(R)$ be the area of $\Sigma_R(\mathbf{x}_0)$.

EXAMPLE 3.2 Using calculus methods, one can show that in $\mathbf{R}^n$ the volume, $V_n(R)$, and the surface area, $A_n(R)$, of any ball of radius R are given by

$$A_n(R) = nR^{-1}V_n(R) = \begin{cases} \dfrac{n\pi^{n/2}}{(n/2)!} R^{n-1} & (n \text{ even}) \\ \dfrac{2n(2\pi)^{(n-1)/2}}{1\cdot 3\cdot 5\cdots n} R^{n-1} & (n \text{ odd}) \end{cases} \tag{1}$$

Definition: A function u has the *mean-value property* at a point $\mathbf{x}_0$ in Ω if

$$u(\mathbf{x}_0) = \frac{1}{A(R)} \int_{\Sigma_R} u(\mathbf{x})\, d\Sigma_R \tag{3.1}$$

for every $R > 0$ such that $B_R(\mathbf{x}_0)$ is contained in Ω.

Theorem 3.1: u is harmonic in an open region Ω if and only if u has the mean-value property at each $\mathbf{x}_0$ in Ω.

By Theorem 3.1, the state function, $u(\mathbf{x})$, for a physical system modeled by Laplace's equation is balanced throughout Ω in the sense that the value of u at any point $\mathbf{x}_0$ is equal to the average of u taken over the surface of any ball in Ω centered at $\mathbf{x}_0$. In other words, *Laplace's equation—and elliptic PDEs in general—are descriptive of physical systems in the equilibrium or steady state.*

Theorem 3.2: Let Ω be a bounded region with boundary S and let u be harmonic in $\bar{\Omega}$. If M and m are, respectively, the maximum and minimum values of $u(\mathbf{x})$ for $\mathbf{x}$ on S, then (*Weak Maximum-Minimum Principle*)

$$m \le u(\mathbf{x}) \le M \quad \text{for all } \mathbf{x} \text{ in } \bar{\Omega}$$

or, more precisely (*Strong Maximum-Minimum Principle*),

$$\textbf{either} \quad m < u(\mathbf{x}) < M \quad \text{for all } \mathbf{x} \text{ in } \Omega$$

$$\textbf{or else} \quad m = u(\mathbf{x}) = M \quad \text{for all } \mathbf{x} \text{ in } \bar{\Omega}$$

EXAMPLE 3.3 If Ω is not bounded, then the (weak) maximum-minimum principle need not hold. In fact,

$$u(x, y) = e^x \sin y$$

satisfies Laplace's equation in $\Omega \equiv \{(x, y): -\infty < x < \infty, 0 < y < \pi\}$, and u is zero on the boundary of Ω, so that $m = M = 0$. But $u(x, y)$ is not identically zero in Ω.

Definition: A function $u(\mathbf{x})$ is *analytic* in Ω if u is in $C^\infty(\Omega)$ and, in a neighborhood of each point $\mathbf{x}$ in Ω, u equals its Taylor series expansion about $\mathbf{x}$.

Theorem 3.3: If u is harmonic in a region Ω, then u is analytic in Ω.

Theorem 3.3 implies that solutions of Laplace's equation cannot exhibit discontinuities in the value of u or of any of its derivatives. This is again characteristic of a physical system in the steady state (any initial disturbances having been smoothed out).

There is a strong connection between harmonic functions in the plane and analytic functions of a complex variable. This connection provides a partial converse to Theorem 3.3.

Theorem 3.4: If $f(z) \equiv f(x + iy) = u(x, y) + iv(x, y)$ is an analytic function of the complex variable z in Ω, then u and v are harmonic in Ω.

Theorem 3.5: A function $u(x, y)$ is harmonic in a simply connected region Ω if and only if, in Ω, u is the real part of some analytic function $f(z)$.

EXAMPLE 3.4 If $f(z) = z^2 = (x + iy)^2 = x^2 - y^2 + i2xy$, then

$$u(x, y) = \operatorname{Re} f(z) = x^2 - y^2 \qquad \text{and} \qquad v(x, y) = \operatorname{Re} -if(z) = 2xy$$

each satisfy Laplace's equation in the plane.

3.2 EXTENDED MAXIMUM-MINIMUM PRINCIPLES

Definition: A continuous function u is *subharmonic* in a region Ω if, for every $\mathbf{x}_0$ in Ω, $u(\mathbf{x}_0)$ is less than or equal to the average of the u-values on the boundary of any ball, $B_R(\mathbf{x}_0)$, in Ω:

$$u(\mathbf{x}_0) \leq \frac{1}{A(R)} \int_{\Sigma_R} u(\mathbf{x})\, d\Sigma_R \tag{3.2}$$

A *superharmonic* function satisfies (*3.2*) with the inequality reversed; it is thus the negative of a subharmonic function.

If u is C^2, then u is subharmonic if and only if $\nabla^2 u \geq 0$, and u is superharmonic if and only if $\nabla^2 u \leq 0$. Clearly, a harmonic function is both subharmonic and superharmonic, and conversely. The maximum-minimum principle, Theorem 3.2, extends to subharmonic and superharmonic functions, as follows:

Theorem 3.6: For Ω, S, m, and M as in Theorem 3.2,

(i) if $\nabla^2 u \geq 0$ in Ω, then $u(\mathbf{x}) < M$ for all $\mathbf{x}$ in Ω or else $u(\mathbf{x}) \equiv M$ in $\bar{\Omega}$;

(ii) if $\nabla^2 u \leq 0$ in Ω, then $u(\mathbf{x}) > m$ for all $\mathbf{x}$ in Ω or else $u(\mathbf{x}) \equiv m$ in $\bar{\Omega}$.

Results similar to Theorem 3.6 hold for elliptic equations more general than Laplace's equation.

Definition: The linear operator

$$L[\;] \equiv \sum_{i,j=1}^{n} a_{ij}(\mathbf{x}) \frac{\partial^2[\;]}{\partial x_i \partial x_j} + \sum_{i=1}^{n} b_i(\mathbf{x}) \frac{\partial[\;]}{\partial x_i} + c(\mathbf{x})[\;] \tag{3.3}$$

is *uniformly elliptic* in Ω if there exists a positive constant λ such that

$$\sum_{i,j=1}^{n} a_{ij}(\mathbf{x})\zeta_i\zeta_j \geq \lambda \sum_{i=1}^{n} \zeta_i^2 \tag{3.4}$$

for all $(\zeta_1, \zeta_2, \ldots, \zeta_n)$ in $\mathbf{R}^n$ and all $\mathbf{x}$ in Ω.

Observe that, if (*3.4*) holds, matrix $\mathbf{A}(\mathbf{x})$ must be positive definite in Ω, which means that $L[\ ]$ is elliptic in Ω, with $Z=0$ and $P=n$ (Section 2.1). On the other hand, assuming that an elliptic operator has all eigenvalues positive (if all are negative, multiply the operator by -1), we have

Theorem 3.7: If $L[\ ]$ is elliptic in $\bar{\Omega}$, it is uniformly elliptic in $\bar{\Omega}$ (*a fortiori*, in Ω).

Theorem 3.8: Let Ω, S, m, and M be as in Theorem 3.2. Suppose in (*3.3*) that $c=0$, $L[\ ]$ is uniformly elliptic in Ω, and a_{ij} and b_i are continuous in $\bar{\Omega}$.

(i) If $L[u] \geq 0$ in Ω, then $u(\mathbf{x}) < M$ for all $\mathbf{x}$ in Ω or else $u(\mathbf{x}) \equiv M$ in $\bar{\Omega}$.

(ii) If $L[u] \leq 0$ in Ω, then $u(\mathbf{x}) > m$ for all $\mathbf{x}$ in Ω or else $u(\mathbf{x}) \equiv m$ in $\bar{\Omega}$.

(iii) If $L[u] = 0$ in Ω, then $m < u(\mathbf{x}) < M$ for all $\mathbf{x}$ in Ω or else $m \equiv u(\mathbf{x}) \equiv M$ in $\bar{\Omega}$.

Theorem 3.9: Let Ω be a bounded region with boundary S. Suppose that $u(\mathbf{x})$ satisfies $L[u]=f$ in Ω, where $L[\ ]$ is uniformly elliptic in Ω and has coefficients a_{ij}, b_i, c which are continuous in $\bar{\Omega}$. Suppose further that, in $\bar{\Omega}$, $c \leq 0$ and f is continuous.

(i) If $f \leq 0$ in $\bar{\Omega}$ and $u(\mathbf{x})$ is nonconstant, then any negative minimum of $u(\mathbf{x})$ must occur on S and not in Ω.

(ii) If $f \geq 0$ in $\bar{\Omega}$ and $u(\mathbf{x})$ is nonconstant, then any positive maximum of $u(\mathbf{x})$ must occur on S and not in Ω.

Theorem 3.10: In the boundary value problem $L[u]=f$ in Ω, $u=g$ on S, suppose that the hypotheses of Theorem 3.9 hold and that g is continuous on S. Let $|a_{ij}|$, $|b_i|$, $|c|$ all be bounded by the constant K, and let λ be as in (*3.4*). If u is C^2 in Ω and C^0 in $\bar{\Omega}$ and if u satisfies the boundary value problem, then, for all $\mathbf{x}_0$ in $\bar{\Omega}$,

$$|u(\mathbf{x}_0)| \leq \max_{\mathbf{x}\in S} |g(\mathbf{x})| + M \max_{\mathbf{x}\in\bar{\Omega}} |f(\mathbf{x})|$$

where $M = M(\lambda, K)$.

3.3 ELLIPTIC BOUNDARY VALUE PROBLEMS

Since elliptic equations in general model physical systems that are not changing with time, the associated auxiliary conditions are typically boundary conditions (Section 1.3).

EXAMPLE 3.5 If Ω is the region $0<x<1, 0<y<1$, then the boundary value problem

$$\begin{aligned} u_{xx} + u_{yy} &= f(x, y) && \text{in } \Omega \\ u(x, 0) = u(x, 1) &= 0 && \text{on } 0<x<1 \\ u_x(0, y) &= 1 && \text{on } 0<y<1 \\ 2yu(1, y) - 5u_x(1, y) &= y^2 && \text{on } 0<y<1 \end{aligned}$$

has a homogeneous Dirichlet condition on the portion of the boundary where $y=0$ or $y=1$. A Neumann condition holds on the part of the boundary where $x=0$. On the edge $x=1$, u satisfies a mixed condition.

A *classical solution* of a (elliptic) boundary value problem satisfies the PDE $L[u]=f$ in Ω, is C^2 in Ω, and is C^0 in $\bar{\Omega}$ (for a Dirichlet condition on S) or C^1 in $\bar{\Omega}$ (for a Neumann or mixed condition on S). It is possible to relax somewhat the smoothness conditions; such *weak solutions* are discussed briefly in Chapter 5. When no qualifier is used, a solution is understood to be a classical solution.

If the region Ω is unbounded, then, in addition to the boundary conditions, a solution is generally required to satisfy a *condition at infinity*, which is frequently dimension dependent.

EXAMPLE 3.6

(*a*) If Ω is the half-plane $y > 0$, then, in the boundary value problem

$$u_{xx} + u_{yy} = f(x, y) \quad \text{in } \Omega$$
$$u(x, 0) = g(x) \quad \text{on } S$$

the usual condition at infinity is that u be bounded,

$$|u(x, y)| < M = \text{const.}$$

for $x^2 + y^2 \to \infty$, $y > 0$.

(*b*) If Ω is the half-space $z > 0$, then, in the boundary value problem

$$u_{xx} + u_{yy} + u_{zz} = f(x, y, z) \quad \text{in } \Omega$$
$$u(x, y, 0) = g(x, y) \quad \text{on } S$$

the typical condition is that u vanish at infinity,

$$|u(x, y, z)| \to 0$$

for $x^2 + y^2 + z^2 \to \infty$, $z > 0$.

The three conditions for a well-posed problem were stated in Section 1.3. For many elliptic boundary value problems, maximum-minimum principles like Theorems 3.8–3.10 or an energy-integral argument can be used to show that conditions (ii) (uniqueness) and (iii) (continuous dependence on data) hold. See Problem 3.14.

EXAMPLE 3.7 Let Ω be a bounded region. The Dirichlet boundary value problem

$$\nabla^2 u = f \quad \text{in } \Omega$$
$$u = g \quad \text{on } S$$

the Neumann problem ($c < 0$)

$$\nabla^2 u + cu = f \quad \text{in } \Omega$$
$$\frac{\partial u}{\partial n} = g \quad \text{on } S$$

and the mixed problem ($\alpha\beta > 0$)

$$\nabla^2 u = f \quad \text{in } \Omega$$
$$\alpha u + \beta \frac{\partial u}{\partial n} = g \quad \text{on } S$$

each have at most one solution and each solution depends continuously on the data functions f and g. Nonetheless, there are mathematically and physically significant elliptic boundary value problems that are ill-posed with regard to conditions (ii) and (iii).

Condition (i), the existence of a solution to a boundary value problem, is generally more difficult to establish. The most satisfactory way to show that a solution exists is to construct it; the solutions to a number of elliptic boundary value problems are constructed in Chapters 7 and 8. A particularly important constructive existence result, for Laplace's equation, is given by

Theorem 3.11 *(Poisson's Integral Formula):* In $\mathbf{R}^n$, if $g(\xi)$ is continuous on Σ_R: $|\xi| = R$ and $A_n(1)$—see (*1*) of Example 3.2—is the area of the unit sphere, then

$$u(\mathbf{x}) = \begin{cases} \dfrac{R^2 - |\mathbf{x}|^2}{RA_n(1)} \displaystyle\int_{\Sigma_R} \frac{g(\boldsymbol{\xi})}{|\mathbf{x} - \boldsymbol{\xi}|^n} d\Sigma_R & |\mathbf{x}| < R \\ g(\mathbf{x}) & |\mathbf{x}| = R \end{cases} \tag{3.5}$$

is a solution to the boundary value problem

$$\nabla^2 u = 0 \quad \text{in } |\mathbf{x}| < R$$
$$u = g \quad \text{on } |\mathbf{x}| = R$$

See Problems 8.7 and 8.38 for a derivation of *(3.5)*. Observe that for $\mathbf{x} = \mathbf{0}$, *(3.5)* coincides with *(3.1)*, the mean-value property.

Sometimes, the *nonexistence* of a solution to a boundary value problem may be demonstrated immediately.

EXAMPLE 3.8 For the Neumann problem

$$\nabla^2 u = f \quad \text{in } \Omega$$
$$\frac{\partial u}{\partial n} = g \quad \text{on } S$$

It follows from the divergence theorem that

$$\int_\Omega f\, d\Omega = \int_\Omega \nabla^2 u\, d\Omega = \int_S \frac{\partial u}{\partial n} dS = \int_S g\, dS$$

Thus, if the *consistency condition* $\int_\Omega f\, d\Omega = \int_S g\, dS$ is not satisfied, the Neumann problem cannot have a solution. Problem 3.20 gives a consistency condition for an elliptic mixed problem.

Solved Problems

3.1 Show that if u is harmonic in an open region Ω of $\mathbf{R}^n$, then u has the mean-value property in Ω.

Suppose that Ω includes the ball $B_\rho(\mathbf{x}_0)$ for $0 \le \rho \le R$. By the divergence theorem,

$$0 = \int_{B_\rho} \nabla^2 u\, dB_\rho = \int_{\Sigma_\rho} \frac{\partial u}{\partial r} d\Sigma_\rho \tag{1}$$

in which we have introduced the radial coordinate $r = |\boldsymbol{\xi} - \mathbf{x}_0|$; $\boldsymbol{\xi}$ being a general point of $\mathbf{R}^n$. Now,

$$\left.\frac{\partial u(r, \ldots)}{\partial r}\right|_{r=\rho} = \frac{\partial u(\rho, \ldots)}{\partial \rho}$$

and (see Example 3.2) $d\Sigma_\rho = \rho^{n-1}\, d\Sigma_1$, where Σ_1 denotes the surface of the unit sphere. Therefore, *(1)* implies

$$0 = \int_{\Sigma_1} \frac{\partial u}{\partial \rho} d\Sigma_1 = \frac{d}{d\rho}\left(\int_{\Sigma_1} u\, d\Sigma_1\right) \tag{2}$$

Integration of *(2)* from $\rho = 0$ to $\rho = R$, where $R = |\mathbf{x} - \mathbf{x}_0|$, yields

$$\begin{aligned}0 &= \int_{\Sigma_1} u(\mathbf{x})\,d\Sigma_1 - \int_{\Sigma_1} u(\mathbf{x}_0)\,d\Sigma_1 \\ &= \frac{1}{R^{n-1}}\int_{\Sigma_R} u(\mathbf{x})\,d\Sigma_R - u(\mathbf{x}_0)A_n(1) \\ &= A_n(1)\left[\frac{1}{A_n(R)}\int_{\Sigma_R} u(\mathbf{x})\,d\Sigma_R - u(\mathbf{x}_0)\right] \end{aligned} \tag{3}$$

since, from Example 3.2, $A_n(R) = A_n(1)R^{n-1}$. The mean-value property of u follows at once.

The converse theorem, that the mean-value property implies harmonicity, can be proved by reversing the above argument *if the prior assumption is made that u is in $C^2(\Omega)$*. A way around such an assumption is shown in Problems 3.2–3.4.

3.2 Suppose that u has the mean-value property in the ball $B_R(\mathbf{x}_0)$. If $u \le M$ in B_R and $u(\mathbf{x}_0) = M$, show that $u = M$ everywhere in B_R.

From the mean-value property and the given conditions on u,

$$M = u(\mathbf{x}_0) = \frac{1}{A(r)}\int_{\Sigma_r} u(\mathbf{x})\,d\Sigma_r \le M \tag{1}$$

for $r \le R$. As equality must hold throughout (*1*), $u(\mathbf{x}) = M$ at every point of Σ_r. Thus, $u(\mathbf{x}) = M$ for all $\mathbf{x}$ in $B_R(\mathbf{x}_0)$.

3.3 Suppose that u has the mean-value property in a bounded region Ω and that u is continuous in $\bar{\Omega}$. Show that if u is nonconstant in Ω, then u attains its maximum and minimum values on the boundary of Ω, not in the interior of Ω.

Since u is continuous in the closed, bounded region $\bar{\Omega}$, u attains its maximum, M, and its minimum, m, somewhere in $\bar{\Omega}$. We will show that if u attains its maximum at an interior point of Ω, then u is constant in $\bar{\Omega}$.

Assume that $u(\mathbf{x}_0) = M$, with $\mathbf{x}_0$ in Ω, and let $\mathbf{x}^*$ be any other point in Ω. Let Γ be a polygonal path in Ω joining $\mathbf{x}_0$ and $\mathbf{x}^*$ and let d be the minimum distance separating Γ and S, the boundary of Ω:

$$d = \min\{|\mathbf{x} - \mathbf{y}| : \mathbf{x} \text{ on } \Gamma,\ \mathbf{y} \text{ on } S\}$$

There exists a sequence of balls $B_R(\mathbf{x}_i)$, $i = 0, 1, \ldots, n$, with $\mathbf{x}_i$ on Γ, satisfying $R \le d$, $\mathbf{x}_{i+1}$ in $B_R(\mathbf{x}_i)$, $\mathbf{x}^*$ in $B_R(\mathbf{x}_n)$. See Fig. 3-1.

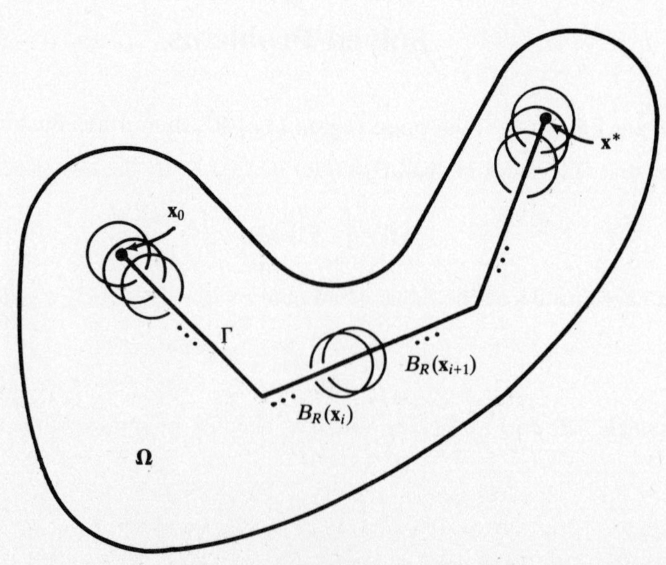

Fig. 3-1

Problem 3.2 shows that u is identically equal to M in each $B_R(\mathbf{x}_i)$, $i = 0, 1, \ldots, n$; hence, $u(x^*) = M$. Since x^* was arbitrary, u must be equal to M throughout Ω and, by continuity of u, throughout $\bar{\Omega}$. This shows that if u is not a constant in Ω, then u can attain its maximum value only on the boundary of Ω.

The above argument, applied to $-u$, establishes that if u is nonconstant, it can attain its minimum value only on S.

3.4 Show that if u has the mean-value property in an open region Ω, then u is harmonic in Ω.

Let $\mathbf{x}_0$ be any point in Ω, and let $B_R(\mathbf{x}_0)$ be wholly contained in Ω. Since Laplace's equation is invariant under a translation of coordinates, we shall suppose $\mathbf{x}_0 = \mathbf{0}$. If v is defined in $B_R(\mathbf{0})$ by Poisson's integral formula, (*3.5*), with $v = u$ on the boundary of B_R, then, by Theorem 3.11, v is harmonic in $\bar{B}_R$. By Problem 3.1, v has the mean-value property in B_R. Because both u and v have the mean-value property in B_R, $w \equiv u - v$ has the mean-value property in B_R. Since $w = 0$ on the boundary of B_R, Problem 3.3 shows that $w = 0$ throughout $\bar{B}_R$. Thus, u is identically equal to the harmonic function v in $\bar{B}_R$, and so u is harmonic at $\mathbf{0}$, which, from the above, represents any point in Ω.

The above proof has an important implication: *Any harmonic function can be expressed in terms of its boundary values on a sphere by Poisson's formula.* Equivalently: *The Dirichlet problem*

$$\nabla^2 u = 0 \qquad \text{in } |\mathbf{x}| < R$$
$$u = g \qquad \text{on } |\mathbf{x}| = R$$

has a unique *solution.*

3.5 Establish the maximum-minimum principle for harmonic functions, Theorem 3.2.

Problem 3.3 establishes Theorem 3.2 for functions having the mean-value property. But, by Problems 3.1 and 3.4, these functions are exactly the harmonic functions.

3.6 Show that if $u(x, y)$ is subharmonic, $\nabla^2 u \geq 0$, in a bounded region Ω and $u \leq M$ on S, the boundary of Ω, then $u \leq M$ everywhere in Ω.

Since Ω is bounded, it can be enclosed in a circle, of radius R, centered at the origin. Let $\epsilon > 0$ be arbitrary and define

$$v(x, y) \equiv u(x, y) + \epsilon(x^2 + y^2) \qquad \text{in } \Omega \tag{1}$$

From (*1*),

$$\nabla^2 v = \nabla^2 u + 2\epsilon > 0 \qquad \text{in } \Omega \tag{2}$$

v can attain a maximum in $\bar{\Omega}$ only on S; for, at an interior maximum, $v_{xx} \leq 0$ and $v_{yy} \leq 0$, which would contradict (*2*). From $u \leq M$ on S, it follows that $v \leq M + \epsilon R^2$ on S and, since v attains its maximum on S, $v \leq M + \epsilon R^2$ everywhere in Ω. By (*1*), $u \leq v$; so

$$u \leq M + \epsilon R^2 \qquad \text{in } \Omega \tag{3}$$

for any $\epsilon > 0$. It follows that u cannot exceed M in Ω; if it did, then, for sufficiently small ϵ, (*3*) would be violated.

The above argument provides an alternate proof that a harmonic function on a bounded region Ω must attain its maximum on the boundary of Ω.

3.7 Show that if $u(x, y)$ is harmonic in a bounded region $\bar{\Omega}$ and u is continuously differentiable in $\bar{\Omega}$, then $|\nabla u|^2$ attains its maximum on S, the boundary of Ω.

Let $w \equiv |\nabla u|^2 = u_x^2 + u_y^2$. Since u is C^1 in $\bar{\Omega}$, w is continuous on S. Therefore, since S is closed and bounded, there is a value M assumed by w on S such that

$$w \leq M \qquad \text{on } S \tag{1}$$

Calculating $\nabla^2 w = \nabla \cdot \nabla w$, we find

$$\nabla^2 w = 2(u_x u_{xxx} + u_{xx}^2 + u_y u_{xxy} + u_{xy}^2 + u_x u_{xyy} + u_{xy}^2 + u_y u_{yyy} + u_{yy}^2)$$

Since u is harmonic in Ω, $u_{xx} + u_{yy} = 0$ in Ω, and

$$u_x u_{xxx} + u_x u_{xyy} = u_x(u_{xx} + u_{yy})_x = 0$$
$$u_y u_{xxy} + u_y u_{yyy} = u_y(u_{xx} + u_{yy})_y = 0$$

Thus,

$$\nabla^2 w = u_{xx}^2 + 2u_{xy}^2 + u_{yy}^2 \geq 0 \qquad \text{in } \Omega \tag{2}$$

or w is subharmonic in Ω. Problem 3.6 and (*1*) now imply that $w = |\nabla u|^2 \leq M$ throughout $\bar{\Omega}$.

This result has a number of interesting physical interpretations; e.g., for steady-state heat flow in a homogeneous medium Ω, the heat flux vector of maximum magnitude must occur on the boundary of Ω.

3.8 Show that the solution to the Dirichlet problem indicated in Fig. 3-2 satisfies

$$0 < u(x, y) < x(2 - x - 2y)$$

in Ω.

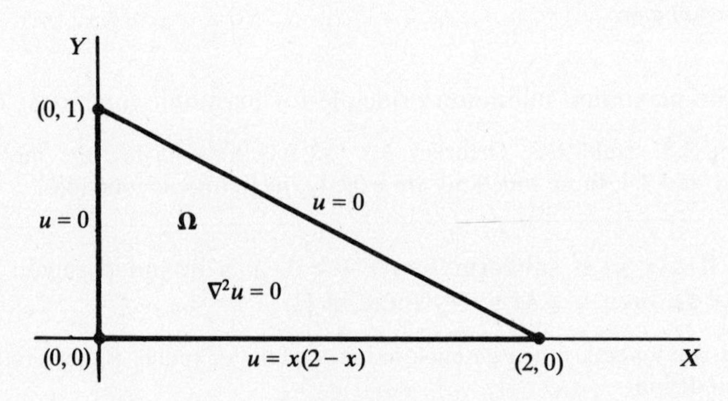

Fig. 3-2

Let $v(x, y) \equiv x(2 - x - 2y)$ and note that $\nabla^2 v = -2$ in Ω. Then, if

$$w(x, y) \equiv u(x, y) - v(x, y)$$

in $\bar{\Omega}$, we have

$$\nabla^2 w = 2 > 0 \qquad \text{in } \Omega$$
$$w = 0 \qquad \text{on } S$$

From Theorem 3.6(i), $w(x, y) < 0$ in Ω; that is,

$$u(x, y) < v(x, y) = x(2 - x - 2y) \qquad \text{in } \Omega$$

Since the minimum of u on S is zero, it follows from Theorem 3.6(ii) that $u(x, y) > 0$ in Ω.

3.9 Let $u(x, y)$ be continuous in the closure of a bounded planar region Ω and let the linear operator

$$L[u] \equiv u_{xx} + u_{yy} + b_1(x, y)u_x + b_2(x, y)u_y$$

be defined in Ω, in which b_1 and b_2 are continuous functions. Prove: (*a*) If $u(x, y)$ satisfies $L[u] = f$ in Ω, with $f > 0$, then u attains its maximum on S, the boundary of Ω, not inside Ω. (*b*) If $u(x, y)$ satisfies $L[u] \geq 0$ in Ω, and if $u \leq M$ on S, then $u \leq M$ in $\bar{\Omega}$.

(*a*) Since u is continuous in the closed bounded region $\bar{\Omega}$, it must attain its maximum, M, somewhere in $\bar{\Omega}$. Let (x_0, y_0) be an interior point of Ω and assume that $u(x_0, y_0) = M$. From calculus, $u_x = 0 = u_y$, $u_{xx} \leq 0$, and $u_{yy} \leq 0$, at (x_0, y_0); hence, b_1 and b_2 being bounded at (x_0, y_0),

$$L[u] = u_{xx} + u_{yy} \leq 0$$

which contradicts $L[u] = f > 0$. We must, then, admit that the maximum is achieved on S.

(*b*) Because b_1 is bounded (below) in Ω, there exists a constant α such that

$$L[e^{\alpha x}] = \alpha e^{\alpha x}(\alpha + b_1) > 0 \qquad \text{in } \Omega \tag{1}$$

For arbitrary $\epsilon > 0$, define $v \equiv u + \epsilon e^{\alpha x}$. Let R be chosen large enough so that the circle centered at the origin with radius R encloses Ω. Since $u \leq M$ on S,

$$v \leq M + \epsilon e^{\alpha R} \qquad \text{on } S \tag{2}$$

From $L(u) \geq 0$ and (*1*), $L[v] > 0$. It now follows from the result of (*a*) that

$$u < v \leq M + \epsilon e^{\alpha R} \qquad \text{in } \bar{\Omega} \tag{3}$$

for arbitrary $\epsilon > 0$, and this implies that $u \leq M$ in $\bar{\Omega}$.

3.10 Under the hypotheses of Problem 3.9, with b_1 and b_2 assumed continuous in $\bar{\Omega}$, show that a solution of $L[u] \geq 0$ cannot attain its maximum, M, at an interior point of Ω unless $u \equiv M$ in $\bar{\Omega}$.

Let M denote the maximum of u on $\bar{\Omega}$ and suppose that u assumes the value M somewhere in Ω. If $u \not\equiv M$ in $\bar{\Omega}$, there exists a disk B_1 in Ω that contains on its boundary an interior point (ξ, η) of Ω where $u(\xi, \eta) = M$ and such that $u(x, y) < M$ inside B_1. Let B_2 be a disk of radius R satisfying $B_2 \subset B_1$, with the boundary of B_2 tangent to the boundary of B_1 at (ξ, η); i.e., $\Sigma_1 \cap \Sigma_2 = (\xi, \eta)$. (See Fig. 3-3.) Since the PDE $L[u] = 0$ is invariant under a translation of coordinates, the origin can be taken to be the center of B_2. Let B_3 be a disk in Ω centered at (ξ, η), of radius less than R. Divide the boundary, Σ_3, of B_3 into the two arcs

$$\sigma_1 \equiv \Sigma_3 \cap \bar{B}_2 \qquad \sigma_2 \equiv \Sigma_3 - \sigma_1$$

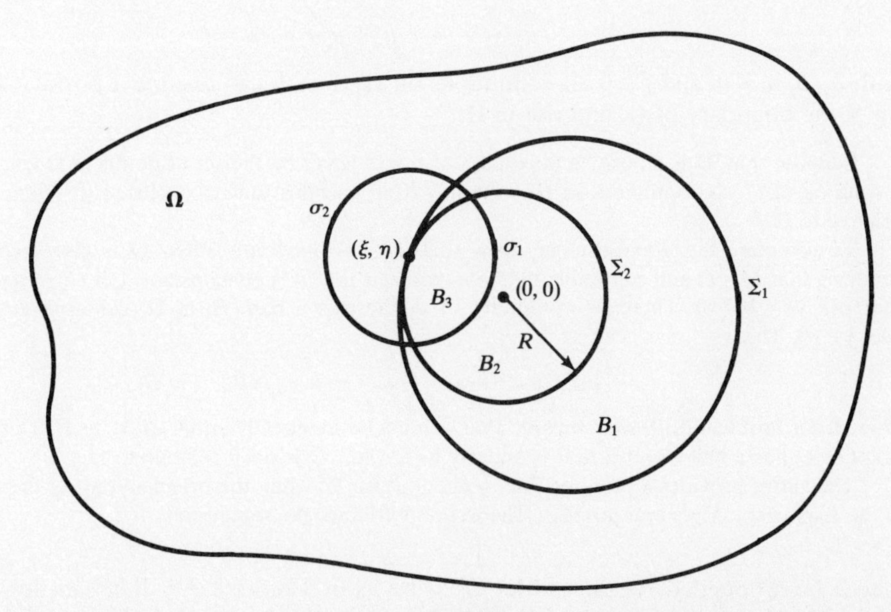

Fig. 3-3

By hypothesis u is continuous on the *closed* arc σ_1, at every point of which $u < M$ (because $\sigma_1 \subset B_1$). It follows that u is bounded away from M on σ_1:

$$u(x, y) \leq M - \epsilon_0 \qquad \text{for some } \epsilon_0 > 0 \text{ and all } (x, y) \text{ on } \sigma_1 \tag{1}$$

Define the comparison function

$$v \equiv e^{-\alpha r^2} - e^{-\alpha R^2}$$

where $r^2 = x^2 + y^2$ and $\alpha > 0$. Note that $v > 0$ in B_2, $v = 0$ on Σ_2, and $v < 0$ outside $\bar{B}_2$. A calculation shows that

$$L[v] = \{4\alpha^2(x^2 + y^2) - 2\alpha(2 + b_1 x + b_2 y)\} e^{-\alpha r^2}$$

Hence, for sufficiently large α, $L[v] > 0$ throughout B_3. From (*1*) it follows that there exists a constant $\beta > 0$ such that

$$u + \beta v < M \qquad \text{on } \Sigma_3 \tag{2}$$

Also, $\beta > 0$, $L[u] \geq 0$, and $L[v] > 0$ in B_3 imply

$$L[u + \beta v] > 0 \qquad \text{in } B_3 \tag{3}$$

Since $u = M$ and $v = 0$ at (ξ, η), (*2*) shows that $u + \beta v$ must assume a maximum greater than or equal to M at some point $(\bar{\xi}, \bar{\eta})$ inside B_3. The necessary conditions for this maximum are

$$u_x + \beta v_x = 0 = u_y + \beta v_y \tag{4}$$

$$u_{xx} + \beta v_{xx} \leq 0 \qquad u_{yy} + \beta v_{yy} \leq 0 \tag{5}$$

However, from (*3*) and (*4*) and the boundedness of the coefficients b_1 and b_2, we conclude that

$$(u_{xx} + \beta v_{xx}) + (u_{yy} + \beta v_{yy}) > 0$$

which contradicts (*5*). Thus, the original assumptions of an interior maximum and a nonconstant function are incompatible.

Problems 3.9 and 3.10 provide a proof of Theorem 3.8(i) for $\mathbf{R}^2$ in the case that the principal part of $L[\ \ ]$ is the Laplacian. The arguments employed contain the essential ingredients of the general proof of Theorem 3.8.

3.11 For a bounded region Ω, show that if u is a nonconstant solution of

$$u_{xx} + u_{yy} + b_1 u_x + b_2 u_y + cu = f \tag{1}$$

with $b_1, b_2, c \leq 0$, and $f \geq 0$ all continuous on $\bar{\Omega}$, then u can assume a positive maximum only on S, the boundary of Ω, and not in Ω.

Suppose that M is a positive maximum of u and let Q be the set of points in Ω where $u = M$. Since a solution of (*1*) is continuous in Ω, we know from calculus that Q is closed (its complement is open) relative to Ω.

To complete the argument, we show that if Q is nonempty, then Q is also open relative to Ω, implying that $Q = \Omega$ and contradicting the hypothesis that u is nonconstant. Let $(\bar{x}, \bar{y})$ be a point in Q, so that $u(\bar{x}, \bar{y}) = M > 0$. Then, by continuity of u, there is a ball, B, in Ω, centered at $(\bar{x}, \bar{y})$, in which $u(x, y) > 0$. Thus,

$$u_{xx} + u_{yy} + b_1 u_x + b_2 u_y = -cu + f \geq 0 \qquad \text{in } B$$

Now, from Problem 3.10, we conclude that u must be identically equal to M in B. This shows that any point in Q has a ball about it that is entirely in Q; i.e., Q is open relative to Ω.

The above provides a proof of Theorem 3.9(ii) for $\mathbf{R}^2$ when the principal part of the elliptic operator is the Laplacian. A general proof of Theorem 3.9 follows the same lines.

3.12 Let Ω be a bounded region and let $L[\ \ ]$ be as in Theorem 3.9. If u satisfies $L[u] = f$ in Ω, $u = g$ on S, and if v satisfies $L[v] \leq f$ in Ω, $v \geq g$ on S, prove that $u \leq v$ in $\bar{\Omega}$.

Since $L[u-v]=f-L[v]\geq 0$ in Ω and $u-v\leq 0$ on S, it follows from Theorem 3.9 that $u-v$ cannot have a positive maximum in $\bar{\Omega}$, which means that $u-v$ cannot assume a positive value in $\bar{\Omega}$.

Similarly, we can show that if $L[w]\geq f$ in Ω and $w\leq g$ on S, then $u\geq w$ in $\bar{\Omega}$.

3.13 In a bounded region Ω, if u satisfies

$$u_{xx}+u_{yy}=f \qquad \text{in } \Omega \tag{1}$$

$$u=g \qquad \text{on } S \tag{2}$$

show that, in $\bar{\Omega}$,

$$|u|\leq \max_{(x,y)\in S}|g(x,y)|+M\max_{(x,y)\in\bar{\Omega}}|f(x,y)| \tag{3}$$

where M is a constant which depends on the size of Ω.

If $\bar{\Omega}$ is not contained in the region $x\geq 0$, a translation of coordinates will leave the forms of (*1*) and (*2*) invariant and will result in $x\geq 0$ throughout $\bar{\Omega}$. Thus, with no loss of generality, we can assume that there exists a positive number a such that $0\leq x\leq a$ for all (x, y) in $\bar{\Omega}$. Define the comparison function

$$v\equiv \max_S|g|+(e^a-e^x)\max_{\bar{\Omega}}|f|$$

Because $e^a-e^x\geq 0$ in $\bar{\Omega}$, $v\geq \max_S |g|$ and

$$v\geq g \qquad \text{on } S \tag{4}$$

Calculating $L[v]=v_{xx}+v_{yy}$, we find, since $e^x\geq e^0=1$ and $f\geq -\max_{\bar{\Omega}}|f|$,

$$L[v]=-e^x\max_{\bar{\Omega}}|f|\leq f \qquad \text{in } \Omega \tag{5}$$

Inequalities (*4*) and (*5*) and Problem 3.12 imply $u\leq v$ in $\bar{\Omega}$. Similarly, with $w=-v$, we have $w\leq g$ on S and $L[w]\geq f$ in Ω, so that $u\geq -v$ in $\bar{\Omega}$. Consequently, in $\bar{\Omega}$,

$$|u|\leq v\leq \max_S|g|+(e^a-1)\max_{\bar{\Omega}}|f|$$

since $e^x\geq 1$ in $\bar{\Omega}$. This establishes (*3*), with $M=e^a-1$.

We term (*3*) an *a priori estimate* of u. When some knowledge of the solution is incorporated, much sharper estimates are possible.

3.14 Let Ω be a bounded region. Show that the Dirichlet problem

$$\nabla^2 u=f \qquad \text{in } \Omega$$

$$u=g \qquad \text{on } S$$

has at most one solution and, if it exists, that the solution depends continuously on the data f and g.

The difference $v\equiv u_1-u_2$ of two solutions would satisfy the homogeneous problem

$$\nabla^2 v=0 \qquad \text{in } \Omega$$

$$v=0 \qquad \text{on } S$$

The maximum-minimum principle, Theorem 3.2, implies $v\equiv 0$, or $u_1\equiv u_2$, in $\bar{\Omega}$.

To establish the continuous dependence of the solution, let $\hat{u}$ satisfy

$$\nabla^2 \hat{u} = \hat{f} \qquad \text{in } \Omega$$
$$\hat{u} = \hat{g} \qquad \text{on } S$$

Theorem 3.10 implies

$$\max_{\bar{\Omega}} |u - \hat{u}| \leq \max_{S} |g - \hat{g}| + M \max_{\bar{\Omega}} |f - \hat{f}| \qquad (M = \text{const.})$$

which shows that small changes in the functions f and g produce small changes in the solution u.

3.15 Let Ω be a bounded region. Show that the Neumann problem

$$\nabla^2 u + cu = f \qquad \text{in } \Omega \tag{1}$$

$$\frac{\partial u}{\partial n} = g \qquad \text{on } S \tag{2}$$

has at most one solution if $c < 0$ in Ω. If $c \equiv 0$ in Ω, show that any two solutions differ by a constant.

If u_1 and u_2 are both solutions of (*1*)–(*2*), $v \equiv u_1 - u_2$ satisfies the homogeneous problem

$$\nabla^2 v + cv = 0 \qquad \text{in } \Omega \tag{3}$$

$$\frac{\partial v}{\partial n} = 0 \qquad \text{on } S \tag{4}$$

Multiply (*3*) by v and integrate over Ω, using (*1.7*) and (*4*):

$$-\int_\Omega |\nabla v|^2 \, d\Omega + \int_\Omega cv^2 \, d\Omega = 0 \tag{5}$$

If $c < 0$ in Ω, the only way (*5*) can hold is for v to be identically zero in Ω, which implies that (*1*)–(*2*) has at most one solution. If $c \equiv 0$ in Ω, (*5*) shows that v is constant in Ω; i.e., the difference of two solutions of (*1*)–(*2*) is a constant.

3.16 Show that if $\alpha\beta > 0$, the mixed problem

$$\nabla^2 u = f \qquad \text{in } \Omega$$

$$\alpha u + \beta \frac{\partial u}{\partial n} = g \qquad \text{on } S$$

has at most one solution in a bounded region Ω.

Again consider the associated homogeneous problem for the difference of two solutions:

$$\nabla^2 v = 0 \qquad \text{in } \Omega \tag{1}$$

$$\alpha v + \beta \frac{\partial v}{\partial n} = 0 \qquad \text{on } S \tag{2}$$

Multiplying (*1*) by v, integrating over Ω, and using (*1.7*), we obtain

$$0 = \int_\Omega v \, \nabla^2 v \, d\Omega = -\int_\Omega |\nabla v|^2 \, d\Omega + \int_S v \frac{\partial v}{\partial n} \, dS$$

or, by (*2*),

$$\int_\Omega |\nabla v|^2 \, d\Omega + \int_S \frac{\alpha}{\beta} v^2 \, dS = 0$$

which is impossible unless $v \equiv 0$ in Ω.

3.17 Let Ω denote the region in $\mathbf{R}^n$ exterior to the unit sphere:

$$\Omega = \{(x_1, x_2, \ldots, x_n) : r^2 \equiv x_1^2 + x_2^2 + \cdots + x_n^2 > 1\}$$

Then S, the boundary of Ω, is the surface of the ball of radius 1 centered at the origin. Show that the exterior Dirichlet problem

$$\nabla^2 u = 0 \qquad \text{in } \Omega \tag{1}$$

$$u = 1 \qquad \text{on } S \tag{2}$$

has infinitely many solutions unless some "behavior at infinity" conditions are imposed.

By use of the chain rule, one easily shows that the one-parameter family of functions

$$v_{(\lambda)}(r) = \begin{cases} \lambda + \dfrac{1-\lambda}{r^{n-2}} & n \neq 2 \\ 1 + (1-\lambda)\log r & n = 2 \end{cases}$$

$-\infty < \lambda < \infty$, all satisfy *(1)*–*(2)*. (Note that only one member of this family, $v_{(1)}(r)$, is harmonic at the origin; it represents the ***unique*** solution of the ***interior*** Dirichlet problem.)

For $n = 2$, only $v_{(1)}$ satisfies the boundedness condition

$$|u| \leq A \qquad \text{in } \Omega \tag{3}$$

and in Problem 3.19 it is shown that $v_{(1)}$ is the unique solution to *(1)*–*(2)*–*(3)* when $n = 2$.

For $n \geq 3$, every $v_{(\lambda)}$ is bounded, but only $v_{(0)}$ satisfies the condition

$$u \to 0, \text{ uniformly, as } r \to \infty \tag{4}$$

In Problem 3.18 it is shown that, for $n \geq 3$, $v_{(0)}$ is the unique solution to *(1)*–*(2)*–*(4)*.

3.18 Show that the exterior boundary value problem in $\mathbf{R}^n$ $(n \geq 3)$

$$\begin{aligned} \nabla^2 u &= f && \text{in } \Omega \ (\Omega \text{ unbounded}) \\ u &= g && \text{on } S \\ u(\mathbf{x}) &\to 0 && \text{uniformly, as } |\mathbf{x}| \to \infty \end{aligned}$$

has at most one solution.

The difference, $v \equiv u_1 - u_2$, of two solutions satisfies

$$\nabla^2 v = 0 \qquad \text{in } \Omega \tag{1}$$

$$v = 0 \qquad \text{on } S \tag{2}$$

$$v(\mathbf{x}) \to 0 \qquad \text{uniformly, as } |\mathbf{x}| \to \infty \tag{3}$$

Let $\mathbf{x}_0$ be an arbitrary point in Ω. From *(3)* it follows that, given $\epsilon > 0$, R can be chosen so that $|\mathbf{x}_0| < R$, and $|v(\mathbf{x})| < \epsilon$ for $|\mathbf{x}| \geq R$. In the bounded region $\hat{\Omega}$ defined by the intersection of Ω and the ball $|\mathbf{x}| < R$, Theorem 3.2 applies. Since the boundary of $\hat{\Omega}$ consists partly of S and partly of the sphere $|\mathbf{x}| = R$, and since $v = 0$ on S while $|v| < \epsilon$ on $|\mathbf{x}| = R$, it follows that $|v(\mathbf{x})| < \epsilon$ throughout $\hat{\Omega}$. In particular, $|v(\mathbf{x}_0)| < \epsilon$. Since both ϵ and $\mathbf{x}_0$ are arbitrary, $v \equiv 0$, or $u_1 \equiv u_2$, in Ω.

3.19 Show that the exterior boundary value problem

$$\begin{aligned} u_{xx} + u_{yy} &= f && \text{in } \Omega \ (\Omega \text{ unbounded}) \\ u &= g && \text{on } S \\ |u(x, y)| &\leq A && \text{in } \Omega \ (A \text{ constant}) \end{aligned}$$

has at most one solution.

The difference, $v \equiv u_1 - u_2$, of two solutions satisfies

$$v_{xx} + v_{yy} = 0 \qquad \text{in } \Omega \tag{1}$$

$$v = 0 \qquad \text{on } S \tag{2}$$

$$|v(x, y)| \leq C \qquad \text{in } \Omega \ (C \text{ constant}) \tag{3}$$

Let Ω' denote the complement of Ω, so that the union of Ω and Ω' is the entire xy-plane. Let $\mathbf{x}_0 = (x_0, y_0)$ be a point in the interior of Ω', let

$$r = [(x - x_0)^2 + (y - y_0)^2]^{1/2}$$

and choose R_1 sufficiently small so that the ball $B_{R_1}(\mathbf{x}_0)$ is in the interior of Ω'. Choose R_2 sufficiently large so that the ball $B_{R_2}(\mathbf{x}_0)$ intersects the region Ω in a nonempty, bounded region $\hat{\Omega}$. If w is defined by

$$w = C \frac{\log (r/R_1)}{\log (R_2/R_1)}$$

then w is harmonic in $\hat{\Omega}$ (by Problem 3.17), w is positive on S, and $w = C$ on the sphere $r = R_2$; hence,

$$-w(x, y) \leq v(x, y) \leq w(x, y) \qquad \text{on the boundary of } \hat{\Omega} \tag{4}$$

Since v and w are both harmonic in the bounded region $\hat{\Omega}$, Theorem 3.2, applied together with (4) to the harmonic functions $w \pm v$, shows that

$$|v(x, y)| \leq w(x, y) \qquad \text{in } \hat{\Omega} \tag{5}$$

To complete the argument, let $(\bar{x}, \bar{y})$ be an arbitrary point in $\hat{\Omega}$. By substituting $(\bar{x}, \bar{y})$ into (5) and allowing $R_2 \to \infty$ with R_1 held fixed, we show that $v(\bar{x}, \bar{y}) = 0$. Thus, $v \equiv 0$ in $\hat{\Omega}$, which implies, by the arbitrariness of the initial R_2, $v \equiv 0$ in Ω.

3.20 Let Ω be a bounded region and consider the mixed problem

$$\nabla^2 u + cu = f \qquad \text{in } \Omega \tag{1}$$

$$\alpha u + \beta \frac{\partial u}{\partial n} = 0 \qquad \text{on } S \tag{2}$$

Show that for (1)–(2) to have a solution it is necessary that the consistency condition

$$\int_\Omega fv \, d\Omega = 0 \tag{3}$$

be satisfied by every solution v to the associated homogeneous problem

$$\nabla^2 v + cv = 0 \qquad \text{in } \Omega \tag{4}$$

$$\alpha v + \beta \frac{\partial v}{\partial n} = 0 \qquad \text{on } S \tag{5}$$

Multiply (1) by v and (4) by u, subtract, and integrate over Ω:

$$\int_\Omega (v \nabla^2 u - u \nabla^2 v) \, d\Omega = \int_\Omega fv \, d\Omega \tag{6}$$

But, by (1.8) and the boundary conditions, the left side of (6) vanishes, yielding the consistency condition (3). The proof includes the special cases $\beta = 0$, $\alpha \neq 0$ (Dirichlet problem) and $\alpha = 0$, $\beta \neq 0$ (Neumann problem).

3.21 Show by example that the initial value problem for Laplace's equation

$$\begin{aligned} u_{xx} + u_{yy} &= 0 && -\infty < x < \infty,\ y > 0 \\ u(x, 0) &= F(x) && -\infty < x < \infty \\ u_y(x, 0) &= G(x) && -\infty < x < \infty \end{aligned}$$

is ill-posed in that the solution does not depend continuously on the data functions F and G.

The following example is due to Hadamard. For $F = F_1 \equiv 0$ and $G = G_1 \equiv 0$, it is clear that $u_1 \equiv 0$ is a solution. For $F = F_2 \equiv 0$ and $G = G_2 = n^{-1} \sin nx$, it is easy to verify that

$$u_2 = \frac{1}{n^2} \sinh ny \sin nx$$

is a solution. The data functions F_1 and F_2 are identical, and

$$\lim_{n\to\infty} |G_1 - G_2| = 0$$

uniformly in x. Therefore, the data pairs F_1, G_1 and F_2, G_2 can be made arbitrarily close by choosing n sufficiently large. Let us compare the solutions u_1 and u_2 at $x = \pi/2$ for an arbitrarily small, fixed, positive y and for n restricted to odd positive integral values:

$$\left| u_1\left(\frac{\pi}{2}, y\right) - u_2\left(\frac{\pi}{2}, y\right) \right| = \frac{1}{n^2} \sinh ny = \frac{e^{ny} - e^{-ny}}{2n^2}$$

Because e^{ny} increases faster than n^2,

$$\lim_{\substack{n\to\infty \\ n \text{ odd}}} \left| u_1\left(\frac{\pi}{2}, y\right) - u_2\left(\frac{\pi}{2}, y\right) \right| = \infty \qquad (y > 0)$$

The conclusion is that, by choosing n sufficiently large, the maximum difference between the data functions can be made arbitrarily small, but the maximum difference between the corresponding solutions is then made arbitrarily large. In general, initial value problems for elliptic PDEs are ill-posed in this fashion.

Supplementary Problems

3.22 Verify that each of the following functions is everywhere harmonic:

(*a*) $x^3 - 3xy^2$ (*b*) $3x^2y - y^3$ (*c*) $e^x \cos y$ (*d*) $e^x \sin y$ (*e*) $6x + y$

3.23 If u and v are solutions of Laplace's equation, show that uv satisfies Laplace's equation if and only if ∇u and ∇v are orthogonal.

3.24 If $u(x, y)$ and $v(x, y)$ are the real and imaginary parts of an analytic function $f(z)$, show that uv satisfies Laplace's equation. [*Hint*: f^2 is analytic.]

3.25 Let the xy- and $\xi\eta$-coordinates be related by a rotation:

$$\xi = x \cos\theta + y \sin\theta \qquad \eta = -x \sin\theta + y \cos\theta$$

where $\theta = \text{const}$. Show that if u is harmonic in x and y, then u is harmonic in ξ and η. [*Hint*: Don't differentiate; appeal to the mean-value property.]

3.26 Show that the surface mean-value property, (*3.1*), is equivalent to the volume mean-value property,

$$u(\mathbf{x}_0) = \frac{1}{V(R)} \int_{B_R(\mathbf{x}_0)} u(\mathbf{x})\, dB_R$$

(see Example 3.2).

3.27 (*a*) Show that if $f(z) \equiv f(x+iy) = u(x, y) + iv(x, y)$ has a continuous derivative in Ω, then u and v satisfy the *Cauchy-Riemann equations*, $u_x = v_y$, $u_y = -v_x$, in Ω. (*b*) Show that if u and v are C^2 and satisfy the Cauchy-Riemann equations, then u and v are harmonic.

3.28 (*a*) In terms of the cylindrical coordinates defined by $x = r\cos\theta$, $y = r\sin\theta$, $z = z$, show that

$$\frac{\partial^2 u}{\partial x^2} + \frac{\partial^2 u}{\partial y^2} + \frac{\partial^2 u}{\partial z^2} = \frac{1}{r}\frac{\partial}{\partial r}\left(r\frac{\partial u}{\partial r}\right) + \frac{1}{r^2}\frac{\partial^2 u}{\partial \theta^2} + \frac{\partial^2 u}{\partial z^2}$$

(*b*) In terms of the spherical coordinates defined by $x = r\sin\theta\cos\phi$, $y = r\sin\theta\sin\phi$, $z = r\cos\theta$, show that

$$\frac{\partial^2 u}{\partial x^2} + \frac{\partial^2 u}{\partial y^2} + \frac{\partial^2 u}{\partial z^2} = \frac{1}{r^2\sin\theta}\left[\frac{\partial}{\partial r}\left(r^2\sin\theta\frac{\partial u}{\partial r}\right) + \frac{\partial}{\partial\theta}\left(\sin\theta\frac{\partial u}{\partial\theta}\right) + \frac{\partial}{\partial\phi}\left(\frac{1}{\sin\theta}\frac{\partial u}{\partial\phi}\right)\right]$$

3.29 (*a*) Show that if $u = u(r, \theta)$ is a harmonic function expressed in polar coordinates and v is defined by $v(\rho, \theta) = u(r, \theta)$, $\rho r = a^2$, then v is a harmonic function of (ρ, θ). (*b*) Let $u(r, \theta, \phi)$ be a harmonic function expressed in the spherical coordinates of Problem 3.28(*b*). Show that if v is defined by the *Kelvin transformation*, $v(\rho, \theta, \phi) = a^{-1}ru(r, \theta, \phi)$, $\rho r = a^2$, then v is a harmonic function of (ρ, θ, ϕ). (*c*) The transformation $\rho r = a^2$ is, in geometrical terms, an inversion in a circle (sphere) of radius a. Show that if the circle (sphere) of inversion remains tangent to a fixed line (plane) as the radius a approaches ∞, the transformation becomes a reflection in the fixed line (plane). Thus, harmonic functions in $\mathbf{R}^2$ and $\mathbf{R}^3$ can be continued by inversion/reflection.

3.30 (*a*) If a harmonic function is positive on the boundary of a bounded region Ω, prove that it is positive throughout Ω. (*b*) Show by example that if Ω is unbounded, the result of (*a*) may not be valid. (*c*) Let u, v, and w be harmonic in a bounded region Ω and let $u \le v \le w$ on the boundary of Ω. Show that $u \le v \le w$ throughout Ω.

3.31 Prove *Harnack's theorem*: A uniformly convergent sequence of harmonic functions converges to a harmonic function. [*Hint*: For "harmonic function" read "function with the mean-value property."]

3.32 Show that a C^2 function u is subharmonic if and only if $\nabla^2 u \ge 0$, and superharmonic if and only if $\nabla^2 u \le 0$.

3.33 (*a*) If u is subharmonic in the ball $B_R(\mathbf{x}_0)$, $u \le M$ in $B_R(\mathbf{x}_0)$, and $u(\mathbf{x}_0) = M$, show that $u = M$ everywhere in $B_R(\mathbf{x}_0)$. (*b*) If u is subharmonic in a bounded region Ω and u attains its maximum value at an interior point of Ω, show that u is constant in Ω.

3.34 Suppose that u is harmonic in a bounded region Ω and v is subharmonic in Ω. Show that if $u = v$ on the boundary of Ω, then $u > v$ throughout Ω.

3.35 If Ω is the region $0 < x < 1$, $0 < y < 1$, use Theorem 3.8 to show that the solution of

$$u_{xx} + u_{yy} - 3u_x = 5 \quad \text{in } \Omega$$
$$u = 0 \quad \text{on } S$$

satisfies $-5/3 < u < 0$ in Ω. [*Hint*: Consider also the Dirichlet problem for $v \equiv u + (5/3)x$.]

3.36 Show that if u satisfies $u_{xx} + e^x u_{yy} - e^y u = 0$ in a bounded region Ω and if $u \le 0$ on the boundary of Ω, then $u \le 0$ throughout Ω.

3.37 If $u(r, \theta)$ satisfies $\nabla^2 u = 0$ in $r<1$, $u(1, \theta) = f(\theta)$, show that Poisson's integral formula in $\mathbf{R}^2$ takes the form

$$u(r, \theta) = \frac{1-r^2}{2\pi} \int_{-\pi}^{\pi} \frac{f(\phi)}{1-2r\cos(\theta-\phi)+r^2}\, d\phi$$

(See Problem 7.13 for a derivation of this version of Poisson's formula.)

3.38 If $u_{xx} + u_{yy} = 0$ in $x^2+y^2<1$, and $u = y^2 x$ on $x^2+y^2=1$, find $u(0,0)$. [*Hint*: The boundary values are antisymmetric about the y-axis.]

3.39 If $u_{xx} + u_{yy} = 0$ in $x^2+y^2<1$, and $u = 3+x+y$ on $x^2+y^2=1$, find $u(\frac{1}{2}, \frac{1}{2})$.

3.40 In the Dirichlet problem $\nabla^2 u(r, \theta) = 0$ in $r<1$, $u(1, \theta) = f(\theta)$ $(-\pi < \theta \le \pi)$, show that a change in the data function $f(\theta)$ over an arbitrarily small interval (θ_1, θ_2) affects the solution value $u(r, \theta)$ for all $r<1$ and all θ.

3.41 Show that the Neumann problem

$$\begin{aligned} \nabla^2 u &= 1 && \text{in } x^2+y^2<1 \\ \frac{\partial u}{\partial n} &= 2 && \text{on } x^2+y^2=1 \end{aligned}$$

does not have a solution.

3.42 Show that

$$\begin{aligned} u_{xx} + u_{yy} + 2u &= f && \text{in } \Omega\text{: } 0<x<\pi, 0<y<\pi \\ u &= 0 && \text{on } S \end{aligned}$$

(*a*) has no solution if $f = 1$; (*b*) has solutions of the form

$$u = \sum_{m,n=1}^{\infty} a_{mn} \sin mx \sin ny$$

for certain constants a_{mn}, if $f = \cos x \cos y$. Determine these constants.

3.43 Show by example that if (i) $c = 0$ or (ii) $c>0$, the boundary value problem of Problem 3.15, with $f = g = 0$, has a nontrivial solution.

3.44 Let Ω be a bounded region whose boundary S consists of nonempty, complementary components S_1 and S_2. Show that the boundary value problem

$$\begin{aligned} \nabla^2 u &= f && \text{in } \Omega \\ u &= g_1 && \text{on } S_1 \\ \frac{\partial u}{\partial n} &= g_2 && \text{on } S_2 \end{aligned}$$

has at most one solution.

3.45 Suppose $L[\ ]$ and Ω satisfy the hypotheses of Theorem 3.9 and consider the boundary value problem $L[u] = f$ in Ω, $u = g$ on S. Let u_1 and u_2 denote, respectively, solutions corresponding to the data (f_1, g_1) and (f_2, g_2). Show that $f_1 \le f_2$ in Ω and $g_1 \le g_2$ on S implies $u_1 \le u_2$ in $\bar{\Omega}$.

3.46 If $u(x, y)$ satisfies

$$\begin{aligned} u_{xx} + u_{yy} &= 0 && \text{in } x^2+y^2<1 \\ (u_x^2 + u_y^2)u &= 0 && \text{on } x^2+y^2=1 \end{aligned}$$

show that u is a constant.

Chapter 4

Qualitative Behavior of Solutions to Evolution Equations

4.1 INITIAL VALUE AND INITIAL–BOUNDARY VALUE PROBLEMS

Unlike elliptic PDEs, which describe a steady state, parabolic or hyperbolic *evolution equations* describe processes that are developing in time. For such an equation, the initial state of the system is part of the auxiliary data for a well-posed problem. If the equation contains time derivatives up to order k, the initial state can be characterized by specifying the initial values of the unknown function and its time derivatives through order $k-1$.

EXAMPLE 4.1 The heat equation serves as the canonical example of a parabolic evolution equation. Problems which are well-posed for the heat equation will be well-posed for more general parabolic equations.

(*a*) **Well-Posed Initial Value Problem** (*Cauchy Problem*)

$$\begin{aligned} u_t &= \kappa\, \nabla^2 u(\mathbf{x}, t) && \text{in } \mathbf{R}^n,\ t>0 \\ u(\mathbf{x}, 0) &= f(\mathbf{x}) && \text{in } \mathbf{R}^n \\ |u(\mathbf{x}, t)| &< M && \text{in } \mathbf{R}^n,\ t>0 \end{aligned}$$

The boundedness condition at infinity (which is not the most general condition possible) is independent of the spatial dimension n.

(*b*) **Well-Posed Initial–Boundary Value Problem**

$$\begin{aligned} u_t &= \kappa\, \nabla^2 u(\mathbf{x}, t) && \text{in } \Omega,\ t>0 \\ u(\mathbf{x}, 0) &= f(\mathbf{x}) && \text{in } \Omega \\ \alpha(\mathbf{x})u(\mathbf{x}, t)+\beta(\mathbf{x})\frac{\partial u}{\partial n}(\mathbf{x}, t) &= g(\mathbf{x}, t) && \text{on } S,\ t>0;\ \alpha\beta \ge 0 \end{aligned}$$

Special values of α and β lead to boundary conditions of Dirichlet or Neumann type (Section 1.3). If Ω is not bounded (e.g., a half-space), then $g(\mathbf{x}, t)$ must be specified over the accessible portion of S and additional behavior-at-infinity conditions may be needed.

EXAMPLE 4.2 The wave equation serves as the prototype for hyperbolic evolution equations.

(*a*) **Well-Posed Initial Value Problem** (*Cauchy Problem*)

$$\begin{aligned} u_{tt} &= a^2\, \nabla^2 u(\mathbf{x}, t) && \text{in } \mathbf{R}^n,\ t>0 \\ u(\mathbf{x}, 0) &= f(\mathbf{x}) && \text{in } \mathbf{R}^n \\ u_t(\mathbf{x}, 0) &= g(\mathbf{x}) && \text{in } \mathbf{R}^n \end{aligned}$$

No behavior-at-infinity conditions are necessary in order to obtain a unique solution to the Cauchy problem for the wave equation.

(*b*) **Well-Posed Initial–Boundary Value Problem**

$$\begin{aligned} u_{tt} &= a^2\, \nabla^2 u(\mathbf{x}, t) && \text{in } \Omega,\ t>0 \\ u(\mathbf{x}, 0) = f(\mathbf{x}) \text{ and } u_t(\mathbf{x}, 0) &= g(\mathbf{x}) && \text{in } \Omega \\ \alpha(\mathbf{x})u(\mathbf{x}, t)+\beta(\mathbf{x})\frac{\partial u}{\partial n}(\mathbf{x}, t) &= h(\mathbf{x}, t) && \text{on } S,\ t>0;\ \alpha\beta \ge 0 \end{aligned}$$

Ω may be unbounded, with no condition at infinity required.

If the initial conditions in a well-posed initial value or initial–boundary value problem for an evolution equation are replaced by conditions on the solution at other than the initial time, the resulting problem may not be well-posed, even when the total *number* of auxiliary conditions is unchanged.

EXAMPLE 4.3

(*a*) **Backward Heat Equation**

$$\begin{aligned} u_t &= \kappa u_{xx}(x,t) && 0<x<1,\ 0<t<T \\ u(x,T) &= f(x) && 0<x<1 \\ u(0,t) = u(1,t) &= 0 && 0<t<T \end{aligned}$$

Here the initial condition of the forward problem has been replaced by a terminal condition specifying the state at a final time $t=T$. The problem is to find previous states $u(x,t)$ $(t<T)$ which will have evolved at time T into the state $f(x)$. For arbitrary $f(x)$, this problem has no solution. Even when the solution exists, it does not depend continuously on the data (see Problem 4.9).

(*b*) **Dirichlet Problem for the Wave Equation**

$$\begin{aligned} u_{tt} &= a^2 u_{xx}(x,t) && 0<x<1,\ 0<t<T \\ u(x,0) &= f(x) && 0<x<1 \\ u(x,T) &= g(x) && 0<x<1 \\ u(0,t) = u(1,t) &= 0 && 0<t<T \end{aligned}$$

Here the initial condition on u_t has been replaced by a terminal condition on u. The solution to this problem does not depend continuously on the data (see Problem 4.20).

4.2 MAXIMUM-MINIMUM PRINCIPLES (PARABOLIC PDEs)

Neither the wave equation nor hyperbolic equations in general satisfy a maximum-minimum principle, but the heat equation and parabolic equations of more general form do so.

Let Ω denote a bounded region in $\mathbf{R}^3$ whose boundary is a smooth closed surface S. Suppose $u(x,y,z,t)$ to be continuous for (x,y,z) in $\bar{\Omega}$ and $0\le t\le T$; for short, in $\bar{\Omega}\times[0,T]$. Let

$$\begin{aligned} M_S &\equiv \max\{u(x,y,z,t)\colon\ (x,y,z) \text{ on } S \text{ and } 0\le t\le T\} \\ M_0 &\equiv \max\{u(x,y,z,t)\colon\ (x,y,z) \text{ in } \Omega \text{ and } t=0\} \\ M &\equiv \max\{M_S, M_0\} \end{aligned}$$

and let m_S, m_0, and m denote the corresponding minimum values for u.

Theorem 4.1 (*Maximum-Minimum Principle for the Heat Equation*): Given that $u(x,y,z,t)$ is continuous in $\bar{\Omega}\times[0,T]$:

(i) If $u_t-\nabla^2 u\le 0$ in $\Omega\times(0,T)$, then $u\le M$ in $\bar{\Omega}\times[0,T]$.

(ii) If $u_t-\nabla^2 u\ge 0$ in $\Omega\times(0,T)$, then $u\ge m$ in $\bar{\Omega}\times[0,T]$.

(iii) If $u_t-\nabla^2 u=0$ in $\Omega\times(0,T)$, then $m\le u\le M$ in $\bar{\Omega}\times[0,T]$.

According to Theorem 4.1(iii), the temperatures inside a heat conductor are bounded by the extreme temperatures attained either inside initially or on the boundary subsequently. Theorem 4.1 is useful in establishing uniqueness and continuous dependence on the data and to obtain various comparison results for the solutions to initial–boundary value problems for the heat equation.

The maximum-minimum principle may be extended in various ways; the next theorem states that if an extreme value of a nonconstant solution of the heat equation occurs on the boundary S, then the normal derivative of the solution (the heat flux) cannot vanish at that point.

Theorem 4.2: Let u be C^1 in $\bar{\Omega} \times [0, T]$ and satisfy

$$u_t - \nabla^2 u = 0 \qquad \text{in } \Omega \times (0, T)$$

Then, either u is constant in $\bar{\Omega} \times [0, T]$ or else

(i) at any point $\boldsymbol{\xi}$ on S such that $u(\boldsymbol{\xi}, \tau) = M$, $\dfrac{\partial u}{\partial n}(\boldsymbol{\xi}, \tau) > 0$;

(ii) at any point $\boldsymbol{\xi}$ on S such that $u(\boldsymbol{\xi}, \tau) = m$, $\dfrac{\partial u}{\partial n}(\boldsymbol{\xi}, \tau) < 0$.

Results analogous to Theorems 4.1 and 4.2 hold for parabolic equations more general than the heat equation.

Definition: If the linear differential operator

$$L[\ ] \equiv \sum_{i,j=1}^{n} a_{ij}(\mathbf{x}, t) \frac{\partial^2[\ ]}{\partial x_i \partial x_j} + \sum_{i=1}^{n} b_i(\mathbf{x}, t) \frac{\partial[\ ]}{\partial x_i} + c(\mathbf{x}, t)[\ ] \tag{4.1}$$

is uniformly elliptic (Section 3.2) in Ω for each t in $[0, T]$, then the operator

$$\frac{\partial[\ ]}{\partial t} - L[\ ]$$

is said to be *uniformly parabolic* in $\Omega \times [0, T]$.

Theorem 4.3: Let Ω be a bounded region in $\mathbf{R}^n$ with smooth boundary S, and suppose that $\partial/\partial t - L$ is uniformly parabolic in $\Omega \times [0, T]$, with coefficients a_{ij} and b_i continuous in $\Omega \times [0, T]$ and coefficient $c \equiv 0$. Suppose also that $u(\mathbf{x}, t)$ is continuous in $\bar{\Omega} \times [0, T]$. Then the conclusions (i), (ii), (iii) of Theorem 4.1 hold, with $\nabla^2[\ ]$ replaced by $L[\ ]$.

Theorem 4.4: Theorem 4.2 remains valid when the operator $\partial/\partial t - \nabla^2$ is replaced by the uniformly parabolic operator $\partial/\partial t - L$ of Theorem 4.3.

The conclusions of Theorems 4.3 and 4.4 regarding the solution to $u_t - L[u] = 0$ continue to hold if $c(\mathbf{x}, t) \leq 0$ and $M \geq 0$—and even in another case (see Problem 4.21).

4.3 DIFFUSIONLIKE EVOLUTION (PARABOLIC PDEs)

Two properties characterize the time-behavior of systems modeled by parabolic PDEs. To describe the second of these, we introduce the notion of an "evolution operator" that takes the initial state $u(\mathbf{x}, 0)$ of the system into the evolved state $u(\mathbf{x}, t)$.

Infinite speed of propagation. At any time $t > 0$ (no matter how small), the solution to a parabolic initial value problem at an arbitrary location $\mathbf{x}$ depends on *all* of the initial data. (See Problem 4.7.) As a consequence, the problem is well-posed only if behavior-at-infinity conditions are imposed.

Smoothing action of the evolution operator. A solution $u(\mathbf{x}, t)$ to the Cauchy problem for the heat equation is, for each $\mathbf{x}$ and all $t > 0$, infinitely differentiable with respect to both $\mathbf{x}$ and t. (See Problem 4.8.)

There is an interesting consequence of the smoothing property of the evolution operator for the heat equation. A sectionally continuous initial state $u(\mathbf{x}, 0)$ can always evolve forward in time in accordance with the heat equation. However, if it is not infinitely differentiable with respect to both $\mathbf{x}$ and t, then it cannot have originated from an earlier state $u(\mathbf{x}, t)$, $t < 0$. Thus the heat equation is irreversible in the mathematical sense that "forward" time is distinguishable from "backward" time. Correspondingly, any physical process for which the heat equation is a mathematical model is irreversible in the sense of the Second Law of Thermodynamics.

In an initial–boundary value problem for a parabolic PDE, the solution will be smooth for all $t > 0$ and all $\mathbf{x}$ *inside* the domain. In order for the solution to be smooth on the boundary and at $t = 0$, it is necessary to impose smoothness and compatibility conditions on the data.

EXAMPLE 4.4 The problem

$$\begin{aligned} u_t(x, t) &= u_{xx}(x, t) && 0 < x < L,\ t > 0 \\ u(x, 0) &= f(x) && 0 < x < L \\ u(0, t) = g(t) \text{ and } u(L, t) &= h(t) && t > 0 \end{aligned}$$

has a solution $u(x, t)$ which is going to be infinitely differentiable with respect to both x and t for $0 < x < L$ and $t > 0$. In addition, the solution will be continuous with respect to x and t for $0 \le x \le L$ and $t \ge 0$, provided (i) $f(x)$ is continuous for $0 \le x \le L$, and $g(t)$ and $h(t)$ are continuous for $t \ge 0$; (ii) $f(0) = g(0)$ and $f(L) = h(0)$ (compatibility of initial and boundary data).

Additional conditions on the smoothness and compatibility of f, g, and h will result in additional smoothness of $u(x, t)$ for $0 \le x \le L$, $t \ge 0$. If such conditions are lacking, the solution may satisfy the initial and boundary conditions in a mean-square but not a pointwise sense.

4.4 WAVELIKE EVOLUTION (HYPERBOLIC PDEs)

The following two properties contrast with those of diffusionlike evolution.

Finite speed of propagation. A solution to an initial value problem for the wave equation corresponding to initial data that vanish outside some bounded region will itself vanish outside a region which is bounded but expands with time. The rate at which this expanding region grows can be interpreted as the (finite) speed of propagation of the effect modeled by the wave equation.

Lack of smoothing action in the evolution operator. The solution of a hyperbolic initial value problem cannot be smoother than the initial data; it may in fact be *less* smooth than the data. When irregularities in the solution to a hyperbolic equation are present, they persist in time and are propagated along characteristics (cf. Problem 2.10).

Finite propagation speed has various consequences. For instance, at any point in the spatial domain at any finite time $t > 0$, the solution to a hyperbolic initial value problem depends on only a portion of the initial data. The set of locations $\mathbf{x}$ such that the solution value assumed at $\mathbf{x}$ at time $t = 0$ affects the value of the solution at $(\mathbf{x}_0, t_0)$ constitutes the *domain of dependence* for $(\mathbf{x}_0, t_0)$. For each $(\mathbf{x}_0, t_0)$ the domain of dependence is of finite extent; consequently, the initial value problem is well-posed without the specification of behavior at infinity.

The lack of smoothing action in wavelike evolution is related to the fact that physical processes modeled by the wave equation are thermodynamically reversible. An initial state $u(\mathbf{x}, 0)$ that is lacking in smoothness can evolve forward in time in accordance with the wave equation and can, as well as not, have originated from an earlier state. Thus we can solve forward in time to find subsequent states into which $u(\mathbf{x}, 0)$ will evolve, or we can solve backward in time to find earlier states from which $u(\mathbf{x}, 0)$ has evolved.

Solved Problems

4.1 Prove Theorem 4.1.

Suppose that $u(x, y, z, t)$ is continuous in $\bar{\Omega} \times [0, T]$ and satisfies

$$u_t - \nabla^2 u \leq 0 \qquad \text{in } \Omega \times (0, T)$$

For $\epsilon > 0$, let $v(x, y, z, t) = u(x, y, z, t) + \epsilon(x^2 + y^2 + z^2)$. Then,

$$v_t - \nabla^2 v = -6\epsilon < 0 \qquad \text{in } \Omega \times (0, T) \tag{1}$$

Suppose now that v assumes its maximum value at (x_0, y_0, z_0, t_0), where (x_0, y_0, z_0) is an interior point of Ω and $0 < t_0 \leq T$. Then, at (x_0, y_0, z_0, t_0),

$$v_t \geq 0 \qquad \text{and} \qquad \nabla^2 v \leq 0$$

which contradicts (*1*). Hence, v must assume its maximum value for (x, y, z) on S and $0 \leq t \leq T$, or else for $t = 0$ and (x, y, z) in Ω.

The definition of v implies that

$$u \leq v \leq \max v = M + \epsilon R^2 \qquad \text{in } \bar{\Omega} \times [0, T]$$

where $R^2 = \max_S (x^2 + y^2 + z^2)$. Since $\epsilon > 0$ is arbitrary, it follows that $u \leq M$ in $\bar{\Omega} \times [0, T]$. Thus Theorem 4.1(i) is proved.

Theorem 4.1(ii) is proved by applying Theorem 4.1(i) to the function $-u$; then Theorems 4.1(i) and 4.1(ii) together imply Theorem 4.1(iii).

4.2 Let Ω, S, T be as previously described and consider the problem

$$u_t - \nabla^2 u = \Phi(x, y, z, t) \qquad \text{in } \Omega \times (0, T)$$
$$u(x, y, z, 0) = F(x, y, z) \qquad \text{in } \Omega$$
$$u(x, y, z, t) = f(x, y, z, t) \qquad \text{in } S \times [0, T]$$

Show that if this problem has a solution u that is continuous in $\bar{\Omega} \times [0, T]$, then this solution is unique.

Method 1

The difference $v(x, y, z, t) \equiv u_1(x, y, z, t) - u_2(x, y, z, t)$ of two continuous solutions is itself continuous and satisfies

$$v_t - \nabla^2 v = 0 \qquad \text{in } \Omega \times (0, T) \tag{1}$$
$$v(x, y, z, 0) = 0 \qquad \text{in } \Omega \tag{2}$$
$$v(x, y, z, t) = 0 \qquad \text{in } S \times [0, T] \tag{3}$$

For the function v, (*2*) implies $m_0 = M_0 = 0$, and (*3*) implies $m_S = M_S = 0$. Hence, $m = M = 0$, and, by Theorem 4.1(iii), $v \equiv 0$.

Method 2

For $0 \leq t \leq T$, define the *energy integral*

$$J(t) \equiv \int_\Omega v^2 \, d\Omega \tag{4}$$

where v is the difference function of Method 1. Clearly, $J(t) \geq 0$ and, by (*2*), $J(0) = 0$. In addition,

$$J'(t) = \int_\Omega 2vv_t \, d\Omega = 2\int_\Omega v \, \nabla^2 v \, d\Omega$$
$$= 2\int_S v \frac{\partial v}{\partial n} \, dS - 2\int_\Omega |\nabla v|^2 \, d\Omega$$

where we have used (*1*) and (*1.7*). By (*3*), the boundary integral vanishes, leaving

$$J'(t) = -2\int_\Omega |\nabla v|^2\, d\Omega \le 0 \qquad (5)$$

Thus $J(t)$ is nonincreasing, which fact, along with $J(0)=0$ and $J(t)\ge 0$, implies that $J(t)\equiv 0$. But then, the integrand $v(x, y, z, t)^2$ in (*4*) being nonnegative and continuous with respect to all arguments, it follows that v is identically zero in Ω for $t\ge 0$.

The energy integral method may be extended to the case of Neumann or mixed boundary conditions, to which Theorem 4.1 is not *directly* applicable. See Problem 4.16.

4.3 For the initial–boundary value problem of Problem 4.2, let u_1 and u_2 denote solutions corresponding to data $\{\Phi_1, F_1, f_1\}$ and $\{\Phi_2, F_2, f_2\}$, respectively. Suppose that $\Phi_1 \le \Phi_2$ in $\Omega\times(0, T)$, $F_1\le F_2$ in Ω, $f_1 \le f_2$ in $S\times[0, T]$. Prove that $u_1\le u_2$ in $\bar\Omega\times[0, T]$.

Letting $v\equiv u_1 - u_2$, we have

$$v_t - \nabla^2 v = \Phi_1 - \Phi_2 \le 0 \qquad \text{in } \Omega\times(0, T)$$

In addition, for the function v,

$$M_0 = \max_\Omega\{F_1 - F_2\}\le 0 \qquad M_S = \max_{S\times[0,T]}\{f_1 - f_2\}\le 0$$

whence $M\le 0$. Then Theorem 4.1(i) implies that $v\le 0$, or $u_1\le u_2$, in $\bar\Omega\times[0, T]$.

4.4 For κ, T positive constants, suppose that $v(x, t)$ satisfies

$$v_t = \kappa v_{xx} \qquad 0<x<1,\ 0<t<T \qquad (1)$$

$$v(x, 0) = 0 \qquad 0<x<1 \qquad (2)$$

$$v_x(0, t) = g(t) \qquad 0<t<T \qquad (3)$$

$$v_x(1, t) = 0 \qquad 0<t<T \qquad (4)$$

If $g(0)=0$ and $g'(t)>0$ for $t>0$ (whence $g(t)>0$ for $t>0$), show that for $0\le x\le 1$, $0\le t\le T$,

(*a*) $v(x, t)\le 0$ (*b*) $v_x(x, t)\ge 0$ (*c*) $v_{xx}(x, t)\le 0$

(*a*) Apply Theorem 4.3 (or, after changing the time variable, Theorem 4.1) to (*1*) in order to conclude that

$$v(x, t)\le M \qquad 0\le x\le 1,\ 0\le t\le T$$

The outward normal derivative of v at $x=0$ is $-v_x(0, t) = -g(t)<0$; hence, by Theorem 4.4 (or Theorem 4.2), the maximum M cannot occur at $x=0$. The outward normal derivative at $x=1$ is $v_x(1, t)=0$; so the maximum cannot occur at $x=1$ either. It follows that $M = M_0 = 0$, whence

$$v(x, t)\le 0 \qquad 0\le x\le 1,\ 0\le t\le T$$

(*b*) Let $u(x, t)\equiv v_x(x, t)$. Differentiate (*1*) with respect to x, to find

$$v_{tx} = \kappa v_{xxx} \qquad 0<x<1,\ 0<t<T \qquad (5)$$

If v is sufficiently smooth, $v_{tx} = v_{xt} = u_t$, and (*5*) becomes

$$u_t = \kappa u_{xx} \qquad 0<x<1,\ 0<t<T \qquad (6)$$

which is the same PDE as is satisfied by v.

Theorem: If a function satisfies a linear PDE with constant coefficients, its derivatives satisfy that same PDE.

Equations (*2*), (*3*), and (*4*) imply the following auxiliary conditions for (*6*):

$$u(x, 0) = 0 \qquad 0 < x < 1 \tag{7}$$

$$u(0, t) = g(t) \qquad 0 < t < T \tag{8}$$

$$u(1, t) = 0 \qquad 0 < t < T \tag{9}$$

For the system (*6*) through (*9*), Theorem 4.3 gives

$$0 = m \leq u(x, t) \qquad 0 \leq x \leq 1,\ 0 \leq t \leq T$$

which is what was to be shown.

(*c*) Define $w(x, t) \equiv v_t(x, t)$. Then, as in (*b*),

$$w_t = k w_{xx} \qquad 0 < x < 1,\ 0 < t < T \tag{10}$$

$$w_x(0, t) = g'(t) \qquad 0 < t < T \tag{11}$$

$$w_x(1, t) = 0 \qquad 0 < t < T \tag{12}$$

In addition, from (*a*), $v(x, t) \leq 0 = v(x, 0)$. Hence,

$$v_t(x, 0) = \lim_{t \to 0^+} \frac{v(x, t) - v(x, 0)}{t} \leq 0$$

or

$$w(x, 0) \leq 0 \qquad 0 < x < 1 \tag{13}$$

Applying Theorem 4.3 to system (*10*) through (*13*), we see that the maximum of w must occur at $t = 0$ in $0 < x < 1$; that is,

$$v_t(x, t) = w(x, t) \leq M \leq 0 \qquad 0 \leq x \leq 1,\ 0 \leq t \leq T$$

Then, since $\kappa > 0$, $v_{xx}(x, t) = \kappa^{-1} v_t(x, t) \leq 0$, for $0 \leq x \leq 1$ and $0 \leq t \leq T$.

4.5 Making use of the results of Problem 4.4, plot $v(x, t)$ versus x for several values of t.

"Profiles" for three time values are plotted in Fig. 4-1. The curves are below the x-axis, in accordance with Problem 4.4(*a*). On $x = 0$ we have $v_t(0, t) \leq 0$ (Problem 4.4(*c*)) and $v_x(0, t) = g(t)$; thus, the starting value is a negative, decreasing function of time, while the starting slope is a positive, increasing function of time. The curves are concave (Problem 4.4(*c*)), and have final slope zero ($v_x(1, t) = 0$).

Inspection of Fig. 4-1 suggests that, for each fixed t in $[0, T]$,

$$v(0, t) \leq v(x, t) \leq v(1, t) \qquad 0 \leq x \leq 1$$

and, indeed, this follows at once from Problem 4.4(*b*).

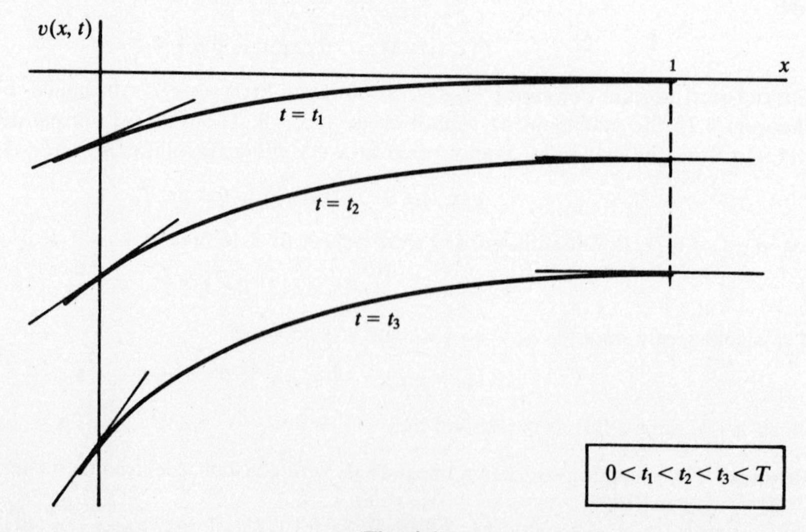

Fig. 4-1

4.6 Let $u(x, t)$ be a solution of the nonlinear problem

$$u_t(x,t) = a(u)u_{xx}(x,t) \qquad 0<x<1,\ 0<t<T \tag{1}$$

$$u(x,0) = 0 \qquad 0<x<1 \tag{2}$$

$$u_x(0,t) = g(t) \qquad 0<t<T \tag{3}$$

$$u_x(1,t) = 0 \qquad 0<t<T \tag{4}$$

Assume that $g(t)$ is continuously differentiable, with $g(0)=0$ and $g'(t)>0$ for $t>0$. In addition, assume that $a(u)$ is continuous and satisfies

$$0<\beta_1 \leq a(u) \leq \beta_2 \qquad \text{for all } u \tag{5}$$

Let $v_i(x, t)$ $(i = 1, 2)$ denote the solutions of the linear problems

$$\begin{aligned} v_{i,t}(x,t) &= \beta_i v_{i,xx}(x,t) && 0<x<1,\ 0<t<T \\ v_i(x,0) &= 0 && 0<x<1 \\ v_{i,x}(0,t) &= g(t) && 0<t<T \\ v_{i,x}(1,t) &= 0 && 0<t<T \end{aligned}$$

Prove that

$$v_2(x,t) \leq u(x,t) \leq v_1(x,t) \qquad 0\leq x\leq 1,\ 0\leq t\leq T \tag{6}$$

Define $h(x,t) \equiv v_2(x,t) - u(x,t)$ $(0<x<1,\ 0<t<T)$. Then $h(x,t)$ must satisfy

$$h_t(x,t) - a(u(x,t))h_{xx}(x,t) = [\beta_2 - a(u(x,t))]v_{2,xx}(x,t) \leq 0 \tag{7}$$

where the inequality follows from (*5*) and Problem 4.4(*c*). Further, we have

$$h(x,0) = 0 \qquad 0<x<1 \tag{8}$$

$$h_x(0,t) = h_x(1,t) = 0 \qquad 0<t<T \tag{9}$$

For the problem (*7*)–(*8*)–(*9*), Theorem 4.4 and (*9*) rule out a boundary maximum for h. Thus, $M = M_0 = 0$, and Theorem 4.3 implies that $h(x,t) \leq 0$; or

$$v_2(x,t) \leq u(x,t) \qquad 0\leq x\leq 1,\ 0\leq t\leq T$$

A similar consideration of the difference $k(x,t) \equiv u(x,t) - v_1(x,t)$ yields the other half of (*6*).

4.7 As is shown in Problem 4.17, the function

$$u(x,t) = \frac{1}{\sqrt{4\pi t}}\int_{-\infty}^{\infty} F(y)\exp\left[\frac{(x-y)^2}{4t}\right]dy \tag{1}$$

solves the initial value problem

$$u_t(x,t) = u_{xx}(x,t) \qquad -\infty<x<\infty,\ t>0 \tag{2}$$

$$u(x,0) = F(x) \qquad -\infty<x<\infty \tag{3}$$

Verify the infinite speed of propagation associated with diffusionlike evolution by showing that (*a*) for each $t>0$ and all x, $u(x,t)$ depends on all the initial data $F(x)$, $-\infty<x<\infty$; (*b*) for the particular data

$$F(x) = \begin{cases} 1 & |x|<\epsilon \\ 0 & |x|>\epsilon \end{cases} \tag{4}$$

$u(x,t)>0$ for all x and every positive t, no matter how small the positive number ϵ.

(*a*) For each $t>0$,

$$\exp\left[-\frac{(x-y)^2}{4t}\right] > 0 \qquad \text{for all } x \text{ and } y$$

It follows from (*1*) that for each $t>0$ and all x, the value $u(x,t)$ incorporates every $F(y)$, $-\infty<y<\infty$.

(*b*) The change of variable

$$z=\frac{x-y}{\sqrt{4t}} \qquad (t>0,\ -\infty<x,\ y<\infty)$$

transforms (*1*) into

$$u(x,t)=\frac{1}{\sqrt{\pi}}\int_{-\infty}^{\infty} e^{-z^2}\,F(x-z\sqrt{4t})\,dz \tag{5}$$

But, by (*4*),

$$F(x-z\sqrt{4t})=\begin{cases}1 & (x-\epsilon)/\sqrt{4t}<z<(x+\epsilon)/\sqrt{4t}\\ 0 & \text{all other } z\end{cases}$$

so that (*5*) becomes

$$u(x,t)=\frac{1}{\sqrt{\pi}}\int_{(x-\epsilon)/\sqrt{4t}}^{(x+\epsilon)/\sqrt{4t}} e^{-z^2}\,dz \qquad -\infty<x<\infty,\ t>0 \tag{6}$$

Since $e^{-z^2}>0$ for all real z, it follows that $u(x,t)>0$ for all x and t. That is, the solution $u(x,t)$ is immediately positive everywhere, even though F is zero everywhere except in the arbitrarily small interval $(-\epsilon,\epsilon)$.

Note, however, that because e^{-z^2} is monotone decreasing, (*6*) implies that

$$u(x,t)<\frac{1}{\sqrt{\pi}}\left(\frac{x+\epsilon}{\sqrt{4t}}-\frac{x-\epsilon}{\sqrt{4t}}\right)\exp\left[-\frac{(x-\epsilon)^2}{4t}\right]=\frac{\epsilon}{\sqrt{\pi t}}\exp\left[-\frac{(x-\epsilon)^2}{4t}\right]$$

Thus, even though the influence of the initial state propagates with infinite speed, the strength of this influence dies out very rapidly (as e^{-r^2}) as the distance ($r=|x-\epsilon|$) from the set where $F\neq 0$ increases. We are therefore able to claim that although solutions to the heat equation exhibit a nonphysical property (infinite speed of propagation), they do behave in a manner that is an acceptable approximation of reality. Practically speaking, effects governed by the heat equation propagate with finite speed. For more on this matter, see Problem 7.8.

4.8 Let $f(x)$ denote a sectionally continuous function in $(0,\pi)$. Then, using separation of variables (Chapter 8; see also Problem 4.18), one shows that

$$u(x,t)=\sum_{n=1}^{\infty} f_n e^{-n^2t}\sin nx \qquad 0<x<\pi,\ t>0 \tag{1}$$

is the solution to

$$u_t(x,t)=u_{xx}(x,t) \qquad 0<x<\pi,\ t>0 \tag{2}$$

$$u(x,0)=f(x) \qquad 0<x<\pi \tag{3}$$

$$u(0,t)=u(\pi,t)=0 \qquad t>0 \tag{4}$$

provided

$$f_n=\frac{2}{\pi}\int_0^{\pi} f(x)\sin nx\,dx \qquad (n=1,2,\dots) \tag{5}$$

Demonstrate the smoothing action of the evolution operator in this case.

At $t=0$, (*1*) reduces to

$$u(x,0)=\sum_{n=1}^{\infty} f_n\sin nx$$

and this series converges pointwise to $\tilde{F}_o(x)$, the odd 2π-periodic extension of $f(x)$, provided $\tilde{F}_o$ and $\tilde{F}'_o$ are sectionally continuous (see Problem 6.1(*b*)).

For each $n = 1, 2, \ldots$ and for each positive t,

$$|f_n e^{-n^2 t} \sin nx| \leq (\text{constant})\, e^{-n^2 t} \leq (\text{constant})\, (e^{-t})^n$$

and the geometric series is convergent. It follows from the Weierstrass M-test that, for each fixed $t > 0$, the series in (*1*) converges absolutely and uniformly to a continuous function of x. The same can be said of the series obtained from (*1*) by term-by-term differentiation any number of times with respect to x and/or t. We conclude that the series in (*1*) represents a function which is not just continuous but is infinitely differentiable with respect to both x and t for $t > 0$ and $0 < x < \pi$.

The evolution operator $\mathscr{E}_t$,

$$\mathscr{E}_t[u(x, 0)] = u(x, t) \qquad t > 0 \tag{6}$$

can be characterized in terms of the Fourier sine-series coefficients of the states, as follows:

$$\mathscr{E}_t[f_n] = f_n e^{-n^2 t} \qquad t > 0 \tag{7}$$

We have seen that whereas the f_n represent a function of x that is not necessarily even C^0, their images under $\mathscr{E}_t$ represent a function that is C^∞. It is characteristic of solutions to the heat equation (and parabolic equations in general) that $u(x, t)$ is an extremely smooth function for $t > 0$, even if $u(x, 0)$ is not particularly smooth.

For parabolic equations having variable coefficients, the smoothing action of the evolution operator may be limited by a lack of smoothness in the coefficients.

4.9 Show that the backward heat problem is ill-posed, as asserted in Example 4.3(*a*). For simplicity, choose $\kappa = 1$.

Write $u_0(x) \equiv u(x, 0)$, the initial state (temperature). Then (cf. Problem 4.8), the function

$$u(x, t) = \sum_{n=1}^{\infty} u_0^{(n)} e^{-n^2\pi^2 t} \sin n\pi x \qquad 0 < x < 1,\ 0 < t < T \tag{1}$$

where

$$u_0^{(n)} = 2 \int_0^1 u_0(x) \sin n\pi x \, dx \qquad (n = 1, 2, \ldots) \tag{2}$$

will solve the problem, provided the $u_0^{(n)}$ are such that

$$f(x) = \sum_{n=1}^{\infty} u_0^{(n)} e^{-n^2\pi^2 T} \sin n\pi x \qquad 0 < x < 1 \tag{3}$$

But the series in (*3*) converges uniformly to an infinitely differentiable function of x, whatever the $u_0^{(n)}$. It follows that no solution exists when $f(x)$ is not infinitely differentiable.

In the case where $f(x)$ *is* infinitely differentiable, the solution does not depend continuously on the data. If, for instance,

$$f(x) = \frac{\sin N\pi x}{N} \qquad (N = \text{integer})$$

the unique solution to the problem is

$$u(x, t) = \frac{1}{N} e^{N^2\pi^2(T-t)} \sin N\pi x \qquad 0 < x < 1,\ 0 < t < T$$

For large N, on the one hand, $|f(x)|$ becomes uniformly small; that is, the data function differs by as little as we wish from the data function $f \equiv 0$, to which corresponds the solution $u \equiv 0$. On the other hand, $|u(x, t)|$ grows with N; i.e., the solution does not remain close to $u \equiv 0$. Thus, there is no continuity of dependence on the data.

4.10 Show that the solution to

$$u_{tt}(x, t) = a^2 u_{xx}(x, t) \qquad -\infty < x < \infty, \ t > 0 \tag{1}$$

$$u(x, 0) = F(x) \qquad -\infty < x < \infty \tag{2}$$

$$u_t(x, 0) = G(x) \qquad -\infty < x < \infty \tag{3}$$

may be given in the *D'Alembert form*

$$u(x, t) = \frac{1}{2}[F(x + at) + F(x - at)] + \frac{1}{2a}\int_{x-at}^{x+at} G(s)\, ds \tag{4}$$

We shall obviously want to apply the theory of Chapter 2. In terms of the characteristic coordinates

$$\xi = x + at \qquad \eta = x - at$$

the problem takes the form

$$u_{\xi\eta}(\xi, \eta) = 0 \qquad -\infty < \eta < \xi < \infty \tag{5}$$

$$u(\xi, \xi) = F(\xi) \qquad -\infty < \xi < \infty \tag{6}$$

$$u_\xi(\xi, \xi) - u_\eta(\xi, \xi) = \frac{1}{a} G(\xi) \qquad -\infty < \xi < \infty \tag{7}$$

Integrating (*5*) in two steps: $u_\xi = \phi(\xi)$ and

$$u(\xi, \eta) = \int \phi(\xi)\, d\xi + \Psi(\eta) = \Phi(\xi) + \Psi(\eta) \tag{8}$$

Applying conditions (*6*) and (*7*) to (*8*):

$$\Phi(\xi) + \Psi(\xi) = F(\xi) \tag{9}$$

$$\Phi'(\xi) - \Psi'(\xi) = \frac{1}{a} G(\xi) \tag{10}$$

Solving (*9*) and the integral of (*10*) for the unknown functions:

$$\Phi(\xi) = \frac{1}{2}\left[F(\xi) + \frac{1}{a}\int^{\xi} G(s)\, ds\right] \tag{11}$$

$$\Psi(\eta) = \frac{1}{2}\left[F(\eta) - \frac{1}{a}\int^{\eta} G(s)\, ds\right] \tag{12}$$

Substitution of (*11*) and (*12*) in (*8*), and transformation back to the variables x and t, gives (*4*). Note that whereas the integral in (*4*) effects one order of smoothing of G (the initial data for u_t), there is *no smoothing of the initial state* F; contrast this with the heat equation.

4.11 For the hyperbolic problem (*1*)–(*2*)–(*3*) of Problem 4.10, (*a*) describe the domain of dependence of a point (x_0, t_0), where $t_0 > 0$; (*b*) if F and G both vanish for $|x| > 1$, show that

$$u(1 + a, t) = 0 \qquad \text{for } 0 \le t \le 1$$

and interpret this result.

(*a*) By (*4*) of Problem 4.10, $u(x_0, t)$ depends on the values of F for the two arguments $x_0 + at_0$ and $x_0 - at_0$, and on the values of G over the interval $(x_0 - at_0, x_0 + at_0)$. Thus, the domain of dependence of (x_0, t_0) is the closed interval $[x_0 - at_0, x_0 + at_0]$, which is precisely the portion of the x-axis cut off by the two characteristics

$$x + at = \xi_0 \qquad x - at = \eta_0$$

that pass through (x_0, t_0). See Fig. 4-2.

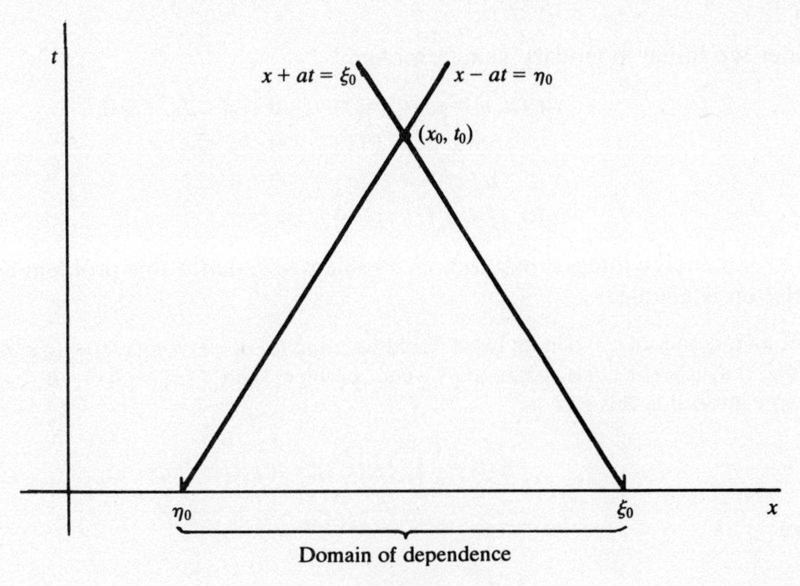

Fig. 4-2

(*b*) Because F and G each vanish outside $|x|<1$, $u(1+a, t)$ must remain zero so long as the domain of dependence of the point $(1+a, t)$, $[1+a-at, 1+a+at]$, remains disjoint from $(-1, 1)$; that is, so long as

$$1+a-at \geq 1 \qquad \text{or} \qquad 0 \leq t \leq 1$$

Now, the distance from the point $(1+a, 0)$ to the interval $(-1, 1)$ is just a units. Consequently, our result may be interpreted to mean that the influence of the initial data requires just 1 unit of time to traverse this distance; i.e., the propagation speed is a units of distance per unit of time.

4.12 Consider the following modification of the n-dimensional wave equation:

$$u_{tt} - \sum_{j=1}^{n} c_j^2 u_{x_j x_j} = 0 \tag{1}$$

where $c_1, \ldots, c_n$ denote real constants. Show that for an arbitrary function F in C^2 and an arbitrary unit vector $\boldsymbol{\alpha} = (\alpha_1, \ldots, \alpha_n)$,

$$u(\mathbf{x}, t) = F(\boldsymbol{\alpha} \cdot \mathbf{x} - \mu t) \tag{2}$$

satisfies (*1*), provided μ satisfies

$$\mu^2 = \sum_{j=1}^{n} c_j^2 \alpha_j^2 \tag{3}$$

Substitute (*2*) in (*1*), to find

$$\left(\mu^2 - \sum_{j=1}^{n} c_j^2 \alpha_j^2\right) F''(\boldsymbol{\alpha} \cdot \mathbf{x} - \mu t) = 0$$

Evidently, if μ satisfies (*3*), then $u(\mathbf{x}, t)$ as given by (*2*) is a solution of (*1*), with no further restrictions on F or on $\boldsymbol{\alpha}$.

For each fixed t, $\boldsymbol{\alpha} \cdot \mathbf{x} = \mu t + \text{const.}$ is the equation of a plane in $\mathbf{R}^n$ having normal vector $\boldsymbol{\alpha}$. For this reason, (*2*) is called a *plane wave solution* to (*1*). The function F is called the *waveform* and $\boldsymbol{\alpha}$ represents the direction in which the wave progresses. While both F and $\boldsymbol{\alpha}$ are arbitrary, the *wave velocity* μ depends on $\boldsymbol{\alpha}$ via (*3*). Evidently, (*1*) models wave propagation in a nonisotropic medium.

4.13 Consider the initial–boundary value problem

$$\begin{aligned} u_{tt}(x,t) &= a^2 u_{xx}(x,t) && 0<x<L,\ t>0 \\ u(x,0) &= F(x) && 0<x<L \\ u_t(x,0) &= G(x) && 0<x<L \\ u_x(0,t) = u_x(L,t) &= 0 && t>0 \end{aligned}$$

Show by the energy integral method (cf. Problem 4.2) that if this problem has a solution, then the solution is unique.

Let $u_1(x,t)$ and $u_2(x,t)$ denote two C^1 solutions and let $v(x,t) \equiv u_1(x,t) - u_2(x,t)$ for $0<x<L$, $t>0$. Then $v(x,t)$ satisfies the initial–boundary value problem with $F(x) = G(x) = 0$ for $0<x<L$. Defining the energy integral in this case as

$$E(t) \equiv \frac{1}{2}\int_0^L [v_t(x,t)^2 + a^2 v_x(x,t)^2]\, dx$$

we have

$$E'(t) = \int_0^L [v_t(x,t)v_{tt}(x,t) + a^2 v_x(x,t)v_{xt}(x,t)]\, dx$$

But

$$v_t v_{tt} + a^2 v_x v_{xt} = v_t(v_{tt} - a^2 v_{xx}) + a^2(v_t v_{xx} + v_x v_{xt}) = a^2 \frac{\partial}{\partial x}(v_x v_t)$$

Therefore,

$$\frac{1}{a^2} E'(t) = v_x(x,t)v_t(x,t)\Big|_{x=0}^{x=L}$$

and the boundary conditions imply that $E'(t) = 0$. The initial conditions imply that $E(0) = 0$; consequently, $E(t) = 0$ for $t \geq 0$. But then the C^0 functions v_t and v_x must be identically zero, so that $v(x,t)$ is a constant. Because $v(x,0) = 0$, this constant must be zero.

If the original boundary conditions are replaced by the conditions $u(0,t) = u(L,t) = 0$, then

$$v_t(0,t) = v_t(L,t) = 0$$

and the uniqueness proof goes through as before.

4.14 Let $f(x)$ and $g(x)$ be defined on $[0, \pi]$, where they are sectionally continuous with sectionally continuous derivatives. Let $\tilde{F}_o$ and $\tilde{G}_o$ denote the odd 2π-periodic extensions of f and g to the entire real axis (Problem 6.1(*b*)). (*a*) Show that the solution of the initial–boundary value problem

$$u_{tt}(x,t) = a^2 u_{xx}(x,t) \qquad 0<x<\pi,\ t>0 \tag{1}$$

$$u(x,0) = f(x) \qquad 0<x<\pi \tag{2}$$

$$u_t(x,0) = g(x) \qquad 0<x<\pi \tag{3}$$

$$u(0,t) = u(\pi,t) = 0 \qquad t>0 \tag{4}$$

is given by

$$u(x,t) = \frac{1}{2}[\tilde{F}_o(x+at) + \tilde{F}_o(x-at)] + \frac{1}{2a}\int_{x-at}^{x+at} \tilde{G}_o(s)\, ds \tag{5}$$

(*b*) Relate the smoothness of the solution $u(x,t)$ to the smoothness of the data f, g and to the compatibility between the initial data and the boundary conditions (*4*).

(*a*) According to Problem 4.10, (*5*) is the solution of the following initial value problem:

$$\begin{aligned} u_{tt}(x, t) &= a^2 u_{xx}(x, t) && -\infty < x < \infty,\ t > 0 \\ u(x, 0) &= \tilde{F}_o(x) && -\infty < x < \infty \\ u_t(x, 0) &= \tilde{G}_o(x) && -\infty < x < \infty \end{aligned}$$

Since $\tilde{F}_o(x) = f(x)$ and $\tilde{G}_o(x) = g(x)$, for $0 < x < \pi$, the expression (*5*) satisfies the PDE (*1*), together with the initial conditions (*2*) and (*3*). Moreover, for $t > 0$,

$$u(0, t) = \frac{1}{2}[\tilde{F}_o(at) + \tilde{F}_o(-at)] + \frac{1}{2a}\int_{-at}^{at} \tilde{G}_o(s)\, ds = 0$$

since $\tilde{F}_o$ and $\tilde{G}_o$ are odd functions. Similarly, using the 2π-periodicity, $u(\pi, t) = 0$ for $t > 0$. Thus $u(x, t)$ as given by (*5*) satisfies (*4*) as well. In Problem 4.13 we proved that the problem (*1*) through (*4*) has at most one solution; therefore, (*5*) is *the* solution.

(*b*) Differentiation of (*5*) gives

$$\frac{\partial^{m+n} u}{\partial x^m \partial t^n} = \frac{a^n}{2}[\tilde{F}_o^{(m+n)}(x + at) + \tilde{F}_o^{(m+n)}(x - at)(-1)^n] + \frac{a^{n-1}}{2}[\tilde{G}_o^{(m+n-1)}(x + at) - (-1)^{n-1}\tilde{G}_o^{(m+n-1)}(x - at)]$$

($m, n = 0, 1, 2, \ldots$). Evidently, the continuity of $u(x, t)$ and its derivatives is determined by the smoothness of $\tilde{F}_o$ and $\tilde{G}_o$, which, in turn, is dependent on the smoothness of f and g in $[0, \pi]$, and the compatibility of f and g with the boundary conditions (*4*).

In Problem 6.4 it will be shown that $\tilde{F}_o(x)$ and all its derivatives through order M are continuous for all x if and only if:

(i) $f(x)$ and all its derivatives through order M are continuous on $[0, \pi]$;

(ii) for all nonnegative integers n such that $2n \le M$, $f^{(2n)}(0) = f^{(2n)}(\pi) = 0$.

Now (ii) is just the condition that f and g and their even-order derivatives satisfy the boundary conditions; this is what is meant by *compatibility* between the initial data and the boundary conditions. If, for some $M > 0$, (i) or (ii) is not satisfied by both f and g, the solution $u(x, t)$ will experience some sort of discontinuity along a characteristic. For example, if $g(0) \ne 0$, then $\tilde{G}_o(x)$ is discontinuous at every integer multiple of π, which means, by (*6*), that $u_t(x, t)$ and $u_x(x, t)$ experience discontinuities for (x, t) such that $x \pm at = k\pi$ (k = integer).

Supplementary Problems

4.15 Determine the most general spherically symmetric solution to the three-dimensional wave equation, $u_{tt} - a^2 \nabla^2 u = 0$. [*Hint*: Find the PDE satisfied by $v(r, t) \equiv r u(r, t)$.]

4.16 In Problem 4.2, let the boundary condition be replaced by

$$\alpha(x, y, z, t) u(x, y, z, t) + \beta(x, y, z, t)\frac{\partial u}{\partial n}(x, y, z, t) = f(x, y, z, t) \qquad \text{in } S \times [0, T]$$

where the continuous functions α and β satisfy

$$\alpha\beta \ge 0 \qquad \alpha^2 + \beta^2 > 0$$

in $S \times [0, T]$. Prove uniqueness by the energy integral method. *Hint*:

$$\alpha v^2 + \beta v \frac{\partial v}{\partial n} = 0 \Rightarrow v \frac{\partial v}{\partial n} = 0$$

4.17 (*a*) Differentiate under the integral sign to verify that

$$u(x, t) = \frac{1}{\sqrt{4\pi t}} \int_{-\infty}^{\infty} F(y)\, e^{-(x-y)^2/4t}\, dy$$

satisfies $u_t = u_{xx}$. (*b*) Infer from (*5*) of Problem 4.7 that if F is continuous, $\lim_{t \to 0^+} u(x, t) = F(x)$.

4.18 For N a positive integer, let

$$u_N(x, t) \equiv \sum_{n=1}^{N} C_n\, e^{-n^2\pi^2 t} \sin n\pi x \qquad 0 < x < 1,\ t > 0$$

Show that u_N satisfies

$$\begin{aligned} u_t &= u_{xx} && 0 < x < 1,\ t > 0 \\ u(0, t) = u(1, t) &= 0 && t > 0 \end{aligned}$$

for all choices of the constants C_n.

4.19 (*a*) Find plane wave solutions for

$$u_{tt} = a_1^2 u_{xx} + a_2^2 u_{yy} + b^2 u \qquad -\infty < x, y < \infty,\ t > 0$$

(*b*) Are there any values of p for which $u(x, y, t) = \sin(x/a_1)\cos(y/a_2)\sin pt$ is a (standing wave) solution of the above equation?

4.20 Consider the problem

$$\begin{aligned} u_{tt} &= u_{xx} && 0 < x < 1,\ 0 < t < T \\ u(x, 0) = u(x, T) &= 0 && 0 < x < 1 \\ u(0, t) = u(1, t) &= 0 && 0 < t < T \end{aligned}$$

Show that if T is irrational, the only solution is $u(x, t) = 0$; whereas if T is rational, the problem has infinitely many nontrivial solutions. Infer that the solution to the Dirichlet problem for the wave equation does not depend continuously on the data.

4.21 Prove that the conclusions of Theorems 4.3 and 4.4 regarding the solution to $u_t - L[u] = 0$ continue to hold if we replace the hypotheses that $M \geq 0$ and $c(\mathbf{x}, t) \leq 0$ with the hypotheses that $M = 0$ and $c(\mathbf{x}, t)$ is bounded above (but may assume positive values). [*Hint:* Let $c(\mathbf{x}, t) \leq A$, and let $u(\mathbf{x}, t) = e^{At} v(\mathbf{x}, t)$.]

4.22 Let $F(\mathbf{x}, t)$ denote a function which is defined and continuous for $\mathbf{x}$ in $\mathbf{R}^n$, $t > 0$. For τ a fixed positive parameter, let $v_F(\mathbf{x}, t; \tau)$ denote the solution of

$$v_{tt}(\mathbf{x}, t) = \nabla^2 v(\mathbf{x}, t) \qquad \mathbf{x} \text{ in } \mathbf{R}^n,\ t > \tau \tag{1}$$

$$v(\mathbf{x}, \tau) = 0 \qquad \mathbf{x} \text{ in } \mathbf{R}^n \tag{2}$$

$$v_t(\mathbf{x}, \tau) = F(\mathbf{x}, \tau) \qquad \mathbf{x} \text{ in } \mathbf{R}^n \tag{3}$$

Show that

$$u(\mathbf{x}, t) \equiv \int_0^t v_F(\mathbf{x}, t; \tau)\, d\tau \tag{4}$$

satisfies

$$u_{tt}(\mathbf{x}, t) = \nabla^2 u(\mathbf{x}, t) + F(\mathbf{x}, t) \qquad \mathbf{x} \text{ in } \mathbf{R}^n,\ t > 0 \tag{5}$$

$$u(\mathbf{x}, 0) = u_t(\mathbf{x}, 0) = 0 \qquad \mathbf{x} \text{ in } \mathbf{R}^n \tag{6}$$

This observation is known as *Duhamel's principle.*

4.23 Derive a version of Duhamel's principle for the heat equation.

4.24 Use Duhamel's principle to solve

$$\begin{aligned} u_{tt} &= a^2 u_{xx} + f(x) && x \text{ in } \mathbf{R}^1,\ t > 0 \\ u(x, 0) = u_t(x, 0) &= 0 && x \text{ in } \mathbf{R}^1 \end{aligned}$$

Chapter 5

First-Order Equations

5.1 INTRODUCTION

First-order PDEs are used to describe a variety of physical phenomena.

EXAMPLE 5.1

(*a*) The first-order system

$$(\rho u)_x + \rho_t = 0$$
$$uu_x + u_t = -\frac{1}{\rho} p_x$$
$$up_x + p_t = -\gamma p u_x$$

governs the one-dimensional adiabatic flow of an ideal gas with velocity u, density ρ, and pressure p.

(*b*) The voltage v and current i in a transmission line satisfy the first-order system

$$\frac{\partial i}{\partial x} + C\frac{\partial v}{\partial t} = -Gv$$
$$\frac{\partial v}{\partial x} + L\frac{\partial i}{\partial t} = -Ri$$

where R, L, C, and G denote respectively resistance, inductance, capacitance, and leakage conductance, all per unit length.

(*c*) Water flow with velocity v and depth u in a slightly inclined, rectangular, open channel is described by the first-order system

$$vu_x + uv_x + u_t = 0$$
$$gu_x + vv_x + v_t = g(S_0 - S_f)$$

where S_0 is the bed slope, S_f measures the frictional resistance to flow, and g is the gravitational acceleration constant. In the equations, the channel width has been taken as the unit of length.

(*d*) Population density u at time t of age-a individuals satisfies the *McKendrick–von Foerster equation*,

$$u_t + u_a = -c(t, a, u)$$

where $c(t, a, u)$ represents the removal rate at time t of age-a individuals.

5.2 CLASSIFICATION

The general *quasilinear system* of n first-order PDEs in n functions of two independent variables is

$$\sum_{j=1}^{n} a_{ij}\frac{\partial u_j}{\partial x} + \sum_{j=1}^{n} b_{ij}\frac{\partial u_j}{\partial y} = c_i \qquad (i = 1, 2, \ldots, n) \tag{5.1}$$

where a_{ij}, b_{ij}, and c_i may depend on $x, y, u_1, u_2, \ldots, u_n$. If each a_{ij} and b_{ij} is independent of $u_1, u_2, \ldots, u_n$, the system (*5.1*) is called *almost linear*. If, in addition, each c_i depends linearly on $u_1, u_2, \ldots, u_n$, the system is said to be *linear*.

EXAMPLE 5.2 The systems of Examples 5.1(*a*) and (*c*) are quasilinear; that of (*b*) is linear; and that of (*d*) is almost linear.

In terms of the $n \times n$ matrices $\mathbf{A} = [a_{ij}]$ and $\mathbf{B} = [b_{ij}]$, and the column vectors $\mathbf{u} = (u_1, u_2, \ldots, u_n)^T$ and $\mathbf{c} = (c_1, c_2, \ldots, c_n)^T$, the system (*5.1*) can be expressed as

$$\mathbf{A}\mathbf{u}_x + \mathbf{B}\mathbf{u}_y = \mathbf{c} \tag{5.2}$$

A system of equations of the form

$$\frac{\partial}{\partial x}\mathbf{F}(\mathbf{u}) + \frac{\partial \mathbf{u}}{\partial y} = \mathbf{0} \tag{5.3}$$

is called a *conservation-law system*; y usually represents a time variable.

EXAMPLE 5.3 For the case $S_0 - S_f = 0$, the system of Example 5.1(*c*) may be written in conservation form as

$$\begin{aligned} (uv)_x + u_t &= 0 \\ (gu + \tfrac{1}{2}v^2)_x + v_t &= 0 \end{aligned}$$

A system of equations of the form

$$\frac{\partial}{\partial x}\mathbf{F}(\mathbf{u}) + \frac{\partial}{\partial y}\mathbf{G}(\mathbf{u}) = \mathbf{0} \tag{5.4}$$

is said to be in *divergence form*. Clearly, any conservation-law system is in divergence form, with $\mathbf{G}(\mathbf{u}) = \mathbf{u}$.

EXAMPLE 5.4 The system of Example 5.1(*a*) is expressible in divergence form as

$$\begin{aligned} (\rho u)_x + \rho_t &= 0 \\ (p + \rho u^2)_x + (\rho u)_t &= 0 \\ \left(\frac{\rho u^3}{2} + \frac{\gamma u p}{\gamma - 1}\right)_x + \left(\frac{\rho u^2}{2} + \frac{p}{\gamma - 1}\right)_t &= 0 \end{aligned}$$

If $\mathbf{A}$ or $\mathbf{B}$ is nonsingular, it is usually possible to classify system (*5.2*) according to type. Suppose $\det(\mathbf{B}) \neq 0$ and define a polynomial of degree n in λ by

$$P_n(\lambda) \equiv \det(\mathbf{A}^T - \lambda \mathbf{B}^T) = \det(\mathbf{A} - \lambda \mathbf{B}) \tag{5.5}$$

System (*5.2*) is classified as

elliptic if $P_n(\lambda)$ has no real zeros.

hyperbolic if $P_n(\lambda)$ has n real, distinct zeros; *or* if $P_n(\lambda)$ has n real zeros, at least one of which is repeated, and the generalized eigenvalue problem $(\mathbf{A}^T - \lambda \mathbf{B}^T)\mathbf{t} = \mathbf{0}$ yields n linearly independent eigenvectors $\mathbf{t}$.

parabolic if $P_n(\lambda)$ has n real zeros, at least one of which is repeated, and the above generalized eigenvalue problem yields fewer than n linearly independent eigenvectors.

An exhaustive classification cannot be carried out when $P_n(\lambda)$ has both real and complex zeros. Since a_{ij} and b_{ij} are allowed to depend on $x, y, u_1, u_2, \ldots, u_n$, the above classification may be position and/or solution dependent.

EXAMPLE 5.5

(*a*) All four systems of Example 5.1 are hyperbolic.

(*b*) If the Cauchy-Riemann equations, $u_x = v_y$, $u_y = -v_x$, are written in the form (*5.2*), then

$$\mathbf{A} = \begin{bmatrix} 1 & 0 \\ 0 & 1 \end{bmatrix} \qquad \mathbf{B} = \begin{bmatrix} 0 & -1 \\ 1 & 0 \end{bmatrix}$$

and $P_2(\lambda) = \lambda^2 + 1$, which has no real zeros. Thus, the Cauchy-Riemann equations are elliptic (as is Laplace's equation for either u or v).

(c) If u satisfies the system of equations $u_t = v_y$, $u_y = v$, then u satisfies the heat equation, $u_t = u_{yy}$. With t playing the role of x in (5.2), we have

$$\mathbf{A} = \begin{bmatrix} 1 & 0 \\ 0 & 0 \end{bmatrix} \qquad \mathbf{B} = \begin{bmatrix} 0 & -1 \\ 1 & 0 \end{bmatrix}$$

and $P_2(\lambda) = \lambda^2$. All eigenvectors corresponding to the real double root $\lambda = 0$ are scalar multiples of $[0, 1]^T$. Hence there is just one linearly independent eigenvector and the first-order system is parabolic.

The method of characteristics for linear second-order PDEs (Chapter 2) may be usefully extended to hyperbolic, but not to elliptic or parabolic, first-order systems. For this reason, *the remainder of this chapter will deal almost exclusively with hyperbolic systems.*

5.3 NORMAL FORM FOR HYPERBOLIC SYSTEMS

If in (5.2), $\mathbf{Au}_x + \mathbf{Bu}_y = \mathbf{c}$, the coefficient matrices $\mathbf{A}$ and $\mathbf{B}$ are such that $\mathbf{A} = \mathbf{DB}$, for some diagonal matrix $\mathbf{D}$, then the system can be written in component form as

$$\sum_{j=1}^{n} b_{ij}\left(d_{ii}\frac{\partial u_j}{\partial x} + \frac{\partial u_j}{\partial y}\right) = c_i \qquad (i = 1, 2, \ldots, n)$$

wherein the ith equation involves differentiation only in a single direction—the direction $dx/dy = d_{ii}$. We say in this case that (5.2) is in *normal form*. When a system is in normal form, techniques of ordinary differential equations become applicable to it.

Suppose that (5.2) is hyperbolic, and let $\lambda_1, \lambda_2, \ldots, \lambda_n$ denote the n real zeros of the polynomial (5.5). The *characteristics* of (5.2) are those curves in the xy-plane along which

$$\frac{dx}{dy} = \lambda_i \qquad (i = 1, 2, \ldots, n) \tag{5.6}$$

EXAMPLE 5.6 For a linear system, the λ_i depend at most on x and y; so the characteristics of (5.2) can be determined by integrating the ordinary equations (5.6). For a quasilinear system, where the λ_i depend on $u_1, u_2, \ldots, u_n$, the characteristics are solution dependent. In the case that (5.2) consists of a single quasilinear PDE, many authors call the plane curves determined by (5.6) the *characteristic base curves*, and use the term "characteristics" or "characteristic curves" to denote the space curves in xyu-space whose projections on the xy-plane are the characteristic base curves. In this Outline we shall use "characteristics" to denote both the plane and the space curves; the context will make it clear which kind of curves is intended.

Theorem 5.1: For (5.2) hyperbolic, let $\mathbf{D}$ denote the $n \times n$ diagonal matrix of the λ_i. Then there exists a nonsingular $n \times n$ matrix $\mathbf{T}$ satisfying

$$\mathbf{TA} = \mathbf{DTB} \tag{5.7}$$

According to Theorem 5.1, if (5.2) is not already in normal form, the transformed system

$$\mathbf{TAu}_x + \mathbf{TBu}_y = \mathbf{Tc} \tag{5.8}$$

is in normal form, with the ith row-equation involving differentiation only in the direction of the tangent to the ith characteristic. Stated otherwise, *a hyperbolic system can always be brought into normal form by taking suitable linear combinations of the equations.*

5.4 THE CAUCHY PROBLEM FOR A HYPERBOLIC SYSTEM

The *Cauchy problem* (or *initial value problem*) for a hyperbolic system (5.2) calls for determining $u_i(x, y)$ $(i = 1, 2, \ldots, n)$ that satisfy (5.2) and take prescribed values (the *initial data*) on some *initial curve*, Γ. If Γ is nowhere tangent to a characteristic of (5.2) and if the coefficients in (5.2) are continuous, the Cauchy problem is well-posed in a neighborhood of Γ. At the other extreme, if Γ coincides with a characteristic, then the Cauchy problem usually will be insoluble (see Problem 5.29).

In illustration, suppose that $n = 3$, and let $\mathscr{C}_1, \mathscr{C}_2, \mathscr{C}_3$ be the characteristics of (*5.2*) that pass through the point R (see Fig. 5-1). The shaded region of the xy-plane enveloped by the characteristics and the initial curve Γ is called the *domain of dependence* of the point R; the portion of Γ between P and Q is called the *interval of dependence* of R. Changes in the initial data exterior to the interval PQ will not affect the solution at R.

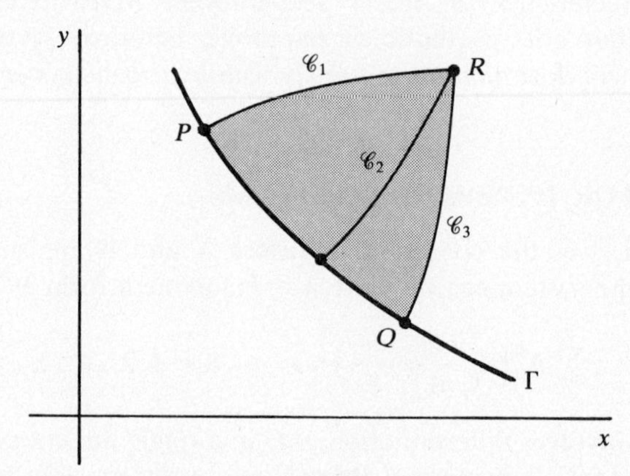

Fig. 5-1

Any discontinuities in the initial data are propagated away from the initial curve along the characteristics defined by (*5.6*). When the system (5.2) is nonlinear, it is possible—even for smooth initial data—for the solution to develop discontinuities some distance from the initial curve. These discontinuities occur when two characteristics carrying contradictory information about the solution intersect. A curve across which one or more of the $u_i(x, y)$ have jump discontinuities is called a *shock*. The position of the shock and the magnitudes of the jumps in the u_i are determined by conservation principles (see Problem 5.17).

Solved Problems

5.1 Show that the open-channel flow equations, Example 5.1(*c*), compose a hyperbolic system and describe the characteristics.

In matrix form the open-channel flow equations are

$$\begin{bmatrix} v & u \\ g & v \end{bmatrix}\begin{bmatrix} u \\ v \end{bmatrix}_x + \begin{bmatrix} 1 & 0 \\ 0 & 1 \end{bmatrix}\begin{bmatrix} u \\ v \end{bmatrix}_t = \begin{bmatrix} 0 \\ g(S_0 - S_f) \end{bmatrix}$$

The characteristic polynomial

$$P_2(\lambda) = \det\left(\begin{bmatrix} v & u \\ g & v \end{bmatrix} - \lambda\begin{bmatrix} 1 & 0 \\ 0 & 1 \end{bmatrix}\right) = (v - \lambda)^2 - gu$$

has two real zeros, $\lambda_1 = v + \sqrt{gu}$ and $\lambda_2 = v - \sqrt{gu}$. Thus, the characteristics are those curves in the xt-plane along which

$$\frac{dx}{dt} = v + \sqrt{gu} \qquad \text{or} \qquad \frac{dx}{dt} = v - \sqrt{gu}$$

The speed (or *celerity*) of a small gravity-wave in shallow water of depth u is given by $c = \sqrt{gu}$. Making the substitution $c^2 = gu$, the open-channel flow equations become

$$2vc_x + cv_x + 2c_t = 0$$
$$2cc_x + vv_x + v_t = g(S_0 - S_f)$$

and the characteristics of this system are the curves along which

$$\frac{dx}{dt} = v + c \qquad \text{or} \qquad \frac{dx}{dt} = v - c$$

5.2 Show that the characteristics of the quasilinear first-order PDE

$$au_x + bu_y = c \tag{1}$$

are the curves along which (*1*) and a knowledge of u are insufficient uniquely to determine u_x and u_y.

First note that, since the only zero of $a - \lambda b$ is $\lambda = a/b$, the characteristics of (*1*) are the curves along which

$$\frac{dx}{dy} = \frac{a}{b}$$

Let $\mathscr{C}$: $x = p(r)$, $y = q(r)$ be a curve along which u is given by $u = f(r)$. From

$$u(p(r), q(r)) = f(r) \qquad \text{along } \mathscr{C}$$

and (*1*), we have

$$p'u_x + q'u_y = f'$$
$$au_x + \ bu_y = c$$

by which u_x and u_y are uniquely determined along $\mathscr{C}$, unless the determinant of the system is zero:

$$\begin{vmatrix} p' & q' \\ a & b \end{vmatrix} = p'b - q'a = 0 \tag{2}$$

But (*2*) holds if and only if

$$\frac{dx}{dy} = \frac{p'}{q'} = \frac{a}{b} \qquad \text{along } \mathscr{C}$$

which is to say, if and only if $\mathscr{C}$ is a characteristic of (*1*).

5.3 (*a*) Show that the first-order quasilinear equation

$$au_x + bu_y = c \qquad (b \neq 0) \tag{1}$$

is in normal form. (*b*) Find the *canonical* or *characteristic equations* for (*1*).

(*a*) We know that a characteristic of (*1*) is defined by

$$\frac{dx}{dy} = \frac{a}{b} \qquad \text{or} \qquad \frac{dx}{a} = \frac{dy}{b} \tag{2}$$

Calculating the derivative of u in the direction $\mathbf{v} = (a, b)$ tangential to the characteristic, we find

$$\mathbf{v} \cdot \nabla u = (a, b) \cdot (u_x, u_y) = au_x + bu_y$$

Therefore, (*1*) involves differentiation in a single direction (along the characteristic); so, by definition, it is in normal form.

(b) Let $\mathscr{C}$: $(x(r), y(r))$ be a characteristic of (1), parameterized by r. Along $\mathscr{C}$,

$$\frac{x'(r)}{y'(r)} = \frac{dx}{dy} = \frac{a}{b} \tag{3}$$

by (2), and $u = u(x(r), y(r))$. By the chain rule, (3), and (1),

$$\frac{du}{dr} = \frac{\partial u}{\partial x}\frac{dx}{dr} + \frac{\partial u}{\partial y}\frac{dy}{dr} = \left(\frac{a}{b}u_x + u_y\right)\frac{dy}{dr} = \frac{c}{b}\frac{dy}{dr}$$

Thus, the canonical or characteristic equations for (1) can be written as

$$\frac{dx}{dr} = \frac{a}{b}\frac{dy}{dr} \qquad \frac{du}{dr} = \frac{c}{b}\frac{dy}{dr} \tag{4a}$$

or symmetrically as

$$\frac{dx}{a} = \frac{dy}{b} = \frac{du}{c} \tag{4b}$$

or unlinked as

$$\frac{\partial x}{\partial r} = a \qquad \frac{\partial y}{\partial r} = b \qquad \frac{\partial u}{\partial r} = c \tag{4c}$$

Form (4c) may be interpreted as indicating a change of coordinates from (x, y) to (r, s): in the new coordinates, the characteristics are the straight lines $s =$ const., and (1) takes the canonical form $u_r = c$.

5.4 Show that a surface $\mathscr{S}$ given by $u = f(x, y)$ defines a solution to the quasilinear first-order equation

$$au_x + bu_y = c \tag{1}$$

if and only if the characteristic equations (4) of Problem 5.3 hold at each point of $\mathscr{S}$. In other words, a solution surface of (1) consists entirely of (space) characteristics.

If $f(x, y) - u = 0$, then

$$0 = d(f(x, y) - u) = f_x dx + f_y dy - du = (f_x, f_y, -1) \cdot (dx, dy, du)$$

Now, if (4b) of Problem 5.3 holds, the vectors (dx, dy, du) and (a, b, c) are parallel, whence

$$0 = (f_x, f_y, -1) \cdot (a, b, c) = af_x + bf_y - c \tag{2}$$

i.e., the function f satisfies (1).

Conversely, if $\mathscr{S}$ is defined by a solution $f(x, y) - u = 0$ of (1), then (2) shows that at any point P of $\mathscr{S}$ the vector (a, b, c) is orthogonal to the surface normal $(f_x, f_y, -1)$. Thus, (a, b, c) represents a direction in the tangent plane at P; a curve $\mathscr{C}$ lying in $\mathscr{S}$ and passing through P in this direction will have, at P, the tangent vector

$$(dx, dy, du) = (\text{const.})(a, b, c)$$

But this relation is just (4b) of Problem 5.3.

5.5 Solve the Cauchy problem

$$a(x, y, u)u_x + b(x, y, u)u_y = c(x, y, u) \tag{1}$$

$$u = u_0(s) \qquad \text{on } \Gamma\text{:} \quad x = F(s), \; y = G(s) \tag{2}$$

where, for all s,

$$\frac{F'(s)}{G'(s)} \neq \frac{a(F(s), G(s), u_0(s))}{b(F(s), G(s), u_0(s))}$$

(which means that Γ is nowhere tangent to a characteristic of (1)).

In the xy-plane, the solution process may be described as the threading of a characteristic through each point of the initial curve Γ (see Fig. 5-2). Thus, for each fixed s, imagine system (*4c*) of Problem 5.6,

$$\frac{\partial x}{\partial r} = a \qquad \frac{\partial y}{\partial r} = b \qquad \frac{\partial u}{\partial r} = c \tag{3}$$

—where the parameter r is chosen so that Γ is represented by $r = 0$—to be solved subject to the initial conditions

$$x(0, s) = F(s) \qquad y(0, s) = G(s) \qquad u(0, s) = u_0(s) \tag{4}$$

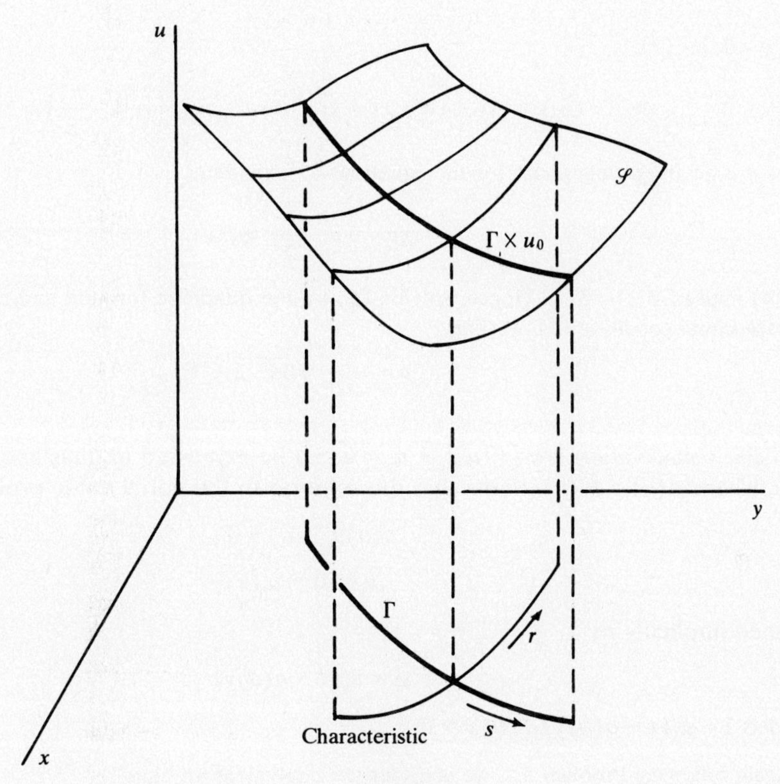

Fig. 5-2

This solution will have the form

$$x = x(r, s) \qquad y = y(r, s) \qquad u = u(r, s)$$

which are the parametric equations of a surface $\mathscr{S}$.

By Problem 5.4, $\mathscr{S}$ is a solution surface for (*1*); and the conditions (*4*) ensure that the curve $\Gamma \times u_0$ lies in $\mathscr{S}$, as required by (*2*). Hence, if we can solve for r and s in terms of x and y, the function

$$u = u(r, s) = u(r(x, y), s(x, y))$$

will solve (*1*)–(*2*). Now, it is in fact possible to invert the transformation $x = x(r, s)$, $y = y(r, s)$ in a neighborhood of Γ, because, along the curve, the Jacobian does not vanish:

$$\begin{aligned}\frac{\partial(x, y)}{\partial(r, s)} &= x_r y_s - y_r x_s = aG'(s) - bF'(s) \\ &= bG'\left(\frac{a}{b} - \frac{F'}{G'}\right) \neq 0\end{aligned}$$

5.6 Solve the quasilinear Cauchy problem

$$xu_x + yuu_y = -xy \tag{1}$$

$$u = 5 \quad \text{on } xy = 1 \quad (x > 0) \tag{2}$$

Following Problem 5.5, we wish to solve

$$x_r = x \qquad y_r = yu \qquad u_r = -xy \tag{3}$$

subject to

$$x(0, s) = s \qquad y(0, s) = \frac{1}{s} \qquad u(0, s) = 5 \tag{4}$$

where $s > 0$. By (3),

$$(xy)_r = x_r y + xy_r = xy + xyu = -u_r - uu_r = \left(-u - \frac{u^2}{2}\right)_r \tag{5}$$

i.e., $1 + u$ is an *integrating factor* for the equations (3), yielding

$$xy = -u - \frac{u^2}{2} + \phi(s) \tag{6}$$

Now (4) implies $\phi(s) = 37/2$. Hence, solving (6) by the quadratic formula and choosing the root that obeys the initial condition (2), we find

$$u = -1 + \sqrt{38 - 2xy}$$

5.7 The scalar conservation law $[F(u)]_x + u_y = 0$ can be expressed in quasilinear form as $a(u)u_x + u_y = 0$, where $a(u) = F'(u)$. Show that the solution to the initial value problem

$$a(u)u_x + u_y = 0 \tag{1}$$

$$u(x, 0) = u_0(x) \tag{2}$$

is defined implicitly by

$$u = u_0(x - a(u)y) \tag{3}$$

provided $1 + u_0'(x - a(u)y)a'(u)y \neq 0$.

Again following Problem 5.5, we consider the equivalent problem

$$x_r = a(u) \qquad y_r = 1 \qquad u_r = 0 \tag{4}$$

$$x(0, s) = s \qquad y(0, s) = 0 \qquad u(0, s) = u_0(s) \tag{5}$$

Integrating the equations (4) in reverse order and applying the conditions (5), we find:

$$u = u_0(s) \qquad y = r \qquad x = a(u_0(s))r + s \tag{6}$$

From (6), there follows

$$u = u_0(s) = u_0(x - a(u_0(s))r) = u_0(x - a(u)y)$$

which is (3).

The expression (3) will actually furnish the solution to (1)–(2) provided the equation

$$\Phi(x, y, u) \equiv u - u_0(x - a(u)y) = 0$$

can be solved for u as a function of x and y. The condition for solvability is

$$\Phi_u(x, y, u) = 1 + u_0'(x - a(u)y)\, a'(u)y \neq 0$$

which certainly holds for $|y|$ sufficiently small.

5.8 One-dimensional, unsteady flow of a compressible fluid at constant pressure, p, is governed by

$$uu_x + u_t = 0 \tag{1}$$

$$(\rho u)_x + \rho_t = 0 \tag{2}$$

$$(eu)_x + e_t + pu_x = 0 \tag{3}$$

where u, ρ, and e are, respectively, the fluid's velocity, density, and internal energy per unit volume. Solve (*1*)–(*3*) subject to the initial conditions

$$u(x,0) = u_0(x) \qquad \rho(x,0) = \rho_0(x) \qquad e(x,0) = e_0(x) \tag{4}$$

According to Problem 5.7, the characteristics of (*1*) are given by

$$x - ut = s = \text{const.}$$

and the solution of (*1*) that obeys (*4*) is given implicitly by $u = u_0(s)$.

Writing (*2*) as a linear equation in ρ,

$$u\rho_x + \rho_t = -u_x\rho \tag{5}$$

we see that (*5*) has the same characteristics, $s = \text{const.}$, as (*1*), and that along a characteristic, on which the running parameter is $r = t$,

$$\frac{d\rho}{dt} = -u_x\rho \tag{6}$$

Now, $u_x = u_0'(s)s_x = u_0'(s)(1 - tu_x)$, or

$$u_x = \frac{u_0'(s)}{1 + u_0'(s)t} \tag{7}$$

Substitute (*7*) in (*6*) and integrate the resulting separable equation, using the initial condition $\rho = \rho_0(s)$ for $t = 0$:

$$\int_{\rho_0(s)}^{\rho} \frac{d\rho}{\rho} = -u_0'(s)\int_0^t \frac{dt}{1 + u_0'(s)t}$$

$$\log\frac{\rho}{\rho_0(s)} = \log\frac{1}{1 + u_0'(s)t}$$

or

$$\rho = \frac{\rho_0(s)}{1 + u_0'(s)t} \tag{8}$$

where $s = x - ut$.

Finally, (*3*) and (*4*) yield the following problem for the new unknown $E \equiv e + p$:

$$uE_x + E_t = -u_xE \qquad E(x,0) = e_0(x) + p$$

This is formally identical to the problem for ρ; hence, by analogy with (*8*),

$$E = \frac{e_0(s) + p}{1 + u_0'(s)t} \qquad \text{or} \qquad e = \frac{e_0(s) - pu_0'(s)t}{1 + u_0'(s)t} \tag{9}$$

5.9 Establish Theorem 5.1.

In component form, (*5.7*) reads

$$\sum_{j=1}^{n} (t_{ij}a_{jk} - \lambda_i t_{ij}b_{jk}) = 0 \qquad (i, k = 1, 2, \ldots, n)$$

which is equivalent to

$$[t_{i1}, t_{i2}, \ldots, t_{in}](\mathbf{A} - \lambda_i\mathbf{B}) = [0, 0, \ldots, 0] \quad (i = 1, 2, \ldots, n) \tag{1}$$

Taking the transpose of each side of (*1*) yields

$$(\mathbf{A}^T - \lambda_i \mathbf{B}^T)\mathbf{t}_i = \mathbf{0} \qquad (i = 1, 2, \ldots, n) \tag{2}$$

where $\mathbf{t}_i$ represents the entries of the ith row of $\mathbf{T}$ arranged as a column vector. Now, because the system (*5.2*) is hyperbolic, there exist n *linearly independent* vectors $\mathbf{t}_i$ satisfying the n matrix equations (*2*). Hence, the matrix $\mathbf{T}$ having these vectors as rows will be nonsingular and will satisfy (*5.7*).

5.10 (*a*) Show that the system

$$2u_x - 2v_x + u_y - 3v_y = v \tag{1}$$

$$u_x - 4v_x \qquad + \ v_y = u \tag{2}$$

is hyperbolic and use Theorem 5.1 to reduce it to normal form. (*b*) Express the system (*1*)–(*2*) in terms of characteristic coordinates.

(*a*) Writing (*1*)–(*2*) in the form (*5.2*), we have

$$\mathbf{A} = \begin{bmatrix} 2 & -2 \\ 1 & -4 \end{bmatrix} \qquad \mathbf{B} = \begin{bmatrix} 1 & -3 \\ 0 & 1 \end{bmatrix}$$

Since $\det(\mathbf{A} - \lambda\mathbf{B}) = \lambda^2 - \lambda - 6$ has distinct real zeros, $\lambda_1 = 3$ and $\lambda_2 = -2$, the system is hyperbolic. According to Theorem 5.1, the rows of the normalizing matrix $\mathbf{T}$ satisfy

$$\begin{bmatrix} 2-\lambda_i & 1 \\ -2+3\lambda_i & -(4+\lambda_i) \end{bmatrix} \begin{bmatrix} t_{i1} \\ t_{i2} \end{bmatrix} = \begin{bmatrix} 0 \\ 0 \end{bmatrix} \qquad (i = 1, 2)$$

For $i = 1$, $\lambda_1 = 3$, we can choose $t_{11} = t_{12} = 1$; for $i = 2$, $\lambda_2 = -2$, we can choose $t_{21} = 1$, $t_{22} = -4$. Thus

$$\mathbf{T} = \begin{bmatrix} 1 & 1 \\ 1 & -4 \end{bmatrix}$$

and the transformed system, (*5.8*), is

$$\begin{bmatrix} 3 & -6 \\ -2 & 14 \end{bmatrix} \begin{bmatrix} u \\ v \end{bmatrix}_x + \begin{bmatrix} 1 & -2 \\ 1 & -7 \end{bmatrix} \begin{bmatrix} u \\ v \end{bmatrix}_y = \begin{bmatrix} v+u \\ v-4u \end{bmatrix}$$

or

$$(3u_x + u_y) - 2(3v_x + v_y) = v + u \tag{3}$$

$$(-2u_x + u_y) - 7(-2v_x + v_y) = v - 4u \tag{4}$$

Equation (*3*) involves differentiation only in the direction

$$\frac{dx}{dy} = 3 = \lambda_1 \tag{5}$$

while (*4*) involves differentiation only in the direction

$$\frac{dx}{dy} = -2 = \lambda_2 \tag{6}$$

(*b*) From (*5*) and (*6*), the characteristics are given by

$$x - 3y = \beta = \text{const.} \tag{7}$$

$$x + 2y = \alpha = \text{const.} \tag{8}$$

The family (*7*), along which α varies, are called the *α-characteristics*; similarly, the family (*8*) are called the *β-characteristics*. Together, (*7*) and (*8*) define an invertible transformation from xy- to $\alpha\beta$-coordinates. We have

$$\frac{\partial}{\partial x} = \frac{\partial}{\partial \alpha} + \frac{\partial}{\partial \beta} \qquad \frac{\partial}{\partial y} = 2\frac{\partial}{\partial \alpha} - 3\frac{\partial}{\partial \beta}$$

so that (*3*)–(*4*) transforms to

$$u_\alpha - 2v_\alpha = \frac{u+v}{5} \tag{9}$$

$$u_\beta - 7v_\beta = \frac{4u-v}{5} \tag{10}$$

The system (9)–(10)–(7)–(8) constitutes the *canonical* or *characteristic form* of (1)–(2).

5.11 With reference to Problem 5.10, solve the initial value problem

$$2u_x - 2v_x + u_y - 3v_y = 0 \tag{1}$$

$$u_x - 4v_x \qquad + \ v_y = 0 \tag{2}$$

$$u(x,0) = u_0(x) \qquad v(x,0) = v_0(x) \tag{3}$$

We know that the characteristics of (1)–(2) are

$$x - 3y = \beta = \text{const.} \qquad \text{and} \qquad x + 2y = \alpha = \text{const.}$$

and that (1)–(2) has the canonical form

$$u_\alpha - 2v_\alpha = 0 \tag{4}$$

$$u_\beta - 7v_\beta = 0 \tag{5}$$

If P, Q, and R are as indicated in Fig. 5-3 and if P has coordinates (x, y), then the coordinates of Q and R are $(x-3y, 0)$ and $(x+2y, 0)$, respectively. By (4), $u - 2v$ is constant on the α-characteristic from Q to P, and, by (5), $u - 7v$ is constant on the β-characteristic from R to P; thus,

$$u(P) - 2v(P) = u_0(Q) - 2v_0(Q) \tag{6}$$

$$u(P) - 7v(P) = u_0(R) - 7v_0(R) \tag{7}$$

Together, (6) and (7) yield the solution to the initial value problem (1)–(2)–(3) as

$$u(P) = u(x,y) = \frac{1}{5}[7u_0(x-3y) - 14v_0(x-3y) - 2u_0(x+2y) + 14v_0(x+2y)]$$

$$v(P) = v(x,y) = \frac{1}{5}[u_0(x-3y) - 2v_0(x-3y) - u_0(x+2y) + 7v_0(x+2y)]$$

For a hyperbolic system, a combination of the variables that remains constant along a characteristic is known as a *Riemann invariant* of the system. By the above, $u - 2v$ and $u - 7v$ are Riemann invariants of (1)–(2).

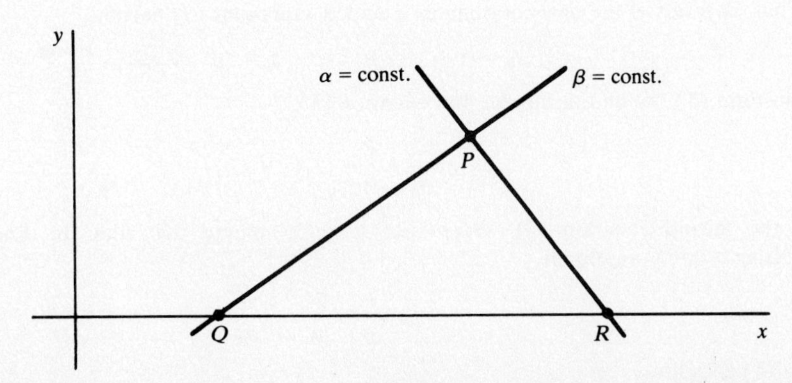

Fig. 5-3

5.12 Bring the open-channel flow equations,

$$vu_x + uv_x + u_t = 0 \tag{1}$$

$$gu_x + vv_x + v_t = g(S_0 - S_f) \tag{2}$$

into canonical form.

In Problem 5.1, the system (*1*)–(*2*) was shown to be hyperbolic, with characteristics given by

$$\frac{dx}{dt} = \lambda_1 = v + \sqrt{gu} \qquad \frac{dx}{dt} = \lambda_2 = v - \sqrt{gu} \tag{3}$$

By (*2*) of Problem 5.9, a matrix $\mathbf{T} = [t_{ij}]$ that will transform (*1*)–(*2*) to normal form satisfies

$$\begin{bmatrix} v - \lambda_i & g \\ u & v - \lambda_i \end{bmatrix} \begin{bmatrix} t_{i1} \\ t_{i2} \end{bmatrix} = \begin{bmatrix} 0 \\ 0 \end{bmatrix} \qquad (i = 1, 2) \tag{4}$$

From (*3*)–(*4*), we may take

$$t_{11} = 1 \qquad t_{12} = \sqrt{u/g} \qquad t_{21} = 1 \qquad t_{22} = -\sqrt{u/g}$$

Now, writing (*1*)–(*2*) in matrix form (*5.2*) and multiplying by $\mathbf{T}$, we find the normal equations

$$(\lambda_1 u_x + u_t) + \sqrt{u/g}(\lambda_1 v_x + v_t) = \sqrt{ug}(S_0 - S_f) \tag{5}$$

$$(\lambda_2 u_x + u_t) - \sqrt{u/g}(\lambda_2 v_x + v_t) = \sqrt{ug}(S_f - S_0) \tag{6}$$

To introduce characteristic coordinates, let the respective solutions to the two ordinary equations (*3*) be

$$F(x, t) = \beta = \text{const.} \qquad G(x, t) = \alpha = \text{const.} \tag{7}$$

i.e., the α-characteristics and the β-characteristics. To show that (*7*) defines a locally invertible coordinate transformation, compute the Jacobian

$$\frac{\partial(\alpha, \beta)}{\partial(x, t)} = \left(\frac{\partial(x, t)}{\partial(\alpha, \beta)}\right)^{-1} = \begin{vmatrix} G_x & G_t \\ F_x & F_t \end{vmatrix} = G_x F_t - F_x G_t$$

But, using (*3*),

$$0 = \frac{dF}{dt} = F_t + F_x(v + \sqrt{gu})$$
$$0 = \frac{dG}{dt} = G_t + G_x(v - \sqrt{gu}) \tag{8}$$

from which it follows that

$$\frac{\partial(\alpha, \beta)}{\partial(x, t)} = -2F_x G_x \sqrt{gu}$$

which is finite and nonzero (recall that u represents the depth of fluid).

Thus, in terms of the new coordinates α and β, equations (*3*) become

$$x_\alpha = (v + \sqrt{gu})t_\alpha \qquad x_\beta = (v - \sqrt{gu})t_\beta \tag{9}$$

To transform (*5*), we find, using the first equation (*8*),

$$\lambda_1 \frac{\partial}{\partial x} + \frac{\partial}{\partial t} = (\lambda_1 G_x + G_t)\frac{\partial}{\partial \alpha} \tag{10}$$

From the second equation (*7*), $G_x x_\alpha + G_t t_\alpha = 1$. Combining this with the first equation (*9*) and substituting in (*10*), we obtain

$$\lambda_1 \frac{\partial}{\partial x} + \frac{\partial}{\partial t} = \frac{1}{t_\alpha}\frac{\partial}{\partial \alpha}$$

so that (*5*) goes into

$$u_\alpha + \sqrt{u/g}\, v_\alpha = \sqrt{gu}(S_0 - S_f)t_\alpha \tag{11}$$

Similarly, we find for the transformation of (6):

$$u_\beta - \sqrt{u/g}\, v_\beta = \sqrt{gu}(S_f - S_0)t_\beta \tag{12}$$

Equations (11), (12), and (9) make up the canonical form of (1)–(2).

5.13 (a) Show that $v+2c$ and $v-2c$, where $c=\sqrt{gu}$, are Riemann invariants (Problem 5.11) of the open-channel flow equations, provided $S_0 - S_f = 0$ (the conservation-law case). (b) Prove that if a single characteristic of the open-channel equations is a straight line, then (i) the entire family that includes that characteristic consists of straight lines; (ii) the Riemann invariant associated with the other family of characteristics is an absolute constant.

(a) In terms of c and v, (11) and (12) of Problem 5.12 read, after cancellation of c/g,

$$2c_\alpha + v_\alpha = 0 \qquad 2c_\beta - v_\beta = 0$$

which imply that $v+2c$ is constant on an α-characteristic and $v-2c$ is constant on a β-characteristic.

(b) Suppose that the particular α-characteristic $F(x,t)=\beta_0$ is a straight line. Then, by (3) of Problem 5.12,

$$\frac{dx}{dt} = v + c = \text{const.}$$

along that characteristic. But, by (a), $v+2c = \text{const.}$ along that same characteristic. Hence v and c must be separately constant along the characteristic $F(x,t)=\beta_0$; i.e., in terms of Fig. 5-4,

$$v(R) = v(S) \qquad \text{and} \qquad c(R) = c(S) \tag{1}$$

On the α-characteristic $F(x,t)=\beta_1$ we have, by (a),

$$v(P) + 2c(P) = v(Q) + 2c(Q) \tag{2}$$

while, on the β-characteristics, we have, by (a),

$$v(P) - 2c(P) = v(R) - 2c(R) \tag{3}$$

$$v(Q) - 2c(Q) = v(S) - 2c(S) \tag{4}$$

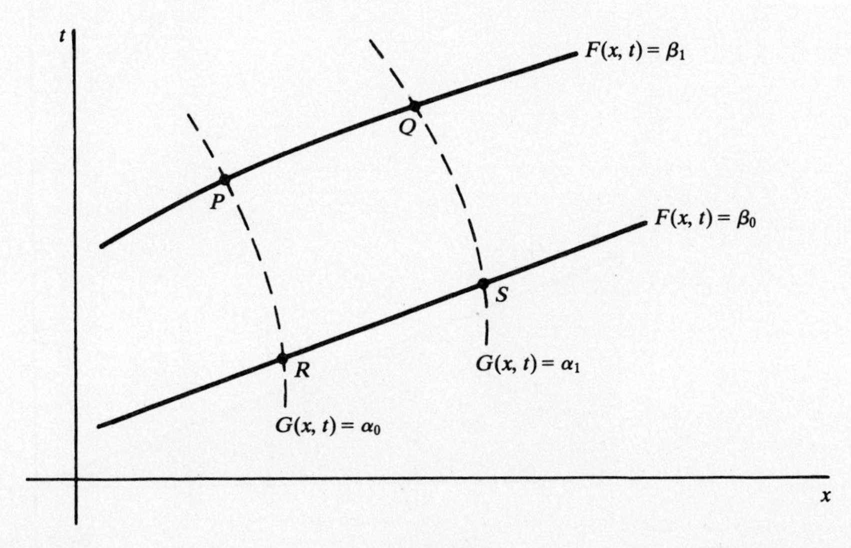

Fig. 5-4

By (*1*), the right side of (*3*) equals the right side of (*4*); so

$$v(P) - 2c(P) = v(Q) - 2c(Q) \tag{5}$$

(i) Together, (*2*) and (*5*) imply $v(P) = v(Q)$ and $c(P) = c(Q)$; hence, on $F(x, t) = \beta_1$,

$$\frac{dx}{dt}(P) = v(P) + c(P) = v(Q) + c(Q) = \frac{dx}{dt}(Q)$$

which shows that characteristic also to be a straight line. (ii) Together, (*3*) and (*5*) imply

$$v(Q) - 2c(Q) = v(R) - 2c(R)$$

i.e., the Riemann invariant $v - 2c$ has the same value at two arbitrary points of the plane, Q and R. This result in effect removes one unknown from the problem.

5.14 A river flows at a uniform depth of 2 meters and a velocity of 1 m/s into an ocean bay. Because of the tide, the water level in the bay, initially the same as the river level, falls at the rate of 0.15 m/h for 8 hours. Neglecting bed slope and frictional resistance, determine (*a*) at what distance upstream the river level is just beginning to fall at the end of the 8-hour period, (*b*) the velocity of the water entering the bay, (*c*) at what time the river level will have fallen 0.6 m at a station 5 km upstream from the bay.

The notation and results of Problems 5.12 and 5.13 will be used. The acceleration of gravity is $g = 9.8\ \text{m/s}^2$.

(*a*) The β-characteristic bordering the zone of quiet (Fig. 5-5) is the straight line

$$\frac{dx}{dt} = v(0, 0) - c(0, 0) = (1 - \sqrt{9.8 \times 2})\ \text{m/s} = -12.3\ \text{km/h} \tag{1}$$

Thus, after 8 hours, the discontinuity in u_t has been propagated $(8\ \text{h})(12.3\ \text{km/h}) = 98.4\ \text{km}$ upstream.

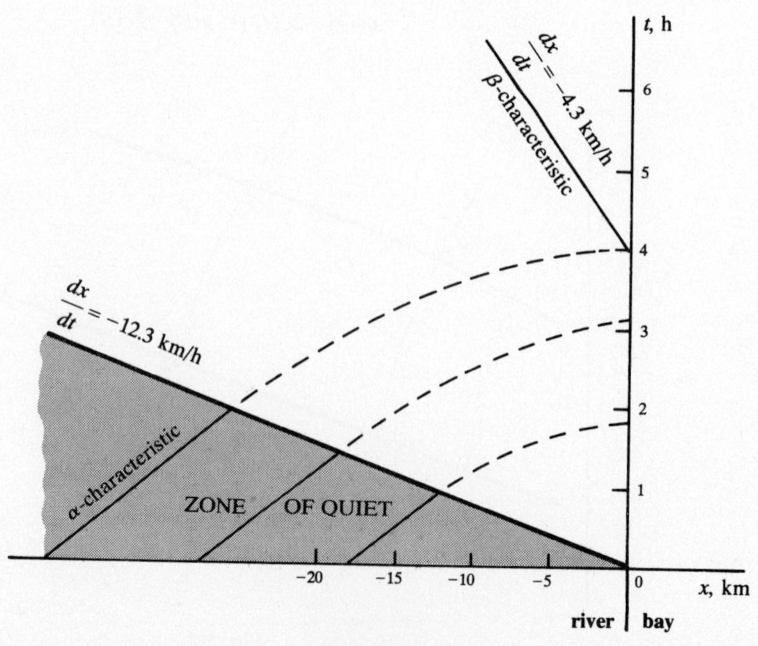

Fig. 5-5

(*b*) In view of (*1*) and Problem 5.13(*b*), all β-characteristics are straight lines, and

$$v(x, t)+2c(x, t)=v(0, 0)+2c(0, 0)=9.9 \text{ m/s} \tag{2}$$

for all x and t. Thus, in units of m/s,

$$v(0, t)=9.9-2c(0, t)=9.9-2\sqrt{9.8(2-0.15t)} \tag{3}$$

(*c*) At the outlet, $x=0$, the water level will have fallen by 0.6 m (from 2 m to 1.4 m) at time

$$t_0=\frac{0.6 \text{ m}}{0.15 \text{ m/h}}=4 \text{ h}$$

We know from Problem 5.13(*b*) that v and c (or u) are separately constant along each β-characteristic. Hence, the β-characteristic through (0, 4), which carries the value $u=1.4$ m, will have slope

$$\begin{aligned}\frac{dx}{dt}&=v(0, 4)-c(0, 4)=[9.9-2\sqrt{9.8(1.4)}]-\sqrt{9.8(1.4)}\\&=-1.2 \text{ m/s}=-4.3 \text{ km/h}\end{aligned}$$

where (*3*) was used to evaluate $v(0, 4)$. It follows that an additional time

$$t_1=\frac{5 \text{ km}}{4.3 \text{ km/h}}=1.2 \text{ h}$$

must pass before the value $u=1.4$ m is felt 5 km upstream. The total time is thus $t_0+t_1=5.2$ h.

5.15 Let Ω be the region $x_1<x<x_2$, $t_1<t<t_2$, and suppose that in Ω

$$\mathbf{u}=(u_1(x, t), u_2(x, t), \ldots, u_n(x, t))$$

solves the divergence-form first-order equation

$$\frac{\partial}{\partial x}F(\mathbf{u})+\frac{\partial}{\partial t}G(\mathbf{u})=\mathbf{0} \tag{1}$$

For any smooth function ϕ on Ω which vanishes on the boundary of Ω, show that

$$\int_{t_1}^{t_2}\int_{x_1}^{x_2}[F(\mathbf{u})\phi_x+G(\mathbf{u})\phi_t]\,dx\,dt=\mathbf{0} \tag{2}$$

From $[F(\mathbf{u})\phi]_x=[F(\mathbf{u})]_x\phi+F(\mathbf{u})\phi_x$ and $\phi=0$ for $x=x_1$ and $x=x_2$, we have

$$\int_{t_1}^{t_2}\int_{x_1}^{x_2}[F(\mathbf{u})]_x\phi\,dx\,dt=-\int_{t_1}^{t_2}\int_{x_1}^{x_2}F(\mathbf{u})\phi_x\,dx\,dt \tag{3}$$

Similarly, since $\phi=0$ for $t=t_1$ and $t=t_2$,

$$\int_{x_1}^{x_2}\int_{t_1}^{t_2}[G(\mathbf{u})]_t\phi\,dt\,dx=-\int_{x_1}^{x_2}\int_{t_1}^{t_2}G(\mathbf{u})\phi_t\,dt\,dx=-\int_{t_1}^{t_2}\int_{x_1}^{x_2}G(\mathbf{u})\phi_t\,dx\,dt \tag{4}$$

Now, multiplying (*1*) by ϕ, integrating over Ω, and applying (*3*) and (*4*), we obtain (*2*).

Smooth (C^∞) functions ϕ which vanish in a neighborhood of (and not merely on) the boundary of Ω are called *test functions* on Ω. We say that $\mathbf{u}$ is a *weak solution* of (*1*) in Ω if (*2*) is valid for all test functions ϕ on Ω. Since (*2*) does not impose any continuity requirements on $\mathbf{u}$, it is possible for a weak solution of (*1*) to have discontinuities.

5.16 Refer to Problem 5.15. Let the rectangular region Ω be partitioned into regions Ω_1 and Ω_2 by the curve $\mathscr{S}$: $x=\sigma(t)$, as indicated in Fig. 5-6. Suppose that $\mathbf{u}$ is a weak solution in Ω, but a continuously differentiable, bounded solution in Ω_1 and in Ω_2. Show that along $\mathscr{S}$ within Ω,

$$(F_1-F_2)=(G_1-G_2)\sigma'(t) \tag{1}$$

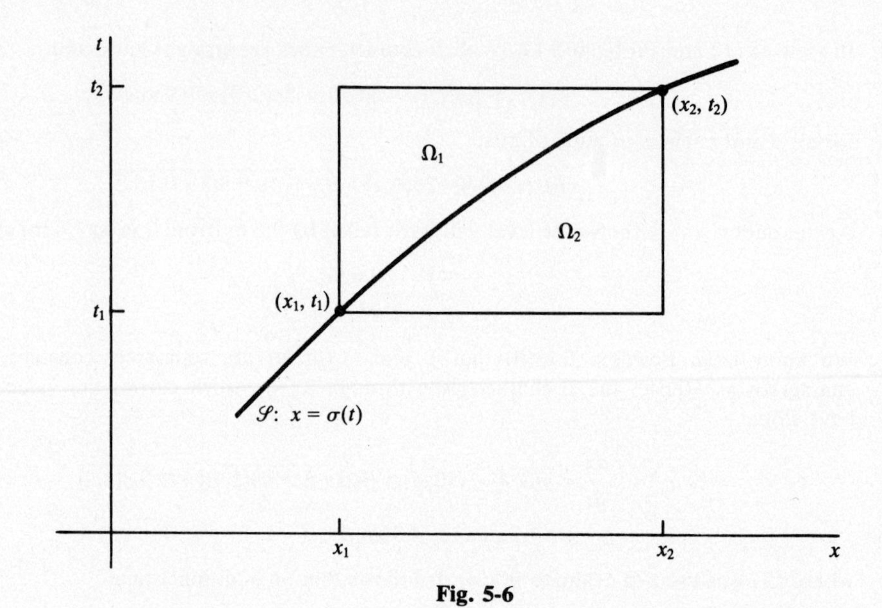

Fig. 5-6

where the subscripts 1 and 2 denote respectively the limits as $(x, t) \to \mathscr{S}$ through regions Ω_1 and Ω_2.

Because $\mathbf{u}$ is a weak solution in Ω, we have, for any test function ϕ on Ω,

$$0 = \iint_{\Omega} (F\phi_x + G\phi_t)\,dx\,dt = \iint_{\Omega_1} (F\phi_x + G\phi_t)\,dx\,dt + \iint_{\Omega_2} (F\phi_x + G\phi_t)\,dx\,dt \tag{2}$$

Since ϕ vanishes on the boundary of Ω,

$$\iint_{\Omega_1} F\phi_x\,dx\,dt = \int_{t_1}^{t_2}\int_{x_1}^{\sigma(t)} [(F\phi)_x - F_x\phi]\,dx\,dt = \int_{\mathscr{S}} F_1\phi\,dt - \iint_{\Omega_1} F_x\phi\,dx\,dt \tag{3}$$

$$\iint_{\Omega_1} G\phi_t\,dx\,dt = \int_{x_1}^{x_2}\int_{\sigma^{-1}(x)}^{t_2} [(G\phi)_t - G_t\phi]\,dt\,dx = -\int_{\mathscr{S}} G_1\phi\,dx - \iint_{\Omega_1} G_t\phi\,dx\,dt \tag{4}$$

Adding (3) and (4), and recalling that (1) of Problem 5.15 holds in Ω_1, we see that

$$\iint_{\Omega_1} (F\phi_x + G\phi_t)\,dx\,dt = \int_{\mathscr{S}} \phi\,(F_1\,dt - G_1\,dx) \tag{5}$$

Similar calculations on Ω_2 show that

$$\iint_{\Omega_2} (F\phi_x + G\phi_t)\,dx\,dt = \int_{\mathscr{S}} \phi\,(-F_2\,dt + G_2\,dx) \tag{6}$$

By (2), the left sides of (5) and (6) sum to zero, whence

$$0 = \int_{\mathscr{S}} \phi\,[(F_1 - F_2)\,dt + (G_2 - G_1)\,dx] = \int_{t_1}^{t_2} \phi\,[(F_1 - F_2) + (G_2 - G_1)\sigma'(t)]\,dt \tag{7}$$

Because ϕ takes arbitrary values along $\mathscr{S}$, (7) implies (1).

If the curve $\mathscr{S}$ represents a shock in the weak solution $\mathbf{u}$, then $\sigma'(t)$ is the velocity of the shock.

5.17 Use Problem 5.16 to derive jump conditions which must hold across a shock in the solution of the conservative open-channel flow equations.

The equations can be put in the divergence form

$$(uv)_x + u_t = 0 \tag{1}$$

$$\left(uv^2+\frac{gu^2}{2}\right)_x+(uv)_t=0 \tag{2}$$

(this is not the conservation form, Example 5.3). Hence, by (*1*) of Problem 5.16, we have across a *bore* or *surge* $\mathcal{S}$: $x=\sigma(t)$

$$u_1v_1-u_2v_2=(u_1-u_2)\sigma'(t) \tag{3}$$

$$\left(u_1v_1^2+\frac{gu_1^2}{2}\right)-\left(u_2v_2^2+\frac{gu_2^2}{2}\right)=(u_1v_1-u_2v_2)\sigma'(t) \tag{4}$$

Assuming unit channel width and constant density, (*3*) asserts that mass is conserved, and (*4*) that momentum is conserved, across $\mathcal{S}$. In the case $\sigma'(t)=0$, the conditions reduce to the well-known *hydraulic jump equations.*

5.18 Water flows in a (one-meter-wide) rectangular channel at a depth of 1 m, with a velocity of 2 m/s. At $x=0$ the depth of the water is suddenly raised to, and subsequently maintained at, 2 m. Neglecting frictional resistance and bed slope, calculate the rate at which the surge moves down the channel and the velocity of the water behind the surge.

In the notation of Problems 5.16 and 5.17, we have the weak solution

behind the surge $(0\le x<\sigma(t))$: $u=u_1=2$ m, $v=v_1$
ahead of the surge $(x>\sigma(t))$: $u=u_2=1$ m, $v=v_2=2$ m/s

The jump conditions (*3*)–(*4*) of Problem 5.17 become two simultaneous equations in the two unknowns v_1 and $\sigma'(t)$. Solving (with $g=9.8$ m/s^2), we find:

$$v_1=4.07 \text{ m/s} \qquad \sigma'(t)=6.14 \text{ m/s}$$

5.19 Use the method of characteristics to solve the initial value problem

$$uu_x+u_t=0 \tag{1}$$

$$u(x,0)=\begin{cases}1 & x\le 0\\ 1-x & 0<x<1\\ 0 & x\ge 1\end{cases} \tag{2}$$

The characteristics are the straight lines

$$\frac{dx}{dt}=u=\text{const.} \tag{3}$$

Using (*2*), the characteristics are constructed as in Fig. 5-7(*a*).

It is seen that points (x,t) below D—that is, in the strip $t<1$—lie on just one characteristic. Thus, on $x\le t<1$, $u=1$; and on $x\ge 1>t$, $u=0$. In between, on the triangular domain isolated in Fig. 5-7(*b*), integration of (*3*) gives

$$x=ut+(1-u) \qquad \text{or} \qquad u=\frac{1-x}{1-t} \tag{4}$$

where the x-intercept was found from (*2*). In summary,

$$\textbf{\textit{for } } t<1 \qquad u=\begin{cases}1 & x\le t\\ \dfrac{1-x}{1-t} & t<x<1\\ 0 & x\ge 1\end{cases}$$

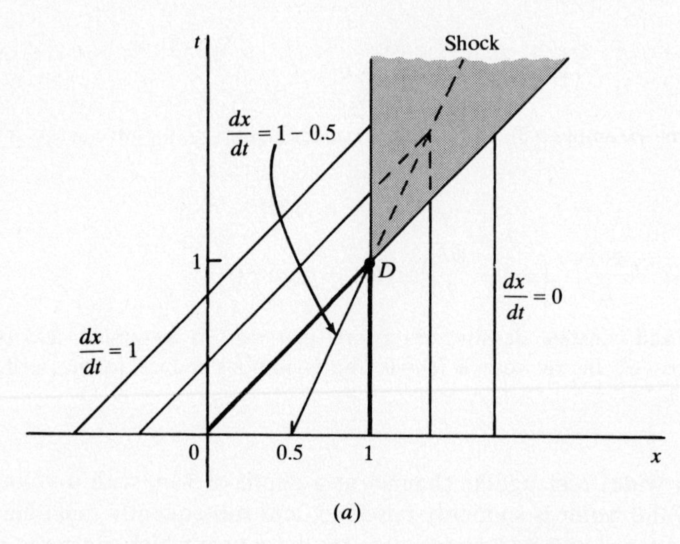

(a)

t
1
D
(x, t)
0
1 − u
1
x

(b)

Fig. 5-7

In the strip $t \geq 1$, points (x, t) of the shaded region in Fig. 5-7(a) lie on two characteristics which bear distinct values of u. The jump condition (1) of Problem 5.16 leads to the equation $x = (t+1)/2$ for the actual shock; i.e., the shock is the prolongation through D of the 45° characteristic. Consequently,

$$\textbf{for } t \geq 1 \qquad u = \begin{cases} 1 & x < (t+1)/2 \\ 0 & x > (t+1)/2 \end{cases}$$

5.20 Show that for $u = f(x/t)$ to be a nonconstant solution of $u_t + a(u)u_x = 0$, f must be the inverse of the function a.

If $u = f(x/t)$,

$$u_t = f'\left(\frac{x}{t}\right) \cdot \frac{-x}{t^2} \qquad \text{and} \qquad u_x = f'\left(\frac{x}{t}\right) \cdot \frac{1}{t}$$

Hence, $u_t + a(u)u_x = 0$ implies that

$$f'\left(\frac{x}{t}\right) \cdot \frac{-x}{t^2} + a\left(f\left(\frac{x}{t}\right)\right) f'\left(\frac{x}{t}\right) \cdot \frac{1}{t} = 0$$

or, assuming $f' \not\equiv 0$ to rule out the constant solution, that

$$a\left(f\left(\frac{x}{t}\right)\right) = \frac{x}{t}$$

This shows the functions a and f to be inverses of each other.

5.21 Solve the initial–boundary value problem

$$u_t + e^u u_x = 0 \qquad x > 0,\ t > 0 \tag{1}$$

$$u(x, 0) = 2 \qquad x > 0 \tag{2}$$

$$u(0, t) = 1 \qquad t > 0 \tag{3}$$

Since the characteristics of (1) are defined by

$$\frac{dx}{dt} = e^u = \text{const.}$$

the characteristics from the positive x-axis have the form $x = e^2t + \text{const.}$, and the characteristics through the positive t-axis are $x = et + \text{const.}$ Therefore (see Fig. 5-8), $u = 2$ for $x \geq e^2t$, and $u = 1$ for $x \leq et$. In the region $et < x < e^2t$, where there are no characteristics, we avail ourselves of the solution $u = \log(x/t)$ found in Problem 5.20. It can be shown that

$$u(x, t) = \begin{cases} 1 & 0 < x \leq et \\ \log(x/t) & et < x < e^2t \\ 2 & e^2t \leq x \end{cases}$$

is the unique continuous weak solution of (1)–(3).

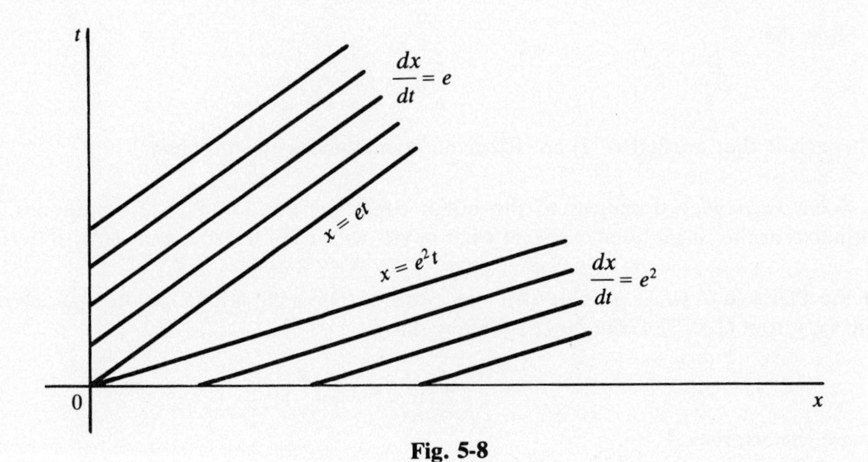

Fig. 5-8

Supplementary Problems

5.22 The Euler equations for steady, isentropic, inviscid, two-dimensional, fluid flow are

$$\rho u_x + u\rho_x + \rho v_y + v\rho_y = 0$$
$$\rho u u_x + \rho v u_y + c^2\rho_x = 0$$
$$\rho u v_x + \rho v v_y + c^2\rho_y = 0$$

Classify this system.

5.23 (*a*) Show that one-way vehicular traffic obeys the continuity equation of fluid dynamics,

$$\rho_t + (v\rho)_x = 0$$

where $\rho \equiv$ vehicles per unit length, $v \equiv$ speed. (*b*) If $v = v(\rho)$, show that ρ is constant for an observer at $x = x(t)$ who moves so that

$$\frac{dx}{dt} = \frac{d(\rho v)}{d\rho}$$

(*c*) Show that if $v = v(\rho)$ and $v'(\rho) \leq 0$, then the rate of propagation of small variations in density cannot exceed the speed of an individual vehicle at that density. (*d*) If $v = v(\rho)$, with what speed is a shock propagated?

5.24 The one-dimensional adiabatic flow equations are

$$(\rho u)_x + \rho_t = 0 \qquad uu_x + u_t = -\frac{1}{\rho}p_x \qquad p = A\rho^\gamma \tag{1}$$

where A and $\gamma > 1$ are constants. (*a*) Letting $c^2 \equiv dp/d\rho$, reduce (*1*) to

$$(\rho u)_x + \rho_t = 0 \qquad \rho u_t + \rho uu_x + c^2\rho_x = 0 \tag{2}$$

and show that (*2*) is a hyperbolic system. (*b*) Transform (*2*) to the canonical form

$$x_\alpha = (u+c)t_\alpha \qquad x_\beta = (u-c)t_\beta \qquad u_\alpha + \frac{c}{\rho}\rho_\alpha = 0 \qquad u_\beta - \frac{c}{\rho}\rho_\beta = 0$$

(*c*) Show that

$$\frac{c}{\rho}\rho_\alpha = \left(\frac{2c}{\gamma - 1}\right)_\alpha$$

and conclude that $u \pm 2c/(\gamma - 1)$ are Riemann invariants for this problem.

5.25 (*a*) Solve $xu_x + yu_y = 0$ subject to the initial condition $u(x, 1) = f(x)$, f continuous. (*b*) If f' is discontinuous at the single point $x = \bar{x}$, at what points will u fail to be continuously differentiable?

5.26 For the PDEs (*a*) $yu_x - xu_y = 0$, (*b*) $yu_x - xu_y = u$, give equations for the characteristic through the point $(x, y, u) = (1, 0, 2)$. [*Hint for* (*b*): Show that

$$\frac{d}{dr}\left(\frac{x}{y}\right) = 1 + \left(\frac{x}{y}\right)^2$$

along a characteristic.]

5.27 Solve the Cauchy problems

(*a*)
$$u_t + u_a = -\frac{cu}{L-a} \qquad t > 0,\ 0 < a < L$$
$$u(t, 0) = b(t) \qquad t > 0$$

where c and L are positive constants

(*b*)
$$xu_x + yu_y = 1 \qquad x > 0,\ y > 0$$
$$u = x^2 + y \qquad 0 < x = 1 - y < 1$$

5.28 Solve by the method of characteristics (c = const.):

$$(a)\quad u_t + cu_x = \phi(x, t),\quad u(x, 0) = \psi(x) \qquad\qquad (b)\quad f(y)u_x + u_y = cu,\quad u(x, 0) = g(x)$$

5.29 Show that the Cauchy problem $u_x + u_y = 1$, $u(x, x) = x^2$, does not have a solution.

5.30 Show that the Cauchy problem $yu_x + xu_y = cu$ (c = const.), $u(x, x) = f(x)$, can have a solution only if $f(x) = bx^c$ (b = const.). If f has the required form, show that

$$u = \left(\frac{x+y}{2}\right)^c g(x^2 - y^2)$$

is a solution for any function g such that $g(0) = b$.

5.31 Solve the initial value problems

(*a*)
$$4u_x - 6v_x + u_t = 0$$
$$u_x - 3v_x + v_t = 0$$
$$u(x, 0) = \sin x \qquad v(x, 0) = \cos x$$

(*b*)
$$3u_y + 2v_x + u_y + v_y = 0$$
$$5u_x + 2v_x - u_y + v_y = 0$$
$$u(x, 0) = \sin x \qquad v(x, 0) = e^x$$

5.32 Refer to Problem 5.23, assuming the speed-density law

$$v = V\left(1 - \frac{\rho}{R}\right)$$

where V and R denote, respectively, the maximum speed and the maximum density. Suppose that cars are traveling along a single-lane road (no passing) at uniform density $R/3$ and uniform speed $2V/3$. At time $t = 0$, a truck enters the road at $x = 0$, inserting itself just behind car A and just ahead of car B. The truck travels at speed $V/3$ until it reaches $x = L$, where it leaves the road (at time $\bar{t} = 3L/V$). Make a graphical determination of $\rho(x, t)$, for $t < \bar{t}$ and all x, by locating the three shocks in the flow and applying the appropriate jump condition across them. Also find the time at which car B catches up with car A.

5.33 In a horizontal, rectangular channel of unit width, water of depth u_0 is held behind a vertical wall. At time $t = 0$ the wall is set in motion with velocity w into the standing water. Show that the shock velocity, W, and the depth behind the shock, U, are determined by

$$\frac{U - u_0}{u_0} = \frac{w}{W - w} \qquad (W - w)^2 = \frac{gu_0}{2}\left(2 - \frac{w}{W}\right)$$

5.34 (*a*) Show that the initial value problem

$$uu_x + u_t = 0$$

$$u(x, 0) = \begin{cases} 1 & 0 < x < 1 \\ 0 & x < 0 \text{ or } x > 1 \end{cases}$$

admits two weak solutions

$$v(x, t) = \begin{cases} 0 & x < t/2 \\ 1 & t/2 < x < 1 + t/2 \\ 0 & x > 1 + t/2 \end{cases}$$

$$w(x, t) = \begin{cases} 0 & x < 0 \\ x/t & 0 < x < t \\ 1 & t < x < 1 + t/2 \\ 0 & x > 1 + t/2 \end{cases}$$

(*b*) For a unique weak solution, the inequalities

$$u_1 > \sigma'(t) > u_2$$

must hold along any shock $x = \sigma(t)$. Verify that w, but not v, satisfies these inequalities.

Chapter 6

Eigenfunction Expansions and Integral Transforms: Theory

6.1 FOURIER SERIES

Let $F(x)$ be an arbitrary function defined in $(-\ell, \ell)$. The infinite trigonometric series

$$\tfrac{1}{2}a_0 + \sum_{n=1}^{\infty} a_n \cos(n\pi x/\ell) + b_n \sin(n\pi x/\ell) \tag{6.1}$$

is called the *Fourier series* for $F(x)$ if the coefficients a_n and b_n are given by

$$a_n = \frac{1}{\ell}\int_{-\ell}^{\ell} F(x)\cos(n\pi x/\ell)\,dx \qquad b_n = \frac{1}{\ell}\int_{-\ell}^{\ell} F(x)\sin(n\pi x/\ell)\,dx \tag{6.2}$$

in which case the coefficients are known as the *Fourier coefficients* for $F(x)$.

Since each of the trigonometric functions in the Fourier series for $F(x)$ is periodic of period 2ℓ, it follows that if the series actually converges to $F(x)$ for $-\ell < x < \ell$, then it converges to the *2ℓ-periodic extension* of $F(x)$,

$$\tilde{F}(x) = F(x) \quad (-\ell < x < \ell) \qquad \text{and} \qquad \tilde{F}(x) = \tilde{F}(x + 2\ell) \tag{6.3}$$

for all x in the domain of $\tilde{F}$; see Problem 6.3.

Theorem 6.1 states sufficient conditions for the convergence of a Fourier series, in terms of properties of $\tilde{F}(x)$. These, of course, derive from properties of $F(x)$, as discussed in Problem 6.3. Recall that a function is *piecewise* or *sectionally continuous* in $(-\infty, \infty)$ if it has at most finitely many finite jump discontinuities in any interval of finite length.

Theorem 6.1: Let $F(x)$ be defined in$(-\ell, \ell)$ and let $\tilde{F}(x)$ denote the 2ℓ-periodic extension of $F(x)$.

(i) If $\tilde{F}(x)$ and $\tilde{F}'(x)$ are both sectionally continuous, the Fourier series for $F(x)$ converges pointwise to $\tilde{F}(x)$ at each point where $\tilde{F}(x)$ is continuous. At each x_0 where $\tilde{F}(x)$ has a jump discontinuity, the series converges to the average of the left- and right-hand limits of $\tilde{F}(x)$ at x_0.

(ii) If $\tilde{F}(x)$ is continuous and $\tilde{F}'(x)$ is sectionally continuous, the Fourier series for $F(x)$ converges uniformly to $\tilde{F}(x)$.

(iii) If $\tilde{F}(x)$ is in C^p and if $\tilde{F}^{(p+1)}(x)$ is sectionally continuous, the series obtained by differentiating the Fourier series for $F(x)$ termwise j times $(j = 0, 1, \ldots, p)$ converges uniformly to $\tilde{F}^{(j)}(x)$.

6.2 GENERALIZED FOURIER SERIES

To extend the notion of Fourier series to other than trigonometric expansions, we first recall the usual definition of the *inner product* of two vectors in $\mathbf{R}^N$:

$$\mathbf{x}\cdot\mathbf{y} \text{ or } \langle \mathbf{x}, \mathbf{y}\rangle \equiv x_1y_1 + x_2y_2 + \cdots + x_Ny_N \tag{6.4}$$

A set of vectors $\{\mathbf{x}_1, \ldots, \mathbf{x}_M\}$ in $\mathbf{R}^N$ is an *orthogonal family* if $\langle \mathbf{x}_i, \mathbf{x}_j\rangle = 0$ for $i \neq j \quad (i, j = 1, \ldots, M)$; it is an *orthonormal family* if

$$\langle \mathbf{x}_i, \mathbf{x}_j\rangle = \delta_{ij} \equiv \begin{cases} 0 & i \neq j \\ 1 & i = j \end{cases} \tag{6.5}$$

Clearly, an orthogonal family of nonzero vectors can always be made into an orthonormal family by dividing each vector $\mathbf{x}_i$ by its *norm*, $\|\mathbf{x}_i\| = \langle \mathbf{x}_i, \mathbf{x}_i \rangle^{1/2}$.

Definition: An orthogonal family is *complete* in $\mathbf{R}^N$ if the only vector orthogonal to every member of the family is the zero vector.

Theorem 6.2: Any complete orthonormal family $\{\mathbf{x}_1, \ldots, \mathbf{x}_M\}$ is a basis of $\mathbf{R}^N$ (i.e., $M = N$) in terms of which an arbitrary vector $\mathbf{v}$ has the representation

$$\mathbf{v} = \sum_{n=1}^{N} \langle \mathbf{v}, \mathbf{x}_n \rangle \mathbf{x}_n \tag{6.6}$$

The coefficients $c_n \equiv \langle \mathbf{v}, \mathbf{x}_n \rangle$ in (*6.6*) are such that the Pythagorean relation

$$\|\mathbf{v}\|^2 = \sum_{n=1}^{N} c_n^2 \tag{6.7}$$

holds when $\mathbf{R}^N$ is referred to the orthonormal basis $\{\mathbf{x}_n\}$. It is the formal resemblance of the right side of (*6.6*) to the Nth partial sum of the Fourier series (*6.1*)–(*6.2*) that serves as the springboard for the generalization that follows.

Let $F(x)$ denote a function which is defined in (a, b) and satisfies

$$\int_a^b F(x)^2\, dx < \infty \tag{6.8}$$

The collection of all such functions will be denoted by $L^2(a, b)$. Two elements, F and G, are said to be *equal in the* $L^2(a, b)$*-sense* if

$$\int_a^b [F(x) - G(x)]^2\, dx = 0$$

This concept of equality is used to define what is meant by the *convergence* of an infinite series of $L^2(a, b)$-functions: $F_1(x) + F_2(x) + \cdots$ *converges* to the limit $F(x)$ in $L^2(a, b)$ if

$$\lim_{N\to\infty} \int_a^b \left[F(x) - \sum_{i=1}^{N} F_i(x)\right]^2 dx = 0 \tag{6.9}$$

(This kind of convergence is frequently referred to as *mean-square convergence.*)

With the introduction of an inner product,

$$\langle F, G \rangle \equiv \int_a^b F(x)G(x)\, dx \tag{6.10}$$

(which is well defined by virtue of Problem 6.16), $L^2(a, b)$ becomes an inner product space. Orthogonality, normality, and completeness are defined exactly as in $\mathbf{R}^N$. In $L^2(a, b)$, a complete orthonormal family is necessarily infinite, but an infinite orthonormal family is not necessarily complete.

EXAMPLE 6.1 In $L^2(-\ell, \ell)$, neither of the infinite orthonormal families

$$\left\{\frac{1}{\sqrt{\ell}}\sin\frac{\pi x}{\ell}, \ \frac{1}{\sqrt{\ell}}\sin\frac{2\pi x}{\ell}, \ \frac{1}{\sqrt{\ell}}\sin\frac{3\pi x}{\ell}, \ \ldots\right\}$$

$$\left\{\frac{1}{\sqrt{2\ell}}, \ \frac{1}{\sqrt{\ell}}\cos\frac{\pi x}{\ell}, \ \frac{1}{\sqrt{\ell}}\cos\frac{2\pi x}{\ell}, \ \ldots\right\}$$

is complete; for instance, for $F(x) \equiv 1$,

$$\left\langle 1, \frac{1}{\sqrt{\ell}}\sin\frac{n\pi x}{\ell}\right\rangle = \frac{1}{\sqrt{\ell}}\int_{-\ell}^{\ell} \sin\frac{n\pi x}{\ell}\, dx = 0 \qquad (n = 1, 2, 3, \ldots)$$

However, the union of these two families *is* a complete orthonormal family, and it generates the Fourier series (*6.1*) for a square-integrable function $F(x)$.

Analogous to Theorem 6.2 we have

Theorem 6.3: If $\{u_n(x)\}$, $n = 1, 2, \ldots$, is a complete orthonormal family in $L^2(a, b)$, then for arbitrary $F(x)$ in $L^2(a, b)$,

$$F(x) \approx \sum_{n=1}^{\infty} \langle F, u_n \rangle u_n(x) \tag{6.11}$$

[mean-square convergence of the series to $F(x)$].

The analog to (*6.7*)

$$\|F(x)\|^2 = \sum_{n=1}^{\infty} F_n^2 \qquad \text{[ordinary convergence]} \tag{6.12}$$

is called the *Parseval relation.*

There remains the problem of finding (nontrigonometric) complete orthonormal families for the construction of generalized Fourier series (*6.11*). The next section will show that such families arise naturally as the solutions to certain boundary value problems for ordinary differential equations.

6.3 STURM–LIOUVILLE PROBLEMS; EIGENFUNCTION EXPANSIONS

Consider the following boundary value problem for the unknown function $w(x)$:

$$\begin{gathered} -(p(x)w'(x))' + q(x)w(x) = \lambda r(x)w(x) \qquad a < x < b \\ C_1 w(a) + C_2 w'(a) = 0 \\ C_3 w(b) + C_4 w'(b) = 0 \end{gathered} \tag{6.13}$$

If (1) $p(x)$, $p'(x)$, $q(x)$, and $r(x)$ are continuous in (a, b); (2) $p(x) > 0$ and $r(x) > 0$ on $[a, b]$; and (3) $C_1^2 + C_2^2 \neq 0$, $C_3^2 + C_4^2 \neq 0$, then (*6.13*) constitutes a *Sturm–Liouville problem.* The function $w(x) \equiv 0$ is a trivial solution to any Sturm–Liouville problem. In addition, for certain values of the parameter λ, there exist nontrivial solutions. Each value of λ for which a nontrivial solution exists is called an *eigenvalue* of the problem, and the corresponding nontrivial solution is called an *eigenfunction.*

Theorem 6.4: The eigenvalues and eigenfunctions of a Sturm–Liouville problem have the following properties.

(i) All eigenvalues are real and compose a countably infinite collection satisfying $\lambda_1 < \lambda_2 < \cdots < \lambda_s \to \infty$.

(ii) To each eigenvalue λ_n there corresponds only one *independent* eigenfunction $w_n(x)$.

(iii) Relative to the inner product (*6.10*), the weighted eigenfunctions

$$u_n(x) \equiv \frac{\sqrt{r(x)}\, w_n(x)}{\|\sqrt{r(x)}\, w_n(x)\|} \qquad (n = 1, 2, \ldots) \tag{6.14}$$

compose a complete orthonormal family in $L^2(a, b)$.

An *eigenfunction expansion*—i.e., a generalized Fourier series for an $L^2(a, b)$-function F based on the family (*6.14*)—not only converges in the mean-square sense (Theorem 6.3), but also:

Theorem 6.5: (i) If F and F' are both sectionally continuous in (a, b), then the series converges pointwise to the value $[F(x+)+F(x-)]/2$ at each x in (a, b).

(ii) If F and F' are continuous in (a, b), if F'' is sectionally continuous, and if F satisfies the boundary conditions of the Sturm–Liouville problem (*6.13*), then the series converges uniformly to $F(x)$ in (a, b).

6.4 FOURIER AND LAPLACE INTEGRAL TRANSFORMS

Theorem 6.4(i) and (iii) also hold for the eigenvalues $n = 0, 1, 2, \ldots$ and eigenfunctions $\{e^{\pm inx}\}$ of the problem

$$\begin{aligned} w''(x) &= \lambda w(x) \qquad -\pi < x < \pi \\ w(-\pi) &= w(\pi) \\ w'(-\pi) &= w'(\pi) \end{aligned}$$

which is of Sturm–Liouville type, except for the boundary conditions, which are periodic instead of separated. Thus, for arbitrary $f(x)$ in $L^2(-\pi, \pi)$, we have (cf. (*6.9*))

$$\lim_{N\to\infty} \int_{-\pi}^{\pi} |f(x) - \tilde{f}_N(x)|^2\, dx = 0 \tag{6.15a}$$

where

$$\tilde{f}_N(x) = \sum_{n=-N}^{N} F_n e^{inx} \tag{6.15b}$$

and

$$F_n = \frac{1}{2\pi}\int_{-\pi}^{\pi} f(x) e^{-inx}\, dx \qquad (n = 0, \pm 1, \pm 2, \ldots) \tag{6.15c}$$

Now suppose that $f(x)$ is in $L^2(-\infty, \infty)$. Unless $f(x)$ is identically zero, it is not periodic and an eigenfunction expansion like (*6.15*) cannot be valid. However, in this case we have

$$\lim_{N\to\infty} \int_{-\infty}^{\infty} |f(x) - f_N(x)|^2\, dx = 0 \tag{6.16a}$$

where

$$f_N(x) = \int_{-N}^{N} F(\alpha)\, e^{i\alpha x}\, d\alpha \tag{6.16b}$$

and

$$F(\alpha) = \frac{1}{2\pi}\int_{-\infty}^{\infty} f(x)\, e^{-i\alpha x}\, dx \tag{6.16c}$$

Note the analogy between (*6.15*) and (*6.16*). The function $F(\alpha)$ defined in (*6.16c*) is called the *Fourier (integral) transform* of $f(x)$; we shall indicate the relationship between the two functions as

$$\mathcal{F}\{f(x)\} = F(\alpha) \qquad \text{or} \qquad \mathcal{F}^{-1}\{F(\alpha)\} = f(x)$$

Operational properties of the Fourier transform are listed in Table 6-1. In addition, Table 6-2 gives a number of specific functions and their Fourier transforms. For our purposes, inversion of the Fourier transform will be carried out by using Table 6-2 as a dictionary, together with certain of the properties from Table 6-1. Note that line 7 of Table 6-1 is equivalent to the inversion formula

$$\int_{-\infty}^{\infty} F(\alpha) e^{ix\alpha}\, d\alpha = f(x) \tag{6.17}$$

The function $f * g$ defined in line 8 is called the *convolution* of the functions f and g. Clearly, the convolution operation is symmetric, associative, and distributive with respect to addition.

Table 6-1. Properties of the Fourier Transform

	$f(x)$	$F(\alpha)=\frac{1}{2\pi}\int_{-\infty}^{\infty} f(x)e^{-i\alpha x}\,dx$
1.	$f^{(n)}(x)$	$(i\alpha)^n F(\alpha)$
2.	$x^n f(x)$	$i^n F^{(n)}(\alpha)$
3.	$f(x-c)$	$e^{-ic\alpha}F(\alpha)$ $(c=\text{const.})$
4.	$e^{icx}f(x)$	$F(\alpha-c)$ $(c=\text{const.})$
5.	$C_1 f_1(x)+C_2 f_2(x)$	$C_1F_1(\alpha)+C_2F_2(\alpha)$
6.	$f(cx)$	$\lvert c\rvert^{-1}F(\alpha/c)$ $(c=\text{const.})$
7.	$F(x)$	$\frac{1}{2\pi}f(-\alpha)$
8.	$f*g(x)\equiv\int_{-\infty}^{\infty} f(x-y)g(y)\,dy$	$2\pi F(\alpha)G(\alpha)$

Table 6-2. Fourier Transform Pairs

	$f(x)$	$F(\alpha)=\frac{1}{2\pi}\int_{-\infty}^{\infty} f(x)e^{-i\alpha x}\,dx$
1.	e^{-cx^2}	$(4\pi c)^{-1/2}e^{-\alpha^2/4c}$ $(c>0)$
2.	$e^{-\lambda\lvert x\rvert}$	$\frac{\lambda/\pi}{\alpha^2+\lambda^2}$ $(\lambda>0)$
3.	$\frac{2\lambda}{x^2+\lambda^2}$	$e^{-\lambda\lvert\alpha\rvert}$ $(\lambda>0)$
4.	$I_A(x)\equiv\begin{cases}1 & \lvert x\rvert<A\\ 0 & \lvert x\rvert>A\end{cases}$	$\frac{\sin A\alpha}{\pi\alpha}$
5.	$\frac{2\sin Ax}{x}$	$I_A(\alpha)$
6.	$E_a(x)\equiv\begin{cases}0 & x<0\\ e^{-ax} & x>0\end{cases}$	$\frac{1}{2\pi}\frac{1}{a+i\alpha}$ $(\text{Re}\,a>0)$

The Fourier transform, as described here, applies to functions $f(x)$ in $L^2(-\infty,\infty)$. A related integral transform, called the *Laplace transform*, is defined by

$$\mathcal{L}\{f(t)\}\equiv\int_0^{\infty} f(t)e^{-st}\,dt\equiv\hat{f}(s) \tag{6.18}$$

This transform may be applied to functions $f(t)$ which are defined for $-\infty<t<\infty$ and satisfy $f(t)=0$

Table 6-3. Properties of the Laplace Transform

$f(t)$	$\hat{f}(s)=\int_0^\infty f(t)e^{-st}\,dt$
1. $C_1f_1(t)+C_2f_2(t)$	$C_1\hat{f}_1(s)+C_2\hat{f}_2(s)$
2. $f(at)$	$a^{-1}\hat{f}(s/a) \quad (a>0)$
3. $f^{(n)}(t)$	$s^n\hat{f}(s)-s^{n-1}f(0)-\cdots-f^{(n-1)}(0)$ $(n=1,2,\ldots)$
4. $t^nf(t)$	$(-1)^n\hat{f}^{(n)}(s) \quad (n=1,2,\ldots)$
5. $e^{ct}f(t)$	$\hat{f}(s-c) \quad (c=\text{const.})$
6. $H(t-b)f(t-b)$, where $H(t)\equiv\begin{cases}0 & t<0\\ 1 & t>0\end{cases}$	$e^{-bs}\hat{f}(s) \quad (b>0)$
7. $f*g(t)\equiv\int_0^t f(t-\tau)g(\tau)\,d\tau$	$\hat{f}(s)\hat{g}(s)$

Table 6-4. Laplace Transform Pairs

$f(t)$	$\hat{f}(s)=\int_0^\infty f(t)e^{-st}\,dt$
1. 1	$\frac{1}{s}$
2. t^n	$\frac{n!}{s^{n+1}} \quad (n=1,2,\ldots)$
3. e^{kt}	$\frac{1}{s-k}$
4. $\sin at$	$\frac{a}{s^2+a^2}$
5. $\cos at$	$\frac{s}{s^2+a^2}$
6. $\frac{1}{\sqrt{\pi t}}$	$\frac{1}{\sqrt{s}}$
7. $\frac{1}{\sqrt{\pi t}}e^{-k^2/4t}$	$\frac{1}{\sqrt{s}}e^{-k\sqrt{s}} \quad (k>0)$
8. $\frac{k}{\sqrt{4\pi t^3}}e^{-k^2/4t}$	$e^{-k\sqrt{s}} \quad (k>0)$
9. $\operatorname{erfc}(k/2\sqrt{t})$, where $\operatorname{erfc} z\equiv\frac{2}{\sqrt{\pi}}\int_z^\infty e^{-u^2}\,du$	$\frac{1}{s}e^{-k\sqrt{s}} \quad (k>0)$

for $t<0$. Note that $f(t)$ need not belong to $L^2(-\infty, \infty)$; it is sufficient that there exist positive constants M and b such that

$$|f(t)| \le Me^{bt} \qquad \text{for } t>0 \tag{6.19}$$

Table 6-3 lists operational formulas for the Laplace transform, and Table 6-4 gives Laplace transforms of specific functions.

Solved Problems

6.1 Let $f(x)$ be a sectionally continuous function in $(0, \ell)$. Determine (a) the *Fourier cosine series*, (b) the *Fourier sine series*, for $f(x)$.

(a) Define

$$F_e(x) \equiv \begin{cases} f(x) & 0<x<\ell \\ f(-x) & -\ell<x<0 \end{cases}$$

an even function in $(-\ell, \ell)$. The Fourier coefficients of $F_e(x)$ are given by *(6.2)* as:

$$a_n = \frac{2}{\ell}\int_0^\ell f(x)\cos(n\pi x/\ell)\,dx \qquad b_n = 0$$

With these coefficients, the series *(6.1)* represents, for all x, the function $\tilde{F}_e(x)$, the *even 2ℓ-periodic extension* of $f(x)$.

(b) Define

$$F_o(x) \equiv \begin{cases} f(x) & 0<x<\ell \\ -f(-x) & -\ell<x<0 \end{cases}$$

an odd function in $(-\ell, \ell)$. The Fourier coefficients of $F_o(x)$ are given by *(6.2)* as:

$$a_n = 0 \qquad b_n = \frac{2}{\ell}\int_0^\ell f(x)\sin(n\pi x/\ell)\,dx$$

With these coefficients, the series *(6.1)* represents, for all x, the function $\tilde{F}_o(x)$, the *odd 2ℓ-periodic extension* of $f(x)$.

6.2 Find all eigenvalues and eigenfunctions for the problem

$$-w''(x) = \lambda w(x) \qquad 0<x<\ell$$
$$w'(0) = w'(\ell) = 0$$

As this problem is of Sturm–Liouville type, Theorem 6.4(i) ensures that the eigenvalues λ are real—negative, zero, or positive. Each of the three possibilities for λ leads to a different form of the general solution to the differential equation, and we must then check to see which solution(s) can satisfy the homogeneous boundary conditions without reducing to the trivial solution.

If $\lambda < 0$, write $\lambda = -\mu^2 < 0$. Then $w(x) = Ae^{\mu x} + Be^{-\mu x}$, and the boundary conditions,

$$w'(0) = \mu(A-B) = 0 \qquad w'(\ell) = \mu(Ae^{\mu\ell} - Be^{-\mu\ell}) = 0$$

are satisfied if and only if $A = B = 0$; i.e., there are no negative eigenvalues.

If $\lambda = 0$, then $w(x) = Ax + B$, and the boundary conditions,

$$w'(0) = A = 0 \qquad w'(\ell) = A = 0$$

are satisfied by $w(x) = B \neq 0$. Thus, $\lambda = 0$ is an eigenvalue, and all corresponding eigenfunctions are constant multiples of $w_0(x) = 1$.

If $\lambda > 0$, write $\lambda = \mu^2 > 0$. Then $w(x) = A \sin \mu x + B \cos \mu x$, and

$$w'(0) = \mu A = 0$$
$$w'(\ell) = \mu A \cos \mu\ell - \mu B \sin \mu\ell = 0$$

The determinant of this system, $-\mu^2 \sin \mu\ell$, vanishes for $\mu_n = n\pi/\ell$ $(n = \pm 1, \pm 2, \ldots)$. Hence, the positive eigenvalues are $\lambda_n = (n\pi/\ell)^2$ $(n = 1, 2, \ldots)$, and the eigenfunctions corresponding to λ_n all are constant multiples of $w_n(x) = \cos n\pi x/\ell$.

It is seen that the eigenfunction expansion on $(0, \ell)$ yielded by the above Sturm–Liouville problem is nothing other than the Fourier cosine series of Problem 6.1(*a*). Changing the boundary conditions to $w(0) = w(\ell) = 0$ would yield the Fourier sine series.

6.3 Given a function $F(x)$, defined on the closed interval $[-\ell, \ell]$, state conditions sufficient to ensure that $\tilde{F}(x)$ is C^p on the whole real axis.

If $F(x)$ were defined merely in $(-\ell, \ell)$, then (*6.3*) would fail to define $\tilde{F}$ at the points $x = \pm\ell, \pm 3\ell, \pm 5\ell, \ldots$, so that questions of continuity would be meaningless. Even if (i) $F(x)$ is defined on $[-\ell, \ell]$, the function $\tilde{F}$ is well defined only if (ii) $F(\ell) = F(-\ell)$. If (i) and (ii) hold and, in addition, $F(x)$ is continuous on $[-\ell, \ell]$, it is apparent that $\tilde{F}(x)$ will be continuous for all x.

Repeating the above argument with respect to the derivatives of $\tilde{F}$, we prove the

Theorem: $\tilde{F}(x)$ is C^p if, for $j = 0, 1, \ldots, p$, $F^{(j)}(x)$ is continuous on $[-\ell, \ell]$ and obeys

$$F^{(j)}(\ell) = F^{(j)}(-\ell)$$

6.4 Given a function $f(x)$, defined on the closed interval $[0, \ell]$, state conditions sufficient to ensure that $\tilde{F}_o(x)$ [Problem 6.1(*b*)] is C^p on the whole real axis.

We apply the result of Problem 6.3 to the function $F_o(x)$ of Problem 6.1(*b*), making two preliminary observations:

(1) For the odd function $F_o(x)$ to be continuous on $[-\ell, \ell]$ and to obey $F_o(\ell) = F_o(-\ell)$, it is sufficient (and necessary) that $f(x)$ be continuous on $[0, \ell]$ and obey $f(0) = f(\ell) = 0$.

(2) For odd j, the jth derivative $F_o^{(j)}(x)$ is an *even* function in $[-\ell, \ell]$. Hence, if it exists, this function automatically satisfies $F_o^{(j)}(\ell) = F_o^{(j)}(-\ell)$.

Theorem: $\tilde{F}_o(x)$ is C^p if $f^{(j)}(x)$ is continuous on $[0, \ell]$ for $j = 0, 1, \ldots, p$, and if

$$f^{(k)}(0) = f^{(k)}(\ell) = 0$$

for $k = 0, 2, 4, \ldots \leq p$.

6.5 Show that if both $f(x)$ and $f'(x)$ belong to $L^2(-\infty, \infty)$, $\lim_{|x|\to\infty} f(x) = 0$.

For all values of x, $[f(x) \pm f'(x)]^2 \geq 0$, from which it follows that

$$\int_a^b f(x)^2\,dx + \int_a^b f'(x)^2\,dx \geq \left| 2\int_a^b f(x)f'(x)\,dx \right| \tag{1}$$

for any real a and b. Now,

$$2\int_a^b f(x)f'(x)\,dx = f(b)^2 - f(a)^2 \tag{2}$$

Moreover, if both $f(x)$ and $f'(x)$ are square integrable,

$$\lim_{\substack{a\to\infty \\ b\to\infty}} \int_a^b f(x)^2\,dx = \lim_{\substack{a\to\infty \\ b\to\infty}} \int_a^b f'(x)^2\,dx = 0 \tag{3}$$

where a and b are allowed to tend to $+\infty$ independently. Together, (*1*), (*2*), and (*3*) imply that $f(x)^2$ approaches a constant as x approaches $+\infty$. Since f is square integrable, this constant must be zero. By similar reasoning, $f(x)$ tends to zero as x tends to $-\infty$.

6.6 Find

$$\mathscr{L}^{-1}\left\{\frac{1}{s}\frac{\cosh a\sqrt{s}}{\cosh b\sqrt{s}}\right\} \qquad (b>a>0)$$

We have:

$$\frac{1}{s}\frac{\cosh a\sqrt{s}}{\cosh b\sqrt{s}} = \frac{1}{s}\frac{e^{a\sqrt{s}}+e^{-a\sqrt{s}}}{e^{b\sqrt{s}}+e^{-b\sqrt{s}}} = \frac{e^{(a-b)\sqrt{s}}+e^{-(a+b)\sqrt{s}}}{s}\,\frac{1}{1+e^{-2b\sqrt{s}}}$$

$$= \frac{e^{(a-b)\sqrt{s}}+e^{-(a+b)\sqrt{s}}}{s}\sum_{n=0}^{\infty}(-1)^n e^{-2nb\sqrt{s}}$$

$$= \sum_{n=0}^{\infty}(-1)^n\frac{1}{s}e^{-[(2n+1)b-a]\sqrt{s}} + \sum_{n=0}^{\infty}(-1)^n\frac{1}{s}e^{-[(2n+1)b+a]\sqrt{s}}$$

Then, by line 9 of Table 6-4 and the linearity of the Laplace transform,

$$\mathscr{L}^{-1}\left\{\frac{1}{s}\frac{\cosh a\sqrt{s}}{\cosh b\sqrt{s}}\right\} = \sum_{n=0}^{\infty}(-1)^n \operatorname{erfc}\left[\frac{(2n+1)b-a}{2\sqrt{t}}\right] + \sum_{n=0}^{\infty}(-1)^n \operatorname{erfc}\left[\frac{(2n+1)b+a}{2\sqrt{t}}\right]$$

Supplementary Problems

6.7 Show that if a series of the form (*6.1*) converges *uniformly* to $F(x)$ in $(-\ell, \ell)$, the coefficients must be given by (*6.2*).

6.8 Compute the Fourier coefficients for:

$$(a)\quad F(x) = \begin{cases} 0 & -\pi<x<0 \\ 1 & 0<x<\pi \end{cases}$$

(b) $G(x)=|x|$ $\quad(-\pi<x<\pi)$ $\qquad$ (c) $H(x)=x$ $\quad(-\pi<x<\pi)$

6.9 Characterize the convergence of the Fourier series from Problem 6.8.

6.10 Write (a) the Fourier sine series, (b) the Fourier cosine series, for the function $F(x)=1$, $0<x<\pi$.

6.11 Find the eigenvalues and corresponding eigenfunctions of

$$-w''(x) = \lambda w(x) \qquad 0<x<\ell$$

under the boundary conditions

(a) $w(0)=w(\ell)=0$ $\qquad$ (d) $w(0)+w'(0)=w(\ell)=0$

(b) $w(0)=w'(\ell)=0$ $\qquad$ (e) $w(0)+w'(0)=w'(\ell)=0$

(c) $w'(0)=w(\ell)=0$ $\qquad$ (f) $w(0)+\alpha w'(0)=w(\ell)+\beta w'(\ell)=0$ $\quad(\alpha>\beta>0)$

6.12 Let $\{u_1(x), u_2(x), \ldots, u_M(x)\}$ be an (incomplete) orthonormal family in $L^2(a, b)$. Given a function $F(x)$ in $L^2(a, b)$, infer from the identity

$$\int_a^b \left[F - \sum_{n=1}^{M} C_n u_n\right]^2 dx \equiv \left[\int_a^b F^2\, dx - \sum_{n=1}^{M}\left(\int_a^b F u_n\, dx\right)^2\right] + \sum_{n=1}^{M}\left(C_n - \int_a^b F u_n\, dx\right)^2$$

where $C_1, \ldots, C_M$ are arbitrary constants, that

(i) Out of all linear fittings of $F(x)$ by the family $\{u_n(x)\}$, the generalized Fourier series yields the smallest mean-square error.

(ii) The generalized Fourier coefficients, F_n, of $F(x)$ obey *Bessel's inequality*,

$$\sum_{n=1}^{M} F_n^2 \le \|F(x)\|^2$$

6.13 Prove a theorem for $\tilde{F}_e(x)$ analogous to that found in Problem 6.4 for $\tilde{F}_o(x)$.

6.14 Let $f(x)$ be defined on $[0, \ell]$ and satisfy, for some $p \ge 2$, the hypotheses of the theorem of Problem 6.4 (6.13). Prove that the Fourier sine (cosine) series converges uniformly. [*Hint*: Integrating by parts p times, show that

$$|b_n|\ (|a_n|) \le \frac{\text{constant}}{n^p}$$

and apply the Weierstrass M-test.

6.15 Find the Fourier series for the following functions:

$$(a)\quad f(x) = \begin{cases} x & 0 < x < \pi/2 \\ \pi - x & \pi/2 < x < 3\pi/2 \\ x - 2\pi & 3\pi/2 < x < 2\pi \end{cases} \qquad (c)\quad f(x) = 3x^2 \qquad -\pi < x < \pi$$

$$(b)\quad f(x) = \begin{cases} 1 & \pi/2 < |x - \pi| < \pi \\ -1 & |x - \pi| < \pi/2 \end{cases} \qquad (d)\quad f(x) = x^3 \qquad -\pi < x < \pi$$

6.16 For f and g in (real) $L^2(a, b)$, show that

$$\left(\int_a^b f(x)g(x)\, dx\right)^2 \le \left(\int_a^b f(x)^2\, dx\right)\left(\int_a^b g(x)^2\, dx\right)$$

6.17 Let $\{u_n(x)\}$ denote a complete orthonormal family of functions in $L^2(a, b)$. For f, g in $L^2(a, b)$, let $f_n \equiv \langle f, u_n\rangle$, $g_n \equiv \langle g, u_n\rangle$ for $n = 1, 2, \ldots$. Prove:

$$(a)\quad \langle f, g\rangle = \sum_{n=1}^{\infty} f_n g_n \qquad (c)\quad \sum_{n=1}^{\infty} f_n^2 < \infty$$

$$(b)\quad \left(\sum_{n=1}^{\infty} f_n g_n\right)^2 \le \left(\sum_{n=1}^{\infty} f_n^2\right)\left(\sum_{n=1}^{\infty} g_n^2\right) \qquad (d)\quad \lim_{n\to\infty} f_n = 0$$

6.18 Refer to Problem 6.17. If f belongs to $L^2(a, b)$, if $\bar{f}$—the $(b - a)$-periodic extension of f—is continuous, and if $\bar{f}'$ is sectionally continuous, prove that

$$\sum_{n=1}^{\infty} |f_n| < \infty$$

6.19 Use the operational properties of the Fourier transform (Table 6-1) to find the Fourier transforms of

$$f(x)=\begin{cases} e^{-2x} & 1<x<4 \\ 0 & \text{otherwise} \end{cases} \qquad g(x)=\begin{cases} 1 & 1<x<3 \\ -4 & 6<x<8 \\ 0 & \text{otherwise} \end{cases}$$

6.20 If $F(\alpha)=\mathscr{F}\{f(x)\}$, use the convolution property of the Fourier transform to find

$$\mathscr{F}^{-1}\left\{\frac{F(\alpha)}{\alpha^2+8\alpha+20}\right\}$$

6.21 Use the operational properties of the Laplace transform (Table 6-3) to find the Laplace transforms of

$$f(t)=t^2\sin at \qquad e(t)=e^{bt}\cos at \qquad g(t)=\begin{cases} 0 & t<1 \\ 1 & 1<t<2 \\ 0 & 2<t \end{cases}$$

6.22 Calculate (*a*) $\mathscr{L}^{-1}\{s^{-4}e^{-3s}\}$, (*b*) $\mathscr{L}^{-1}\{\sqrt{s}\hat{f}(s)\}$ if $\hat{f}(s)=\mathscr{L}\{f(t)\}$, (*c*) $\mathscr{L}^{-1}\{\exp(-k\sqrt{s+h})\}$.

6.23 Find the inverse Laplace transforms of

$$\hat{f}(s)=\frac{\cosh as}{\sinh bs}\frac{1}{s} \qquad \hat{g}(s)=\frac{\sinh a\sqrt{s}}{\sinh b\sqrt{s}}\frac{1}{\sqrt{s}}$$

Chapter 7

Eigenfunction Expansions and Integral Transforms: Applications

The techniques of Chapter 6 can yield exact solutions to certain PDEs, by reducing them to ordinary differential equations or even to algebraic equations. For success, it is essential that the PDE be linear and hence allow superposition of solutions.

7.1 THE PRINCIPLE OF SUPERPOSITION

Let $L[\ ]$ denote any *linear* partial differential operator; e.g., (*3.3*). Then, for arbitrary, sufficiently smooth functions $u_1, \ldots, u_N$ and arbitrary constants $c_1, \ldots, c_N$,

$$L[c_1u_1 + \cdots + c_Nu_N] = c_1L[u_1] + \cdots + c_NL[u_N] \tag{7.1}$$

and so

$$L[u_j] = 0 \quad (j = 1, \ldots, N) \Rightarrow L[c_1u_1 + \cdots + c_Nu_N] = 0 \tag{7.2}$$

(*7.2*) is one statement of the *principle of superposition.*

For infinite linear combinations such that

$$\sum_{k=1}^{\infty} c_k u_k \qquad \text{and} \qquad \sum_{k=1}^{\infty} c_k L[u_k]$$

both converge, we have

$$L\left[\sum_{k=1}^{\infty} c_k u_k\right] = \sum_{k=1}^{\infty} c_k L[u_k] \tag{7.3}$$

and the superposition principle reads:

$$L[u_k] = 0 \quad (\text{all } k) \Rightarrow L\left[\sum_{k=1}^{\infty} c_k u_k\right] = 0 \tag{7.4}$$

For a third version, suppose $u(\mathbf{x}, \lambda)$ to be a function of $\mathbf{x}$ in $\mathbf{R}^n$ depending on parameter λ, $a < \lambda < b$, and $g(\lambda)$ to be an integrable function of λ on (a, b). Then, if

$$\int_a^b g(\lambda)u(\mathbf{x}, \lambda)\, d\lambda \qquad \text{and} \qquad \int_a^b g(\lambda)L[u(\mathbf{x}, \lambda)]\, d\lambda$$

both exist, we have

$$L\left[\int_a^b g(\lambda)u(\mathbf{x}, \lambda)\, d\lambda\right] = \int_a^b g(\lambda)L[u(\mathbf{x}, \lambda)]\, d\lambda \tag{7.5}$$

and

$$L[u(\mathbf{x}, \lambda)] = 0 \quad (a < \lambda < b) \Rightarrow L\left[\int_a^b g(\lambda)u(\mathbf{x}, \lambda)\, d\lambda\right] = 0 \tag{7.6}$$

7.2 SEPARATION OF VARIABLES

If $u(x, y)$ satisfies a linear PDE in x and y, then the method of *separation of variables* for this problem begins with the assumption that $u(x, y)$ is of the form $X(x)\,Y(y)$. This has the effect of replacing the single PDE with two ordinary differential equations. The theory of eigenfunction expansions enters into the treatment of any inhomogeneous aspects of the problem.

EXAMPLE 7.1 By examining the Solved Problems, where numerous applications of the method of eigenfunction expansion (separation of variables) are made, we see that for the method to be successful, the problem must have the following attributes:

(1) At least one of the independent variables in the problem must be restricted to a finite interval. Moreover, the domain of the problem must be a coordinate cell in the coordinate system in which the PDE is expressed (e.g., in Cartesian coordinates, a rectangle; in polar coordinates, a sector). See Problem 7.17(*b*).

(2) The PDE must separate; see Problem 7.17(*a*).

(3) In general, homogeneous boundary conditions must be arranged such that at least one of the separated problems is a Sturm–Liouville problem. (If this is not the case, it can often be made so by reduction to subproblems or by a change of dependent variable.) See Problems 7.18 and 7.4.

7.3 INTEGRAL TRANSFORMS

The integral transforms which are most generally applicable are the Fourier and Laplace transforms. Others, such as the Hankel and Mellin transforms, are sometimes useful, but they will not be considered here.

EXAMPLE 7.2 Examination of the Solved Problems reveals that integral transforms apply in the following situations:

(1) The PDE has constant coefficients (otherwise the Fourier or Laplace transform would not produce an ordinary differential equation in the transform space).

(2) The independent variable ranges over an unbounded interval. If the interval is $(-\infty, \infty)$, then the Fourier transform is the likely transform to use. If the interval is $(0, \infty)$ and, in addition, if the initial conditions are appropriate, then the Laplace transform is indicated.

Solved Problems

7.1 For $f(x)$ in $L^2(0, \ell)$, find $u(x, t)$ satisfying

$$u_t = \kappa u_{xx} \qquad 0 < x < \ell,\ t > 0 \tag{1}$$

$$u(x, 0) = f(x) \qquad 0 < x < \ell \tag{2}$$

$$u(0, t) = u(\ell, t) = 0 \qquad t > 0 \tag{3}$$

Assume that $u(x, t) = X(x)\,T(t)$. Then

$$u_t(x, t) = X(x)\,T'(t) \qquad u_{xx}(x, t) = X''(x)\,T(t)$$

and it follows from (*1*) that, for $0 < x < \ell$ and $t > 0$,

$$\frac{T'(t)}{\kappa\, T(t)} = \frac{X''(x)}{X(x)} \tag{4}$$

Since the left side of (*4*) is a function of t alone and the right side is a function of x alone, equality holds for all $0 < x < \ell$ and every $t > 0$ if and only if there exists a constant, $-\lambda$, such that

$$\frac{T'(t)}{\kappa T(t)} = -\lambda = \frac{X''(x)}{X(x)}$$

for $0 < x < \ell$, $t > 0$. This is equivalent to the two separate equations,

$$T'(t) = -\lambda\kappa T(t) \qquad \text{and} \qquad -X''(x) = \lambda X(x)$$

In addition, the boundary conditions (*3*) imply that $X(0) = X(\ell) = 0$. Hence,

$$-X''(x) = \lambda X(x)$$
$$X(0) = X(\ell) = 0$$

is a Sturm–Liouville problem, with eigenvalues $\lambda_n = (n\pi/\ell)^2$ and corresponding eigenfunctions $X_n(x) = \sin(n\pi x/\ell)$ $(n = 1, 2, \ldots)$.

A solution of

$$T'(t) = -\lambda_n\kappa T(t) \qquad (t > 0,\ n = 1, 2, \ldots)$$

is easily found to be $T_n(t) = e^{-\lambda_n\kappa t}$. Thus, for each n, $u_n(x, t) \equiv e^{-\lambda_n\kappa t}\sin(n\pi x/\ell)$ satisfies the PDE (*1*) and the homogeneous boundary conditions (*3*). By the principle of superposition, the function

$$u(x, t) = \sum_{n=1}^{\infty} c_n u_n(x, t)$$

has these same properties, for any set of constants c_n for which the series converges. Finally, the initial condition (*2*) will be satisfied if

$$f(x) = \sum_{n=1}^{\infty} c_n u_n(x, 0) = \sum_{n=1}^{\infty} c_n \sin(n\pi x/\ell)$$

which determines the c_n as the coefficients of the Fourier sine series for $f(x)$ (see Problem 6.1(*b*)):

$$c_n = \frac{2}{\ell}\int_0^{\ell} f(x)\sin(n\pi x/\ell)\,dx$$

7.2 Solve

$$\begin{aligned} u_t(x, t) &= \kappa u_{xx}(x, t) + F(x, t) && 0 < x < \ell,\ t > 0 \\ u(x, 0) &= f(x) && 0 < x < \ell \\ u(0, t) = u(\ell, t) &= 0 && t > 0 \end{aligned}$$

Because the equation here is inhomogeneous, we must use a modified separation of variables procedure. If $F(x, t)$ were zero, Problem 7.1 would give the solution as

$$u(x, t) = \sum_{n=1}^{\infty} c_n e^{-\kappa(n\pi/\ell)^2 t}\sin(n\pi x/\ell)$$

Therefore, borrowing the idea of "variation of parameters," we assume a solution of the form

$$u(x, t) = \sum_{n=1}^{\infty} u_n(t)\sin(n\pi x/\ell) \tag{1}$$

for certain unknown functions $u_n(t)$; in addition, we write

$$F(x, t) = \sum_{n=1}^{\infty} F_n(t)\sin(n\pi x/\ell) \qquad f(x) = \sum_{n=1}^{\infty} f_n \sin(n\pi x/\ell) \tag{2}$$

where the Fourier coefficients $F_n(t)$ and f_n are given by the usual integral formulas. Substituting (*1*) and (*2*) into the PDE yields

$$\sum_{n=1}^{\infty} [u_n'(t) + \kappa(n\pi/\ell)^2 u_n(t) - F_n(t)]\sin(n\pi x/\ell) = 0 \tag{3}$$

and the initial condition becomes

$$\sum_{n=1}^{\infty} [u_n(0) - f_n] \sin(n\pi x/\ell) = 0 \tag{4}$$

Because $\{\sin(n\pi x/\ell)\}$ is a complete orthogonal family in $L^2(0, \ell)$, (3) and (4) imply that for $n = 1, 2, \ldots,$

$$u_n'(t) + \kappa(n\pi/\ell)^2 u_n(t) = F_n(t) \qquad t > 0 \tag{5}$$

$$u_n(0) = f_n \tag{6}$$

There are several standard techniques for solving (5) subject to (6); we choose to take the Laplace transform of (5), applying line 3 of Table 6-3:

$$\hat{u}_n(s) = \frac{f_n}{s + \kappa(n\pi/\ell)^2} + \frac{\hat{F}_n(s)}{s + \kappa(n\pi/\ell)^2} \tag{7}$$

Then, inverting the transform with the aid of line 7 of Table 6-3,

$$u_n(t) = f_n e^{-\kappa(n\pi/\ell)^2 t} + \int_0^t e^{-\kappa(n\pi/\ell)^2(t-\tau)} F_n(\tau)\, d\tau$$

and (1) becomes

$$u(x, t) = \sum_{n=1}^{\infty} f_n e^{-\kappa(n\pi/\ell)^2 t} \sin(n\pi x/\ell) + \sum_{n=1}^{\infty} \left[\int_0^t e^{-\kappa(n\pi/\ell)^2(t-\tau)} F_n(\tau)\, d\tau\right] \sin(n\pi x/\ell) \tag{8}$$

The first series on the right of (8) reflects the influence of the initial state, $u(x, 0)$; the second series reflects the influence of the forcing term, $F(x, t)$.

7.3 Exhibit the steady-state solution to Problem 7.2 if $f(x) \equiv 0$ and (a) $F(x, t) = \phi(x)$ (i.e., time-independent forcing), (b) $F(x, t) = \phi(x) \sin t$. For simplicity, take $\ell = 1$.

By Problem 7.2,

$$u(x, t) = \sum_{n=1}^{\infty} \left[\int_0^t e^{-\kappa(n\pi)^2(t-\tau)} F_n(\tau)\, d\tau\right] \sin n\pi x \tag{1}$$

in which

$$F_n(\tau) = 2\int_0^1 F(x, \tau) \sin n\pi x\, dx \qquad (n = 1, 2, \ldots) \tag{2}$$

(a)
$$F_n(\tau) = \phi_n = 2\int_0^1 \phi(x) \sin n\pi x\, dx \qquad (n = 1, 2, \ldots)$$

and since

$$\int_0^t e^{-\kappa(n\pi)^2(t-\tau)}\, d\tau = \frac{1}{\kappa}\frac{1}{(n\pi)^2}(1 - e^{-\kappa(n\pi)^2 t})$$

(1) reduces to

$$u(x, t) = \frac{1}{\kappa}\sum_{n=1}^{\infty} \frac{\phi_n}{(n\pi)^2}(1 - e^{-\kappa(n\pi)^2 t}) \sin n\pi x$$

Letting $t \to \infty$, we obtain as the steady-state solution

$$u_\infty(x) = \frac{1}{\kappa}\sum_{n=1}^{\infty} \frac{\phi_n}{(n\pi)^2} \sin n\pi x \tag{3}$$

Differentiating (3) twice with respect to x,

$$u_\infty''(x) = -\frac{1}{\kappa}\sum_{n=1}^{\infty} \phi_n \sin n\pi x = -\frac{1}{\kappa}\phi(x)$$

That is, $u_\infty(x)$ satisfies the original inhomogeneous heat equation with all time dependence suppressed. In this sense, $u_\infty(x)$ is an equilibrium solution.

(*b*)
$$F_n(\tau) = (\sin \tau)\left(2\int_0^1 \phi(x) \sin n\pi x \, dx\right) = \phi_n \sin \tau \qquad (n = 1, 2, \ldots)$$

and
$$\int_0^t e^{-\kappa(n\pi)^2(t-\tau)} \sin \tau \, d\tau = \frac{k(n\pi)^2 \sin t - \cos t + e^{-\kappa(n\pi)^2 t}}{1+\kappa^2(n\pi)^4}$$

Hence
$$u(x, t) = \sum_{n=1}^{\infty} \frac{\kappa(n\pi)^2 \sin t - \cos t + e^{-\kappa(n\pi)^2 t}}{1+\kappa^2(n\pi)^4} \phi_n \sin n\pi x$$

For large $t > 0$, $u(x, t)$ approaches
$$u_\infty(x, t) = \sum_{n=1}^{\infty} \frac{\kappa(n\pi)^2 \sin t - \cos t}{1+\kappa^2(n\pi)^4} \phi_n \sin n\pi x$$

Evidently, when the forcing is time dependent, the steady-state solution is also time dependent and cannot be obtained as the solution of a time-independent heat equation.

7.4 Solve
$$\begin{aligned} u_t(x, t) &= \kappa u_{xx}(x, t) & 0 < x < \ell,\ t > 0 \\ u(x, 0) &= 0 & 0 < x < \ell \\ u(0, t) = f_0(t),\ u(\ell, t) &= f_1(t) & t > 0 \end{aligned}$$

if $f_0(0) = f_1(0) = 0$.

If we attempt to separate variables directly, we shall be led to the following consequences of the inhomogeneous boundary conditions:
$$X(0)\,T(t) = f_0(t) \qquad \text{and} \qquad X(\ell)\,T(t) = f_1(t)$$

for $t > 0$. Neither of these implies anything directly about $X(0)$ or $X(\ell)$, and as a result we do not obtain a Sturm–Liouville problem for $X(x)$.

To reduce the problem to one with homogeneous boundary conditions, let us write
$$u(x, t) = v(x, t) + \left(1 - \frac{x}{\ell}\right) f_0(t) + \frac{x}{\ell} f_1(t)$$

The problem for $v(x, t)$ is then
$$\begin{aligned} v_t(x, t) - \kappa v_{xx}(x, t) &= F(x, t) & 0 < x < \ell,\ t > 0 \\ v(x, 0) &= 0 & 0 < x < \ell \\ v(0, t) = v(\ell, t) &= 0 & t > 0 \end{aligned}$$

where
$$F(x, t) \equiv -\left(1 - \frac{x}{\ell}\right) f_0'(t) - \frac{x}{\ell} f_1'(t)$$

We obtain the solution at once by setting $f(x) \equiv 0$ in Problem 7.2:
$$v(x, t) = \sum_{n=1}^{\infty} \left[\int_0^t e^{-\kappa(n\pi/\ell)^2(t-\tau)} F_n(\tau)\, d\tau\right] \sin(n\pi x/\ell) \tag{1}$$

where the Fourier coefficients F_n are given by
$$\begin{aligned} F_n(t) &= -\frac{2}{\ell}\int_0^\ell \left[\left(1 - \frac{x}{\ell}\right) f_0'(t) + \frac{x}{\ell} f_1'(t)\right] \sin(n\pi x/\ell)\, dx \\ &= \frac{2}{n\pi}[(\cos n\pi) f_1'(t) - f_0'(t)] \end{aligned} \tag{2}$$

7.5 Rework Problem 7.4 by the Laplace transform method.

Let $\hat{u}(x, s)$ denote the Laplace transform of $u(x, t)$ with respect to t. Then:

$$s\hat{u}(x,s) - 0 = \kappa \frac{d^2}{dx^2}\hat{u}(x,s) \qquad 0 < x < \ell \tag{1}$$

$$\hat{u}(0,s) = \hat{f}_0(s), \quad \hat{u}(\ell, s) = \hat{f}_1(s) \tag{2}$$

With $\sigma^2 \equiv s/\kappa$, sinh $x\sigma$ and cosh $x\sigma$ are two linearly independent solutions of (*1*). Also, sinh $x\sigma$ and sinh $(\ell - x)\sigma$ are linearly independent solutions, and these will be more convenient for our purposes, as one of them vanishes at $x = 0$, and the other at $x = \ell$. In fact, the linear combination

$$\hat{u}(x,s) = \frac{\hat{f}_1(s)}{\sinh \ell\sigma}\sinh x\sigma + \frac{\hat{f}_0(s)}{\sinh \ell\sigma}\sinh(\ell - x)\sigma \tag{3}$$

satisfies (*1*) and (*2*).

Using the approach of Problem 6.6, we obtain, for $b > a > 0$,

$$\frac{\sinh a\sqrt{s}}{\sinh b\sqrt{s}} = \sum_{n=0}^{\infty} e^{-[(2n+1)b-a]\sqrt{s}} - \sum_{n=0}^{\infty} e^{-[(2n+1)b+a]\sqrt{s}}$$

so that, by line 8 of Table 6-4,

$$\mathcal{L}^{-1}\left\{\frac{\sinh a\sqrt{s}}{\sinh b\sqrt{s}}\right\} = \sum_{n=0}^{\infty} \frac{(2n+1)b-a}{\sqrt{4\pi t^3}} e^{-[(2n+1)b-a]^2/4t} - \sum_{n=0}^{\infty} \frac{(2n+1)b+a}{\sqrt{4\pi t^3}} e^{-[(2n+1)b+a]^2/4t}$$

$$= -\sum_{n=-\infty}^{\infty} \frac{(2n+1)b+a}{\sqrt{4\pi t^3}} e^{-[(2n+1)b+a]^2/4t}$$

Choosing $b = \ell/\sqrt{\kappa}$ and $a = x/\sqrt{\kappa}$ or $(\ell - x)/\sqrt{\kappa}$, we obtain

$$\mathcal{L}^{-1}\left\{\frac{\sinh x\sigma}{\sinh \ell\sigma}\right\} = -\sum_{n=-\infty}^{\infty} \frac{(2n+1)\ell + x}{\sqrt{4\pi\kappa t^3}} e^{-[(2n+1)\ell+x]^2/4\kappa t}$$

$$\mathcal{L}^{-1}\left\{\frac{\sinh(\ell - x)\sigma}{\sinh \ell\sigma}\right\} = -\sum_{n=-\infty}^{\infty} \frac{(2n+2)\ell - x}{\sqrt{4\pi\kappa t^3}} e^{-[(2n+2)\ell-x]^2/4\kappa t}$$

From line 7 of Table 6-3, it follows that if we define

$$M(x,t) \equiv -\sqrt{\frac{\kappa}{\pi t}} \sum_{n=-\infty}^{\infty} e^{-(2n\ell - x)^2/4\kappa t}$$

then $u(x, t)$ can be expressed in the form

$$u(x,t) = \int_0^t M_x(x, t-\tau) f_0(\tau)\,d\tau + \int_0^t M_x(\ell - x, t - \tau) f_1(\tau)\,d\tau$$

This form of the solution involves a series that converges rapidly for small values of t, whereas the form obtained by separating variables is to be preferred for large values of t.

7.6 Solve

$$u_t(x,t) = \kappa u_{xx}(x,t) \qquad -\infty < x < \infty, \ t > 0 \tag{1}$$

$$u(x,0) = f(x) \qquad -\infty < x < \infty \tag{2}$$

$$u(x,t) \text{ of exponential growth in } x \tag{3}$$

This is a well-posed problem for the heat equation, with (*3*) playing the role of boundary conditions on the variable x. Under the Fourier transform with respect to x, the problem becomes:

$$\frac{d}{dt}U(\alpha,t) = -\kappa\alpha^2 U(\alpha,t) \qquad t > 0$$

$$U(\alpha, 0) = F(\alpha)$$

of which the solution is $U(\alpha, t) = F(\alpha)\, e^{-\kappa\alpha^2 t}$. Inverting by use of Table 6-2, line 1, and Table 6-1, line 8, we obtain

$$u(x, t) = \frac{1}{\sqrt{4\pi\kappa t}} \int_{-\infty}^{\infty} e^{-(x-y)^2/4\kappa t} f(y)\, dy \tag{4}$$

It can be directly verified (see Problem 4.17) that (*4*) is a solution to (*1*)–(*2*)–(*3*), independent of the validity of the steps used above to construct it. In fact, it is the unique solution (cf. Problem 4.2).

7.7 Solve

$$\begin{aligned} u_t(x, t) &= \kappa u_{xx}(x, t) && x>0,\ t>0 \\ u(x, 0) &= f(x) && x>0 \\ u(0, t) &= g(t) && t>0 \\ |u(x, t)| &< M && x>0,\ t>0 \end{aligned}$$

Reduce the problem to subproblems for $u_1(x, t)$ and $u_2(x, t)$ such that $u = u_1 + u_2$.

Subproblem 1

$$\begin{aligned} u_{1,t} &= \kappa u_{1,xx} && x>0,\ t>0 \\ u_1(x, 0) &= 0 && x>0 \\ u_1(0, t) &= g(t) && t>0 \\ |u_1(x, t)| &< M && x>0,\ t>0 \end{aligned}$$

Taking the Laplace transform with respect to t, we obtain the problem

$$s\hat{u}_1(x, s) - 0 = \kappa \frac{d^2}{dx^2} \hat{u}_1(x, s) \qquad x>0 \tag{1}$$

$$\hat{u}_1(0, s) = \hat{g}(s) \tag{2}$$

$$|\hat{u}_1(x, s)| < \frac{M}{s} \qquad x>0,\ s>0 \tag{3}$$

The solution of (*1*) that obeys (*2*) and (*3*) is $\hat{u}_1(x, s) = \hat{g}(s) \exp(-x\sqrt{s/\kappa})$. Then, by line 8 of Table 6-4,

$$u_1(x, t) = \frac{x}{\sqrt{4\pi\kappa}} \int_0^t \frac{1}{(t-\tau)^{3/2}} \exp\left[-\frac{x^2}{4\kappa(t-\tau)}\right] g(\tau)\, d\tau$$

Subproblem 2

$$\begin{aligned} u_{2,t}(x, t) &= \kappa u_{2,xx}(x, t) && x>0,\ t>0 \\ u_2(x, 0) &= f(x) && x>0 \\ u_2(0, t) &= 0 && t>0 \\ |u_2(x, t)| &< M && x>0,\ t>0 \end{aligned}$$

Let $F_o(x)$ denote the extension of $f(x)$ as an odd function over the whole x-axis, and consider the problem

$$v_t(x, t) = \kappa v_{xx}(x, t) \qquad -\infty < x < \infty,\ t>0 \tag{1}$$

$$v(x, 0) = F_o(x) \qquad -\infty < x < \infty \tag{2}$$

It is obvious physically that an initially antisymmetric temperature distribution must evolve antisymmetrically; that is, the solution $v(x, t)$ of (*1*)–(*2*) ought to be odd in x. If it is, and in addition is continuous and bounded for all x and all positive t, then its restriction to $x>0$ provides the solution $u_2(x, t)$ of subproblem 2. Now, by Problem 7.6, the unique solution of (*1*)–(*2*) is

$$\begin{aligned} v(x, t) &= \frac{1}{\sqrt{4\pi\kappa t}} \int_{-\infty}^{\infty} e^{-(x-y)^2/4\kappa t} F_o(y)\, dy \\ &= \frac{1}{\sqrt{4\pi\kappa t}} \int_0^{\infty} [e^{-(x-y)^2/4\kappa t} - e^{-(x+y)^2/4\kappa t}] f(y)\, dy \end{aligned}$$

It is easy to see that, under mild conditions on $f(y)$, this function possesses all the desired properties.

7.8 Consider the following special case of Problem 7.7: $f(x) \equiv 0$, $g(t) = g_0 = \text{const}$. Show that (*a*) $u(x, t) = g_0 \operatorname{erfc}(x/\sqrt{4\kappa t})$; (*b*) the "front" $u(x, t) = \alpha g_0$ $(0 < \alpha < 1)$ propagates into the region $x > 0$ at speed $z_\alpha \sqrt{\kappa / t}$, where $\operatorname{erfc} z_\alpha = \alpha$.

(*a*) From Problem 7.7,

$$u(x, t) = u_1(x, t) = \frac{x}{\sqrt{4\pi\kappa}} \int_0^t \frac{1}{(t-\tau)^{3/2}} \exp\left[-\frac{x^2}{4\kappa(t-\tau)}\right] g_0 \, d\tau$$

The transformation $\lambda^2 = x^2/4\kappa(t-\tau)$ changes this to

$$u(x, t) = g_0 \frac{2}{\sqrt{\pi}} \int_{x/\sqrt{4\kappa t}}^{\infty} e^{-\lambda^2} \, d\lambda = g_0 \operatorname{erfc}(x/\sqrt{4\kappa t})$$

(*b*) For $0 < \alpha < 1$, let z_α denote the unique solution of $\operatorname{erfc} z_\alpha = \alpha$. Then $u(x, t) = \alpha g_0$ for all x and t that satisfy $x = z_\alpha \sqrt{4\kappa t}$. Therefore, at time $t > 0$, the point $x_\alpha(t)$ at which $u = \alpha g_0$ moves with speed

$$\frac{dx}{dt} = z_\alpha \frac{\sqrt{\kappa}}{\sqrt{t}}$$

7.9 Solve

$$u_{tt}(x, t) = a^2 u_{xx}(x, t) \qquad 0 < x < \ell, \; t > 0 \tag{1}$$

$$u(x, 0) = f(x), \; u_t(x, 0) = g(x) \qquad 0 < x < \ell \tag{2}$$

$$u(0, t) = u(\ell, t) = 0 \qquad t > 0 \tag{3}$$

We suppose $u(x, t) = X(x)T(t)$ and are led to

$$\frac{X''(x)}{X(x)} = \frac{T''(t)}{a^2 T(t)} = -\lambda^2 \qquad X(0) = X(\ell) = 0$$

This yields two separated problems:

$$X''(x) + \lambda^2 X(x) = 0 \qquad 0 < x < \ell$$
$$X(0) = X(\ell) = 0$$

and

$$T''(t) + a^2\lambda^2 T(t) = 0 \qquad t > 0$$

The respective solutions are, with $\lambda^2 = \lambda_n^2 = (n\pi/\ell)^2$ $(n = 1, 2, \ldots)$,

$$X_n(x) = \sin(n\pi x/\ell)$$
$$T_n(t) = A_n \cos(n\pi a t/\ell) + B_n \sin(n\pi a t/\ell)$$

Hence,

$$u(x, t) = \sum_{n=1}^{\infty} A_n \sin(n\pi x/\ell) \cos(n\pi a t/\ell) + \sum_{n=1}^{\infty} B_n \sin(n\pi x/\ell) \sin(n\pi a t/\ell)$$

Now the initial conditions (2) require

$$f(x) = \sum_{n=1}^{\infty} A_n \sin(n\pi x/\ell) \qquad (0 < x < \ell)$$

$$g(x) = \sum_{n=1}^{\infty} B_n (n\pi a/\ell) \sin(n\pi x/\ell) \qquad (0 < x < \ell)$$

These conditions will be satisfied if A_n and $(n\pi a/\ell)B_n$ respectively equal f_n and g_n, the Fourier sine-series coefficients of the functions $f(x)$ and $g(x)$. Therefore,

$$u(x, t) = \sum_{n=1}^{\infty} f_n \sin(n\pi x/\ell) \cos(n\pi a t/\ell) + \sum_{n=1}^{\infty} \frac{\ell}{n\pi a} g_n \sin(n\pi x/\ell) \sin(n\pi a t/\ell) \tag{4}$$

By use of the relations

$$\sin(n\pi x/\ell)\cos(n\pi at/\ell) = \frac{1}{2}\left[\sin\frac{n\pi}{\ell}(x+at) + \sin\frac{n\pi}{\ell}(x-at)\right]$$

$$\sin(n\pi x/\ell)\sin(n\pi at/\ell) = \frac{1}{2}\left[\cos\frac{n\pi}{\ell}(x-at) - \cos\frac{n\pi}{\ell}(x+at)\right]$$

$$= \frac{n\pi}{2\ell}\int_{x-at}^{x+at}\sin\frac{n\pi z}{\ell}\,dz$$

we can rewrite (*4*) as

$$u(x,t) = \frac{1}{2}\sum_{n=1}^{\infty} f_n\left[\sin\frac{n\pi}{\ell}(x+at) + \sin\frac{n\pi}{\ell}(x-at)\right] + \frac{1}{2a}\sum_{n=1}^{\infty}\int_{x-at}^{x+at} g_n \sin\frac{n\pi z}{\ell}\,dz$$

$$= \frac{1}{2}[\tilde{F}_o(x+at) + \tilde{F}_o(x-at)] + \frac{1}{2a}\int_{x-at}^{x+at}\tilde{G}_o(z)\,dz \tag{5}$$

Evidently, as was suggested in Problem 4.14(*a*), the initial–boundary value problem (*1*)–(*2*)–(*3*) is equivalent to the pure initial value problem in which the initial data are the *periodic* functions $\tilde{F}_o$ and $\tilde{G}_o$. Indeed, for such data, separation of variables in a half-period strip has led to the same D'Alembert solution of the wave equation as is furnished by the method of characteristics when applied over the entire *xt*-plane. See Problem 4.10.

7.10 Solve

$$u_{tt}(x,t) = a^2 u_{xx}(x,t) \qquad -\infty < x < \infty,\ t > 0$$
$$u(x,0) = f(x) \qquad -\infty < x < \infty$$
$$u_t(x,0) = g(x) \qquad -\infty < x < \infty$$

For special, periodic f and g, the solution has already been found in Problem 7.9. Here again we shall retrieve the D'Alembert formula. Apply the Fourier transform in x, to get

$$\frac{d^2U(\alpha,t)}{dt^2} = -a^2\alpha^2 U(\alpha,t) \qquad t>0 \tag{1}$$

$$U(\alpha,0) = F(\alpha) \tag{2}$$

$$\frac{dU}{dt}(\alpha,0) = G(\alpha) \tag{3}$$

Solving (*1*), $U(\alpha,t) = A_1 \sin a\alpha t + A_2 \cos a\alpha t$; then (*2*) and (*3*) imply $A_2 = F(\alpha)$, $A_1 = G(\alpha)/a\alpha$. Hence,

$$U(\alpha,t) = G(\alpha)\frac{\sin a\alpha t}{a\alpha} + F(\alpha)\cos a\alpha t = G(\alpha)\frac{\sin a\alpha t}{a\alpha} + \frac{1}{2}F(\alpha)(e^{ia\alpha t} + e^{-ia\alpha t})$$

where we have chosen to write $\cos a\alpha t$ in the exponential form for convenience in inverting. Using Tables 6-1 and 6-2, we obtain:

$$u(x,t) = g * \frac{\pi}{a} I_{at}(x) + \frac{1}{2}[f(x+at) + f(x-at)]$$

$$= \frac{1}{2a}\int_{-at}^{at} g(x-y)\,dy + \frac{1}{2}[f(x+at) + f(x-at)]$$

$$= \frac{1}{2a}\int_{x-at}^{x+at} g(z)\,dz + \frac{1}{2}[f(x+at) + f(x-at)]$$

7.11 Solve

$$u_{tt}(x,t) = a^2 u_{xx}(x,t) \qquad x>0,\ t>0$$
$$u(x,0) = f(x),\ u_t(x,0) = g(x) \qquad x>0$$
$$u(0,t) = h(t) \qquad t>0$$

The approach of Problem 7.7 is perhaps best here.

Subproblem 1

$$\begin{aligned} u_{1,tt}(x,t) &= a^2 u_{1,xx}(x,t) & x>0,\ t>0 \\ u_1(x,0)=f(x),\ u_{1,t}(x,0) &= g(x) & x>0 \\ u_1(0,t) &= 0 & t>0 \end{aligned}$$

In the usual way, we extend the problem to $x<0$ by making the initial data odd functions of x; this (plus continuity) forces $u_1(0,t)=0$:

$$\begin{aligned} v_{tt}(x,t) &= a^2 v_{xx}(x,t) & -\infty<x<\infty,\ t>0 \\ v(x,0)=F_o(x),\ v_t(x,0) &= G_o(x) & -\infty<x<\infty \end{aligned}$$

Problem 7.10 gives

$$v(x,t)=\frac{1}{2}[F_o(x+at)+F_o(x-at)]+\frac{1}{2a}\int_{x-at}^{x+at} G_o(z)\,dz$$

and so

$$u_1(x,t)=\begin{cases} \dfrac{1}{2}[f(x+at)+f(x-at)]+\dfrac{1}{2a}\displaystyle\int_{x-at}^{x+at} g(z)\,dz & 0<t<x/a \\ \dfrac{1}{2}[f(x+at)-f(at-x)]+\dfrac{1}{2a}\displaystyle\int_{at-x}^{at+x} g(z)\,dz & 0<x/a<t \end{cases}$$

Subproblem 2

$$\begin{aligned} u_{2,tt}(x,t) &= a^2 u_{2,xx}(x,t) & x>0,\ t>0 \\ u^2(x,0)=u_{2,t}(x,0) &= 0 & x>0 \\ u_2(0,t) &= h(t) & t>0 \end{aligned}$$

Apply the Laplace transform in t, to get

$$s^2\hat{u}_2(x,s)-0=a^2\frac{d^2}{dx^2}\hat{u}_2(x,s) \qquad x>0$$

$$\hat{u}_2(0,s)=\hat{h}(s)$$

Then, $\hat{u}_2(x,s)=C_1 e^{-xs/a}+C_2 e^{xs/a}$. In order that $u_2(x,t)$ remain bounded for all positive x and t, we require that $C_2=0$. The initial condition then implies $C_1=\hat{h}(s)$, and we have

$$\hat{u}_2(x,s)=\hat{h}(s)\,e^{-xs/a}$$

It then follows from line 6 of Table 6-3 that

$$u_2(x,t)=\begin{cases} 0 & 0<t<x/a \\ h(t-x/a) & 0<x/a<t \end{cases}$$

Our solution,

$$u=u_1+u_2=\begin{cases} \dfrac{1}{2}[f(x+at)+f(x-at)]+\dfrac{1}{2a}\displaystyle\int_{x-at}^{x+at} g(z)\,dz & 0<t<x/a \\ \dfrac{1}{2}[f(x+at)-f(at-x)]+\dfrac{1}{2a}\displaystyle\int_{at-x}^{at+x} g(z)\,dz+h(t-x/a) & 0<x/a<t \end{cases}$$

should be compared with Fig. 7-1, the characteristic diagram.

At $x=x_0$ at time $t=t_0<x_0/a$, the domain of dependence is the interval $[x_0-at_0, x_0+at_0]$. Moreover, the backward characteristics through (x_0, t_0) do not meet the line $x=0$ in the half-plane $t>0$. Therefore,

$$u(x_0,t_0)=\frac{1}{2}[f(x_0+at_0)+f(x_0-at_0)]+\frac{1}{2a}\int_{x_0-at_0}^{x_0+at_0} g(z)\,dz$$

For $t=t_1>x_0/a$, the domain of dependence of (x_0, t_1) is the interval $[at_1-x_0, at_1+x_0]$, and one of the backward characteristics through (x_0, t_1) cuts the line $x=0$ at $t=t_1-(x_0/a)$. Then,

$$u(x_0,t_1)=\frac{1}{2}[f(at_1+x_0)-f(at_1-x_0)]+\frac{1}{2a}\int_{at_1-x_0}^{at_1+x_0} g(z)\,dz+h(t_1-x_0/a)$$

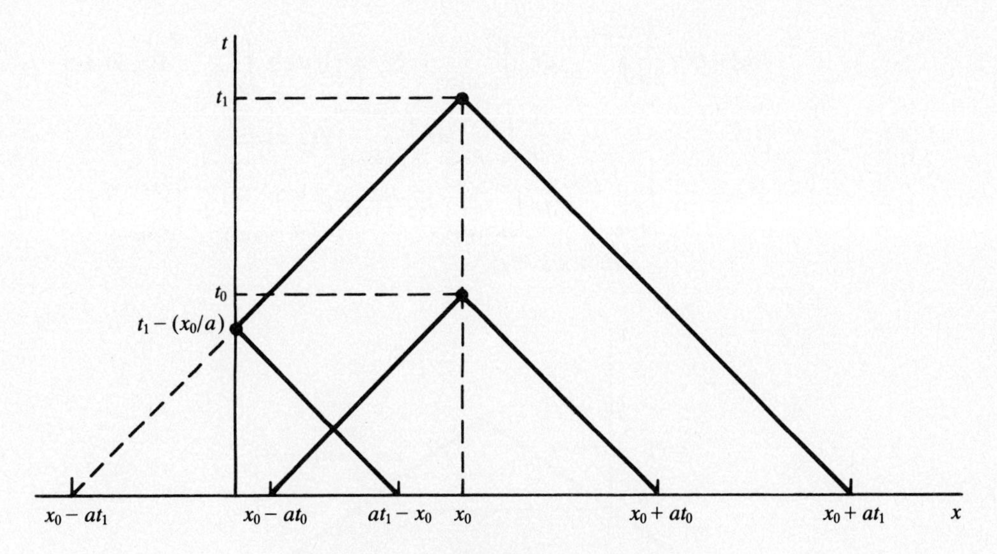

Fig. 7-1

7.12 Solve

$$\begin{aligned} u_{tt}(x,t) &= a^2u_{xx}(x,t)+f(x,t) && x>0,\ t>0 \\ u(x,0) &= u_t(x,0)=0 && x>0 \\ u(0,t) &= h(t) && t>0 \end{aligned}$$

In the usual way, we reduce this to two simpler subproblems with solutions such that $u=u_1+u_2$.

Subproblem 1

$$\begin{aligned} u_{1,tt}(x,t) &= a^2u_{1,xx}(x,t) && x>0,\ t>0 \\ u_1(x,0) &= u_{1,t}(x,0)=0 && x>0 \\ u_1(0,t) &= h(t) && t>0 \end{aligned}$$

This is just subproblem 2 of Problem 7.11; hence,

$$u_1(x,t)=\begin{cases} 0 & 0<t<x/a \\ h(t-x/a) & 0<x/a<t \end{cases}$$

Subproblem 2

$$\begin{aligned} u_{2,tt}(x,t) &= a^2u_{2,xx}(x,t)+f(x,t) && x>0,\ t>0 \\ u_2(x,0) &= u_{2,t}(x,0)=0 && x>0 \\ u_2(0,t) &= 0 && t>0 \end{aligned}$$

or, extending f and u as odd functions of x,

$$\begin{aligned} v_{tt}(x,t) &= a^2v_{xx}(x,t)+F_o(x,t) && -\infty<x<\infty,\ t>0 \\ v(x,0) &= v_t(x,0)=0 && -\infty<x<\infty \end{aligned}$$

The solution for v may be obtained at once from Duhamel's principle (Problem 4.22) and the D'Alembert solution of the wave equation (Problem 7.10) for initial data prescribed at $t=\tau$. Thus,

$$v(x,t)=\int_0^t\left[\frac{1}{2a}\int_{x-a(t-\tau)}^{x+a(t-\tau)}F_o(z,\tau)\,dz\right]d\tau \tag{1}$$

Finally, restricting v to $x>0$, we obtain from (*1*), for $0\le t\le x/a$,

$$u_2(x,t)=\frac{1}{2a}\int_0^t d\tau\int_{x-a(t-\tau)}^{x+a(t-\tau)}f(z,\tau)\,dz \qquad (0<t<x/a)$$

For $0<x/a<t$, the triangle of integration in (*1*) must be decomposed into two regions, as indicated in Fig. 7-2. We find:

$$u_2(x,t)=\frac{1}{2a}\int_0^{t-(x/a)}d\tau\left\{\int_{x-a(t-\tau)}^{0}[-f(-z,\tau)]\,dz+\int_0^{x+a(t-\tau)}f(z,\tau)\,dz\right\}$$
$$+\frac{1}{2a}\int_{t-(x/a)}^{t}d\tau\int_{x-a(t-\tau)}^{x+a(t-\tau)}f(z,\tau)\,dz$$
$$=\frac{1}{2a}\int_0^{t-(x/a)}d\tau\int_{a(t-\tau)-x}^{a(t-\tau)+x}f(z,\tau)\,dz+\frac{1}{2a}\int_{t-(x/a)}^{t}d\tau\int_{x-a(t-\tau)}^{x+a(t-\tau)}f(z,\tau)\,dz$$
$$(0<x/a<t)$$

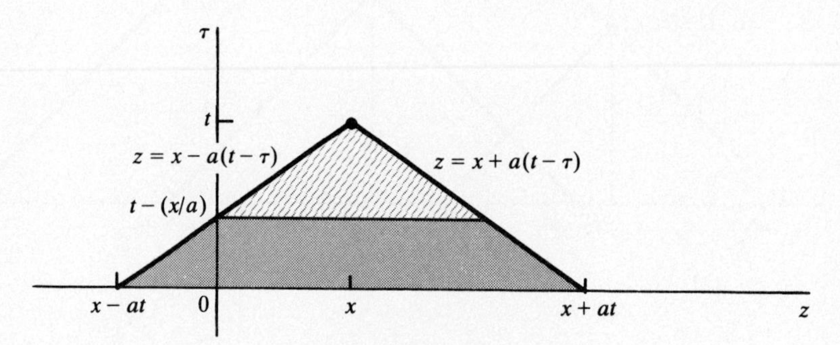

Fig. 7-2

7.13 Solve the following Dirichlet problem for Laplace's equation:

$$\frac{1}{r}\frac{\partial}{\partial r}(ru_r)+\frac{1}{r^2}u_{\theta\theta}=0 \qquad r<1,\ -\pi<\theta<\pi$$
$$u(1,\theta)=f(\theta) \qquad -\pi<\theta<\pi$$
$$u(r,-\pi)=u(r,\pi),\ u_\theta(r,-\pi)=u_\theta(r,\pi) \qquad r<1$$

If we suppose that $u(r,\theta)=R(r)\Theta(\theta)$, then, in the usual manner, we find:

$$\frac{r(rR')'}{R}=-\frac{\Theta''}{\Theta}=\lambda$$

subject to $\Theta(-\pi)=\Theta(\pi)$ and $\Theta'(-\pi)=\Theta'(\pi)$. Thus the separated problems are

$$-\Theta''=\lambda\Theta \qquad -\pi<\theta<\pi$$
$$\Theta(-\pi)=\Theta(\pi),\ \Theta'(-\pi)=\Theta'(\pi)$$

and

$$r^2R''+rR'-\lambda R=0 \qquad r<1$$
$$R(0)\text{ finite}$$

As found in Section 6.4, the eigenvalues of the θ-problem are $\lambda_n=n^2$ $(n=0,1,2,\ldots)$, with corresponding eigenfunctions

$$\Theta_n(\theta)=e^{in\theta}\text{ and }\Theta_{-n}(\theta)=e^{-in\theta} \qquad (n=1,2,\ldots)$$

and $\Theta_0(\theta)=1$. Now the r-problem may be solved to give

$$R_n(r)=r^n \qquad (n=0,1,2,\ldots)$$

The superposition for $u(r,\theta)$ is therefore

$$u(r,\theta)=c_0+\sum_{n=1}^{\infty}(c_nr^n e^{in\theta}+c_{-n}r^n e^{-in\theta})$$
$$=\sum_{n=-\infty}^{\infty}c_nr^{|n|}e^{in\theta} \tag{1}$$

The boundary condition $u(1, \theta) = f(\theta)$ now determines the c_n as the coefficients in the complex Fourier series for $f(\theta)$ over $(-\pi, \pi)$:

$$c_n = \frac{1}{2\pi}\int_{-\pi}^{\pi} f(\phi)\, e^{-in\phi}\, d\phi \qquad (n = 0, \pm 1, \ldots) \tag{2}$$

[cf. (6.15c)]. Substitution of (2) in (1), and transposition of summation and integration, yields

$$u(r, \theta) = \frac{1}{2\pi}\int_{-\pi}^{\pi}\left[\sum_{n=-\infty}^{\infty} r^{|n|}\, e^{in(\theta-\phi)}\right] f(\phi)\, d\phi$$

But, by the formula for the sum of a geometric series,

$$\sum_{n=-\infty}^{\infty} r^{|n|}\, e^{in(\theta-\phi)} = \frac{1-r^2}{1-2r\cos(\theta-\phi)+r^2}$$

and

$$u(r, \theta) = \frac{1-r^2}{2\pi}\int_{-\pi}^{\pi} \frac{f(\phi)}{1-2r\cos(\theta-\phi)+r^2}\, d\phi$$

which is the Poisson integral formula in the unit circle of $\mathbf{R}^2$.

7.14 Solve

$$u_{xx}(x, y) + u_{yy}(x, y) = 0 \qquad -\infty < x < \infty,\ y > 0$$
$$u(x, 0) = f(x) \qquad -\infty < x < \infty$$

The use of the Fourier transform in x is indicated:

$$-\alpha^2 U(\alpha, y) + \frac{d^2}{dy^2} U(\alpha, y) = 0 \qquad y > 0$$
$$U(\alpha, 0) = F(\alpha)$$

The solution of the transformed problem which remains bounded for large y is

$$U(\alpha, y) = F(\alpha)\, e^{-|\alpha| y}$$

Inverting by means of Table 6-2, line 3, and Table 6-1, line 8, we find:

$$u(x, y) = \frac{y}{\pi}\int_{-\infty}^{\infty} \frac{f(z)}{y^2 + (x-z)^2}\, dz = \frac{y}{\pi}\int_{-\infty}^{\infty} \frac{f(x-z)}{y^2 + z^2}\, dz \tag{1}$$

Note that the change of variable $z = y \tan \eta$ takes the second integral (1) into

$$u(x, y) = \frac{1}{\pi}\int_{-\pi/2}^{\pi/2} f(x - y \tan \eta)\, d\eta \qquad (y > 0) \tag{2}$$

Hence, for $f(x)$ continuous,

$$\lim_{y \to 0+} u(x, y) = f(x) \qquad (-\infty < x < \infty)$$

7.15 Solve

$$u_{xx}(x, y) + u_{yy}(x, y) = 0 \qquad -\infty < x < \infty,\ y > 0$$
$$u_y(x, 0) = g(x) \qquad -\infty < x < \infty$$

Let $w(x, y) \equiv u_y(x, y)$. Then $w(x, y)$ satisfies

$$w_{xx}(x, y) + w_{yy}(x, y) = 0 \qquad -\infty < x < \infty,\ y > 0$$
$$w(x, 0) = g(x) \qquad -\infty < x < \infty$$

from which, by (1) of Problem 7.14,

$$w(x, y) = \frac{1}{2\pi}\int_{-\infty}^{\infty} \frac{2y}{y^2 + (x - z)^2} g(z)\, dz \tag{1}$$

Now, $u(x, y) = \int^y w(x, \zeta)\, d\zeta + C(x)$. A y-antiderivative of w is obtained by integrating on y under the integral sign in (*1*):

$$\int^y w(x, \zeta)\, d\zeta = \frac{1}{2\pi}\int_{-\infty}^{\infty} \left[\int^y \frac{2\zeta}{\zeta^2 + (x - z)^2}\, d\zeta\right] g(z)\, dz$$
$$= \frac{1}{2\pi}\int_{-\infty}^{\infty} \log\,(y^2 + (x - z)^2)\, g(z)\, dz$$

As for $C(x)$, it must be bounded and harmonic; hence, a constant. (If u is steady-state temperature, C is an arbitrary reference temperature.)

7.16 Solve

$$\begin{aligned} u_{xx}(x, y) + u_{yy}(x, y) &= 0 \qquad x > 0,\ y > 0 \\ u(x, 0) &= f(x) \qquad x > 0 \\ u(0, y) &= g(y) \qquad y > 0 \end{aligned}$$

Neither variable is restricted to a bounded interval; so separation of variables is not indicated. Neither variable ranges over the whole real line; so the Fourier transform does not seem to apply. Finally, the Laplace transform does not apply, since the equation is second-order in either variable but there is only one initial condition for either variable. However, a reduction of the problem to two subproblems,

$$u(x, y) = u_1(x, y) + u_2(x, y)$$

permits application of the Fourier transform.

Subproblem 1

$$\begin{aligned} u_{1,xx} + u_{1,yy} &= 0 \qquad x > 0,\ y > 0 \\ u_1(x, 0) &= f(x) \qquad x > 0 \\ u_1(0, y) &= 0 \qquad y > 0 \end{aligned}$$

As previously, we obtain u_1 as the restriction to the first quadrant of a function v that satisfies

$$\begin{aligned} v_{xx}(x, y) + v_{yy}(x, y) &= 0 \qquad -\infty < x < \infty,\ y > 0 \\ v(x, 0) &= F_o(x) \qquad -\infty < x < \infty \end{aligned}$$

where $F_o(x)$ denotes the odd extension of $f(x)$. By (*1*) of Problem 7.14,

$$\begin{aligned} v(x, y) &= \frac{y}{\pi}\int_{-\infty}^{\infty} \frac{F_o(z)}{y^2 + (x - z)^2}\, dz \\ &= \frac{y}{\pi}\left[-\int_{-\infty}^{0} \frac{f(-z)}{y^2 + (x - z)^2}\, dz + \int_0^{\infty} \frac{f(z)}{y^2 + (x - z)^2}\, dz\right] \\ &= \frac{y}{\pi}\int_0^{\infty}\left[\frac{1}{y^2 + (x - z)^2} - \frac{1}{y^2 + (x + z)^2}\right] f(z)\, dz \\ &= \frac{4xy}{\pi}\int_0^{\infty} \frac{z}{(x^2 + y^2 + z^2)^2 - 4x^2z^2} f(z)\, dz \end{aligned}$$

Subproblem 2

$$\begin{aligned} u_{2,xx} + u_{2,yy} &= 0 \qquad x > 0,\ y > 0 \\ u_2(x, 0) &= 0 \qquad x > 0 \\ u_2(0, y) &= g(y) \qquad y > 0 \end{aligned}$$

This is just subproblem 1 with x and y interchanged and f replaced by g; hence u_2 is the restriction to the first quadrant of

$$w(x, y) = \frac{4xy}{\pi}\int_0^{\infty} \frac{z}{(x^2 + y^2 + z^2)^2 - 4y^2z^2} g(z)\, dz$$

7.17 Give two examples of boundary value problems for linear PDEs where separation of variables fails.

(*a*)

$$\begin{aligned} u_{xx}(x, y) + u_{xy}(x, y) + u_{yy}(x, y) &= 0 && 0<x<1,\ 0<y<1 \\ u(x, 0) = u(x, 1) &= 0 && 0<x<1 \\ u(0, y) = F(y),\ u(1, y) &= 0 && 0<y<1 \end{aligned}$$

If we suppose that $u(x, y) = X(x)Y(y)$, then $X''(x)Y(y) + X'(x)Y'(y) + X(x)Y''(y) = 0$ and there is no way to separate this expression so as to have a function of x alone on one side of the equation and a function of y alone on the other side.

(*b*)

$$\begin{aligned} u_{xx}(x, y) + u_{yy}(x, y) &= 0 && 0<x<1,\ 0<y<x \\ u(x, 0) &= f(x) && 0<x<1 \\ u(1, y) &= 0 && 0<y<1 \\ u(x, x) &= 0 && 0<x<1 \end{aligned}$$

If we suppose that $u(x, y) = X(x)Y(y)$, then the equation separates into two ordinary differential equations. However, the boundary conditions do not lead to a problem of Sturm–Liouville type either for $X(x)$ or $Y(y)$. The difficulty lies in the fact that the region $\{0<x<1, 0<y<x\}$ is not a coordinate cell. This problem may, in fact, be solved by means of a clever transformation (see Problem 7.18). In general, when Ω is not a coordinate cell, no such transformation is possible.

7.18 Transform Problem 7.17(*b*) so that it becomes separable, and carry out the solution. For convergence of the series, assume that $f(0) = f(1) = 0$.

Extend the problem to the square $0<x<1, 0<y<1$, making the boundary data antisymmetric with respect to the diagonal $y = x$ (see Fig. 7-3). This forces $v(x, x) = 0$. Now, decompose the v-problem into two subproblems such that $v = v_1 + v_2$.

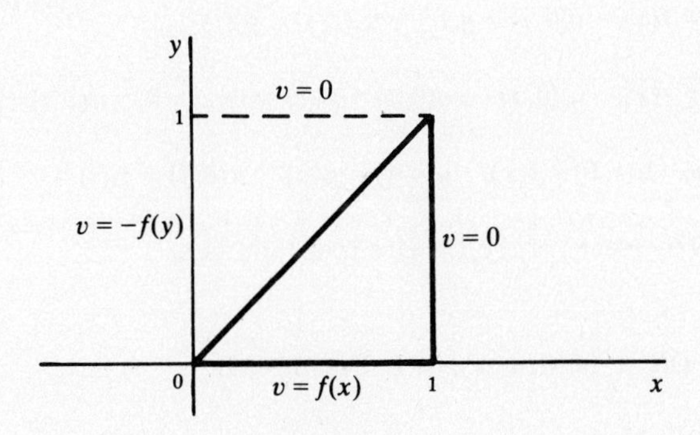

Fig. 7-3

Subproblem 1

$$\begin{aligned} v_{1,xx}(x, y) + v_{1,yy}(x, y) &= 0 && 0<x<1,\ 0<y<1 \\ v_1(x, 0) = f(x),\ v_1(x, 1) &= 0 && 0<x<1 \\ v_1(0, y) = v_1(1, y) &= 0 && 0<y<1 \end{aligned}$$

Upon separation of variables, the three homogeneous boundary conditions give the eigenfunctions

$$\sin n\pi x \sinh n\pi(1-y) \qquad (n = 1, 2, \ldots)$$

The remaining boundary condition then determines the superposition coefficients c_n through

$$f(x) = \sum_{n=1}^{\infty} (c_n \sinh n\pi) \sin n\pi x \qquad (0<x<1)$$

or

$$c_n \sinh n\pi = f_n \equiv 2\int_0^1 f(x) \sin n\pi x \, dx$$

(the nth coefficient in the Fourier sine series for f). Therefore,

$$v_1(x, g) = \sum_{n=1}^{\infty} \frac{f_n}{\sinh n\pi} \sin n\pi x \sinh n\pi(1-y) \qquad (0<x<1,\ 0<y<1)$$

Subproblem 2 is just subproblem 1 with x and y interchanged and f replaced by $-f$. Consequently,

$$v_2(x, y) = \sum_{n=1}^{\infty} \frac{-f_n}{\sinh n\pi} \sin n\pi y \sinh n\pi(1-x) \qquad (0<x<1,\ 0<y<1)$$

We conclude that the solution to Problem 7.17(b) is

$$u(x, y) = \sum_{n=1}^{\infty} \frac{f_n}{\sinh n\pi} [\sin n\pi x \sinh n\pi(1-y) - \sin n\pi y \sinh n\pi(1-x)]$$

$$(0<x<1, 0<y<x)$$

Supplementary Problems

THE HEAT EQUATION ON A FINITE INTERVAL

Solve for $u(x, t)$ $(0<x<\ell, t>0)$. Use eigenfunction expansions in Problems 7.19–7.22.

7.19 $u_t = \kappa u_{xx}$, $u(x, 0) = f(x)$, $u_x(0, t) = g_0(t)$, $u_x(\ell, t) = g_1(t)$.

7.20 $u_t = \kappa u_{xx}$, $u(x, 0) = f(x)$, $u(0, t) = g_0(t)$, $u_x(\ell, t) = g_1(t)$.

7.21 $u_t = \kappa u_{xx}$, $u(x, 0) = f(x)$, $u_x(0, t) - pu(0, t) = g_0(t)$ with $p>0$, $u(\ell, t) = g_1(t)$.

7.22 $u_t = \kappa u_{xx} + bu_x + cu$, $u(x, 0) = f(x)$, $u(0, t) = g_0(t)$, $u(\ell, t) = g_1(t)$.

7.23 Solve Problem 7.19 by means of the Laplace transform in t.

THE HEAT EQUATION ON A SEMI-INFINITE INTERVAL

Solve for $u(x, t)$ $(x>0,\ t>0)$.

7.24 $u_t = \kappa u_{xx}$, $u(x, 0) = 0$, $u_x(0, t) = g(t)$.

7.25 $u_t = \kappa u_{xx}$, $u(x, 0) = 0$, $u_x(0, t) - pu(0, t) = g(t)$ with $p>0$. [*Hint*: Let $v(x, t) = u_x(x, t) - pu(x, t)$.]

7.26 $u_t = \kappa u_{xx} + bu_x + cu$, $u(x, 0) = 0$, $u(0, t) = f(t)$.

THE WAVE EQUATION ON A FINITE INTERVAL

Solve for $u(x, t)$ $(0<x<\ell, t>0)$. Use eigenfunction expansions or the D'Alembert formula in Problems 7.27–7.31, along with Duhamel's principle where appropriate. In Problems 7.32–7.34, apply the Laplace transform.

7.27 $u_{tt} = a^2 u_{xx}, \quad u(x, 0) = f(x), \quad u_t(x, 0) = g(x), \quad u_x(0, t) = u_x(\ell, t) = 0.$

7.28 $u_{tt} = a^2 u_{xx}, \quad u(x, 0) = f(x), \quad u_t(x, 0) = g(x), \quad u_x(0, t) = u(\ell, t) = 0.$

7.29 $u_{tt} = a^2 u_{xx} + f(x, t), \quad u(x, 0) = u_t(x, 0) = 0, \quad u(0, t) = u(\ell, t) = 0.$

7.30 $u_{tt} = a^2 u_{xx} - 2cu_t, \quad u(x, 0) = 0, \quad u_t(x, 0) = g(x), \quad u(0, t) = u(\ell, t) = 0.$

7.31 $u_{tt} = a^2 u_{xx} - 2cu_t + g(x), \quad u(x, 0) = u_t(x, 0) = 0, \quad u(0, t) = u(\ell, t) = 0.$

7.32 $u_{tt} = a^2 u_{xx}, \quad u(x, 0) = u_t(x, 0) = 0, \quad u(0, t) = f(t), \quad u(\ell, t) = g(t).$

7.33 $u_{tt} = a^2 u_{xx}, \quad u(x, 0) = u_t(x, 0) = 0, \quad u_x(0, t) = f(t), \quad u_x(\ell, t) = g(t).$

7.34 $u_{tt} = a^2 u_{xx}, \quad u(x, 0) = u_t(x, 0) = 0, \quad u_x(0, t) = f(t), \quad u(\ell, t) = g(t).$

THE WAVE EQUATION ON A SEMI-INFINITE INTERVAL

Solve for $u(x, t)$ $(x > 0, \ t > 0)$.

7.35 $u_{tt} = a^2 u_{xx}, \quad u(x, 0) = u_t(x, 0) = 0, \quad u_x(0, t) = f(t).$

7.36 $u_{tt} = a^2 u_{xx}, \quad u(x, 0) = u_t(x, 0) = 0, \quad u_x(0, t) - pu(0, t) = f(t)$ with $p > 0$.

LAPLACE'S EQUATION ON BOUNDED DOMAINS

7.37

$$\begin{aligned} u_{xx}(x, y) + u_{yy}(x, y) &= 0 & & 0 < x < 1, \ 0 < y < 1 \\ u_x(0, y) = f(y), \ u_x(1, y) &= g(y) & & 0 < y < 1 \\ u_y(x, 0) = p(x), \ u_y(x, 1) &= q(x) & & 0 < x < 1 \end{aligned}$$

7.38

$$\begin{aligned} \nabla^2 u(r, \theta) &= 0 & & 0 \le r \le 1, \ -\pi < \theta < \pi \\ u_r(1, \theta) &= f(\theta) & & -\pi < \theta < \pi \end{aligned}$$

where $\int_{-\pi}^{\pi} f(\theta) \, d\theta = 0$.

7.39

$$\begin{aligned} \nabla^2 u(r, \theta) &= 0 & & a < r < b, \ 0 < \theta < \pi \\ u(r, 0) = u(r, \pi) &= 0 & & a < r < b \\ u(a, \theta) = 0, \ u(b, \theta) &= 1 & & 0 < \theta < \pi \end{aligned}$$

LAPLACE'S EQUATION ON UNBOUNDED DOMAINS

7.40

$$\begin{aligned} u_{xx}(x, y) + u_{yy}(x, y) &= 0 & & x > 0, \ y > 0 \\ u_x(0, y) &= f(y) & & y > 0 \\ u_y(x, 0) &= g(x) & & x > 0 \end{aligned}$$

7.41

$$\begin{aligned} u_{xx}(x, y) + u_{yy}(x, y) &= 0 & & -\infty < x < \infty, \ 0 < y < 1 \\ u_y(x, 0) = f(x), \ u_y(x, 1) &= g(x) & & -\infty < x < \infty \end{aligned}$$

7.42

$$\begin{aligned} \nabla^2 u(r, \theta) &= 0 & & r > 1, \ -\pi < \theta < \pi \\ u(1, \theta) &= f(\theta) & & -\pi < \theta < \pi \\ |u(r, \theta)| &< M & & r > 1, \ -\pi < \theta < \pi \end{aligned}$$

[*Hint*: Use the results of Problems 7.13 and 3.29(*a*).]

7.43

$$\begin{aligned} \nabla^2 u(r, \theta) &= 0 & & r > 1, \ -\pi < \theta < \pi \\ u_r(1, \theta) &= f(\theta) & & -\pi < \theta < \pi \\ |u(r, \theta)| &< M & & r > 1, \ -\pi < \theta < \pi \end{aligned}$$

where $\int_{-\pi}^{\pi} f(\theta) \, d\theta = 0$. [*Hint*: Use the results of Problems 7.38 and 3.29(*a*), bearing in mind that the inversion will reverse the sign of the boundary derivative.]

Chapter 8

Green's Functions

8.1 INTRODUCTION

In a region Ω with boundary S, let

$$L[u] = f(\mathbf{x}) \quad \text{in } \Omega \tag{8.1}$$

$$B[u] = 0 \quad \text{on } S \tag{8.2}$$

represent, respectively, a linear, second-order PDE and linear boundary–initial conditions, such that for each continuous f, the problem (8.1)–(8.2) has a unique solution. Then $G(\mathbf{x}; \boldsymbol{\xi})$ is the *Green's function* for the problem if this unique solution is given by

$$u(\mathbf{x}) = \int_\Omega G(\mathbf{x}; \boldsymbol{\xi}) f(\boldsymbol{\xi})\, d_\xi \Omega$$

(We attach a subscript to the volume element to emphasize that the integration is with respect to the ξ-variables.)

EXAMPLE 8.1 We know (Problems 4.2, 4.7, 4.17, 7.6) that the initial value problem for the heat equation,

$$v_t(x, t) - v_{xx}(x, t) = 0 \qquad -\infty < x < \infty,\ t > 0$$

$$v(x, 0) = f(x) \qquad -\infty < x < \infty$$

has the unique solution

$$v(x, t) = \frac{1}{\sqrt{4\pi t}} \int_{-\infty}^{\infty} \exp\left[-\frac{(x-\xi)^2}{4t}\right] f(\xi)\, d\xi$$

It then follows from Duhamel's principle (Problem 4.23) that the problem

$$u_t(x, t) - u_{xx}(x, t) = f(x) \qquad -\infty < x < \infty,\ t > 0 \tag{1}$$

$$u(x, 0) = 0 \qquad -\infty < x < \infty \tag{2}$$

has the unique solution

$$u(x, t) = \int_0^t v(x, t - \tau)\, d\tau = \int_0^t \int_{-\infty}^{\infty} \frac{1}{\sqrt{4\pi(t-\tau)}} \exp\left[-\frac{(x-\xi)^2}{4(t-\tau)}\right] f(\xi)\, d\xi\, d\tau \tag{3}$$

From (3) we infer that the Green's function for the problem (1)–(2) is

$$G(x, t; \xi, \tau) = \frac{1}{\sqrt{4\pi(t-\tau)}} \exp\left[-\frac{(x-\xi)^2}{4(t-\tau)}\right] \qquad (t > \tau > 0) \tag{4}$$

It is seen that the Green's function (4) exhibits singular behavior as $\mathbf{x} = (x, t)$ approaches $\boldsymbol{\xi} = (\xi, \tau)$. This holds true for Green's functions in general, and is reflected in the fact that $G(\mathbf{x}; \boldsymbol{\xi})$ for (8.1)–(8.2) satisfies, as a function of $\mathbf{x}$, the PDE

$$L[\sigma] = \delta(\mathbf{x} - \boldsymbol{\xi}) \tag{5}$$

which is (8.1) with $f(\mathbf{x})$ replaced by the "function" $\delta(\mathbf{x} - \boldsymbol{\xi})$ (see Problem 8.1). We call a solution of (5) a *singularity solution for* $L[\ ]$. The essence, then, of the Green's function method is to represent $f(\mathbf{x})$ in (8.1) as a "sum" of delta functions, thereby obtaining $u(\mathbf{x})$ as the "sum" of the corresponding singularity solutions (adjusted to obey (8.2)).

8.2 LAPLACE'S EQUATION

It follows from Problem 3.17 that the function

$$\sigma(\mathbf{x};\boldsymbol{\xi}) = \begin{cases} \dfrac{1}{2}|x-\xi| & \text{for } n=1 \\ \dfrac{1}{2\pi}\log|\mathbf{x}-\boldsymbol{\xi}| & \text{for } n=2 \\ \dfrac{1}{(2-n)A_n(1)}|\mathbf{x}-\boldsymbol{\xi}|^{2-n} & \text{for } n\geq 3 \end{cases} \tag{8.3}$$

where $\mathbf{x}$ and $\boldsymbol{\xi}$ represent distinct points and $A_n(1)$ denotes the area of the unit sphere (see Example 3.2), is the *singularity solution* for the Laplacian operator $\nabla^2[\ \ ]$ in $\mathbf{R}^n$.

EXAMPLE 8.2 For Laplace's equation in three-space, $u_{xx}+u_{yy}+u_{zz}=0$, with $\mathbf{x}=(x,y,z)$ and $\boldsymbol{\xi}=(\xi,\eta,\zeta)$, the singularity solution is

$$\sigma(x,y,z;\xi,\eta,\zeta) = -\frac{1}{4\pi[(x-\xi)^2+(y-\eta)^2+(z-\zeta)^2]^{1/2}}$$

Theorem 8.1: If $\boldsymbol{\xi}$ is a fixed point of $\mathbf{R}^n$, then:

(i) for $n\geq 1$, $\sigma(\mathbf{x};\boldsymbol{\xi})$ is dependent only on the distance $r\equiv|\mathbf{x}-\boldsymbol{\xi}|$;

(ii) for $n\geq 1$, $\nabla^2_{\mathbf{x}}\sigma=0$ for all $\mathbf{x}\neq\boldsymbol{\xi}$;

(iii) for $n>1$, the integral of the normal derivative of σ over any sphere centered at $\mathbf{x}=\boldsymbol{\xi}$ is equal to one. (This extends in the obvious way to $\mathbf{R}^1$.)

Since $\sigma(\mathbf{x};\boldsymbol{\xi})=\sigma(\boldsymbol{\xi};\mathbf{x})$, Theorem 8.1 is valid with the roles of $\mathbf{x}$ and $\boldsymbol{\xi}$ reversed.

Because Poisson's equation, $\nabla^2 u=\rho(\mathbf{x})$, is solved in unbounded $\mathbf{R}^n$ by

$$u(\mathbf{x}) = \int_{\mathbf{R}^n} \sigma(\mathbf{x};\boldsymbol{\xi})\rho(\boldsymbol{\xi})\,d_{\boldsymbol{\xi}}\Omega$$

the singularity solution serves as the Green's function for $\nabla^2[\ \ ]$ when no boundaries are present; we call it the *free-space Green's function.* We now show how to modify this function so that it gives the Green's function when boundary conditions have to be satisfied.

Let Ω be a region with boundary S, and take $\boldsymbol{\xi}$ to be a fixed point inside Ω. If, as a function of $\mathbf{x}$, $\nabla^2 h=0$ in Ω, then the function

$$\phi(\mathbf{x},\boldsymbol{\xi}) = \sigma(\mathbf{x};\boldsymbol{\xi}) + h$$

where h may depend on $\boldsymbol{\xi}$ as well as on $\mathbf{x}$, is called a *fundamental solution* of Laplace's equation in Ω.

EXAMPLE 8.3 Let Ω be the upper half of the xy-plane and let $\boldsymbol{\xi}=(\xi,\eta)$ be a fixed point in Ω. Both

$$\phi_1(x,y;\xi,\eta) = \sigma(x,y;\xi,\eta)+x^2-y^2$$

and

$$\phi_2(x,y;\xi,\eta) = \sigma(x,y;\xi,\eta)+\sigma(x,y;\xi,-\eta)$$

are fundamental solutions of Laplace's equation in Ω.

Theorem 8.2: Let Ω be a bounded region to which the divergence theorem applies. If $\phi(\mathbf{x};\boldsymbol{\xi})$ is a fundamental solution of Laplace's equation in Ω and if $\nabla^2 u=f(\mathbf{x})$ in Ω, then

$$u(\boldsymbol{\xi}) = \int_\Omega \phi(\mathbf{x};\boldsymbol{\xi})f(\mathbf{x})\,d\Omega + \int_S \left(u\frac{\partial\phi}{\partial n}-\phi\frac{\partial u}{\partial n}\right)dS \tag{8.4}$$

where

$$\frac{\partial}{\partial n} \equiv \mathbf{n}\cdot\nabla_{\mathbf{x}}$$

If f and u are sufficiently well-behaved at infinity, then Theorem 8.2 is valid in unbounded regions. In *(8.4)* the roles of $\mathbf{x}$ and $\boldsymbol{\xi}$ can be interchanged to obtain an expression for $u(\mathbf{x})$ ($\mathbf{x}$ in Ω).

Suppose that for each continuous f and g, the mixed problem

$$\nabla^2 u = f(\mathbf{x}) \qquad \text{in } \Omega$$

$$\alpha u + \beta \frac{\partial u}{\partial n} = g(\mathbf{x}) \qquad \text{on } S$$

has a unique solution. The *Green's function*, $G(\mathbf{x}; \boldsymbol{\xi})$, for this problem is the fundamental solution of Laplace's equation in Ω that satisfies

$$\alpha G + \beta \frac{\partial G}{\partial n} = 0 \qquad \text{on } S$$

This homogeneous boundary condition ensures that, for $\phi = G$, the boundary integral in *(8.4)* depends only on the known functions G, α, β, and g, and not on u. (It does not, however, ensure that the boundary integral will vanish; that can be enforced only when the boundary condition on u is homogeneous, as in *(8.2)*.)

The remainder of this section assumes $\beta = 0$; i.e., it treats the Dirichlet problem

$$\nabla^2 u = f(\mathbf{x}) \qquad \text{in } \Omega \tag{8.5}$$

$$u = g(\mathbf{x}) \qquad \text{on } S \tag{8.6}$$

Theorem 8.3: The Green's function for *(8.5)*–*(8.6)* is unique and is given by

$$G(\mathbf{x}; \boldsymbol{\xi}) = \sigma(\mathbf{x}; \boldsymbol{\xi}) + h(\mathbf{x}; \boldsymbol{\xi})$$

where, for each fixed $\boldsymbol{\xi}$ in Ω, h satisfies

$$\nabla_{\mathbf{x}}^2 h = 0 \qquad \text{in } \Omega$$

$$h = -\sigma \qquad \text{on } S$$

Theorem 8.4: For each fixed $\boldsymbol{\xi}$ in Ω, the Green's function for *(8.5)*–*(8.6)* satisfies

$$\nabla_{\mathbf{x}}^2 G(\mathbf{x}; \boldsymbol{\xi}) = \delta(\mathbf{x} - \boldsymbol{\xi}) \qquad \text{in } \Omega$$

$$G = 0 \qquad \text{on } S$$

where $\delta(\mathbf{x} - \boldsymbol{\xi})$ is the n-dimensional Dirac delta function.

Theorem 8.5: The Green's function for *(8.5)*–*(8.6)* is symmetric, $G(\mathbf{x}; \boldsymbol{\xi}) = G(\boldsymbol{\xi}; \mathbf{x})$, and it is negative for all distinct $\mathbf{x}$ and $\boldsymbol{\xi}$ in Ω.

Theorem 8.6: If Ω is bounded, the Green's function for *(8.5)*–*(8.6)* has the eigenfunction expansion

$$G(\mathbf{x}; \boldsymbol{\xi}) = \sum_{r=1}^{\infty} \frac{u_r(\mathbf{x})\, u_r(\boldsymbol{\xi})}{\lambda_r}$$

where $\nabla^2 u_r(\mathbf{x}) = \lambda_r u_r(\mathbf{x})$ in Ω, $u_r(\mathbf{x}) = 0$ on S, $u_r \not\equiv 0$.

Theorem 8.7: The solution of *(8.5)*–*(8.6)* is

$$u(\mathbf{x}) = \int_{\Omega} G(\mathbf{x}; \boldsymbol{\xi}) f(\boldsymbol{\xi})\, d_{\boldsymbol{\xi}}\Omega + \int_S g(\boldsymbol{\xi}) \frac{\partial G}{\partial n}(\mathbf{x}; \boldsymbol{\xi})\, d_{\boldsymbol{\xi}} S$$

where

$$\frac{\partial G}{\partial n}(\mathbf{x}; \boldsymbol{\xi}) \equiv \mathbf{n} \cdot \nabla_{\boldsymbol{\xi}} G(\mathbf{x}; \boldsymbol{\xi})$$

The techniques for constructing Green's functions for Laplace's equation include the method of images (see Problem 8.5), eigenfunction expansions, and integral transforms. In two dimensions, conformal mappings of the complex plane provide a powerful means of constructing Green's functions for Laplace's equation.

Theorem 8.8: (i) Let $w = F(z)$ be an analytic function which maps the region Ω in the z-plane onto the upper half of the w-plane, with $F'(z) \neq 0$ in Ω. Then, if $z = x + iy$ and $\zeta = \xi + i\eta$ are any two points in Ω, the Green's function for (8.5)–(8.6) in $\mathbf{R}^2$ is given by

$$G(x, y; \xi, \eta) = \frac{1}{2\pi} \log \left| \frac{F(z) - F(\zeta)}{F(z) - \overline{F(\zeta)}} \right|$$

where the overbar denotes the complex conjugate. (ii) Let $w = f(z)$ be an analytic function which maps the region Ω in the z-plane onto the unit circle in the w-plane, with $f(\zeta) = 0$ and $f'(z) \neq 0$ in Ω. Then the Green's function for (8.5)–(8.6) in $\mathbf{R}^2$ is given by

$$G(x, y; \xi, \eta) = \frac{1}{2\pi} \log |f(z)|$$

8.3 ELLIPTIC BOUNDARY VALUE PROBLEMS

Given a second-order, linear, partial differential operator, $L[\ ]$, defined by

$$L[u] \equiv \sum_{i,j=1}^{n} a_{ij} \frac{\partial^2 u}{\partial x_i \partial x_j} + \sum_{i=1}^{n} b_i \frac{\partial u}{\partial x_i} + cu \tag{8.7}$$

the *adjoint operator*, $L^*[\ ]$, is defined by

$$L^*[v] \equiv \sum_{i,j=1}^{n} \frac{\partial^2}{\partial x_i \partial x_j} (a_{ij} v) - \sum_{i=1}^{n} \frac{\partial}{\partial x_i} (b_i v) + cv \tag{8.8}$$

It is assumed that the a_{ij} are in C^2 and the b_i are in C^1. For any pair of C^2 functions u and v, *Lagrange's identity*,

$$vL[u] - uL^*[v] = \sum_{i=1}^{n} \frac{\partial}{\partial x_i} \left[\sum_{j=1}^{n} a_{ij} \left(v \frac{\partial u}{\partial x_j} - u \frac{\partial v}{\partial x_j} \right) + uv \left(b_i - \sum_{j=1}^{n} \frac{\partial a_{ij}}{\partial x_j} \right) \right] \tag{8.9}$$

holds. If M_i denotes the expression in square brackets on the right side of (8.9) and $\mathbf{M} \equiv (M_1, M_2, \ldots, M_n)$, then Lagrange's identity takes the form

$$vL[u] - uL^*[v] = \nabla \cdot \mathbf{M} \tag{8.10}$$

If (8.10) is integrated over a region Ω with boundary S, then the divergence theorem shows that

$$\int_\Omega vL[u]\, d\Omega = \int_\Omega uL^*[v]\, d\Omega + \int_S \mathbf{M} \cdot \mathbf{n}\, dS \tag{8.11}$$

where, as ever, $\mathbf{n}$ is a unit outward normal to S.

Consider the linear boundary value problem

$$L[u] = f \quad \text{in } \Omega \tag{8.12}$$

$$B[u] = 0 \quad \text{on } S \tag{8.13}$$

where $L[\ ]$ is an *elliptic* operator of the form (8.7) and

$$B[u] = \alpha u + \beta \frac{\partial u}{\partial n}$$

The *adjoint boundary conditions*, $B^*[v] = 0$ on S, are a minimal set of homogeneous conditions on v such that $B[u] = B^*[v] = 0$ on S implies $\mathbf{M} \cdot \mathbf{n} = 0$ on S. The PDE (8.12) is called *self-adjoint* if $L^*[\ ] = L[\ ]$; problem (8.12)–(8.13) is self-adjoint if $L^*[\ ] = L[\ ]$ and $B^*[\ ] = B[\ ]$.

Theorem 8.9: Let $\mathbf{x} = (x_1, x_2, \ldots, x_n)$ and $\boldsymbol{\xi} = (\xi_1, \xi_2, \ldots, \xi_n)$. If (*8.12*)–(*8.13*) has a Green's function, $G(\mathbf{x}; \boldsymbol{\xi})$, then, as a function of the x-variables, G satisfies

$$L_{\mathbf{x}}[G] = \delta(\mathbf{x} - \boldsymbol{\xi}) \quad \text{in } \Omega \tag{8.14}$$

$$B_{\mathbf{x}}[G] = 0 \quad \text{on } S \tag{8.15}$$

As a function of the ξ-variables, G satisfies

$$L_{\boldsymbol{\xi}}^*[G] = \delta(\mathbf{x} - \boldsymbol{\xi}) \quad \text{in } \Omega \tag{8.16}$$

$$B_{\boldsymbol{\xi}}^*[G] = 0 \quad \text{on } S \tag{8.17}$$

Theorem 8.10: If $G(\mathbf{x}; \boldsymbol{\xi})$ is the Green's function for (*8.12*)–(*8.13*), then G is symmetric in $\mathbf{x}$ and $\boldsymbol{\xi}$ if and only if the problem is self-adjoint.

Theorem 8.11: For (*8.12*)–(*8.13*) to have a Green's function, it is necessary that $u \equiv 0$ be the only solution to $L[u] = 0$ in Ω, $B[u] = 0$ on S.

8.4 DIFFUSION EQUATION

For the diffusion equation in n space-variables, $u_t - \kappa \nabla^2 u = 0$, the singularity solution is

$$K(\mathbf{x} - \boldsymbol{\xi}, t - \tau) = H(t - \tau)\,[4\pi\kappa(t - \tau)]^{-n/2} \exp\left[\frac{-|\mathbf{x} - \boldsymbol{\xi}|^2}{4\kappa(t - \tau)}\right] \tag{8.18}$$

Theorem 8.12: (i) As a function of $\mathbf{x}$ and t, the singularity solution (*8.18*) satisfies

$$K_t - \kappa \nabla_{\mathbf{x}}^2 K = 0 \qquad (\mathbf{x}, t) \neq (\boldsymbol{\xi}, \tau)$$

$$\lim_{t \to \tau+} K(\mathbf{x} - \boldsymbol{\xi}, t - \tau) = \delta(\mathbf{x} - \boldsymbol{\xi})$$

(ii) As a function of $\boldsymbol{\xi}$ and τ, the singularity solution (*8.18*) satisfies

$$-K_\tau - \kappa \nabla_{\boldsymbol{\xi}}^2 K = 0 \qquad (\boldsymbol{\xi}, \tau) \neq (\mathbf{x}, t)$$

$$\lim_{\tau \to t-} K(\mathbf{x} - \boldsymbol{\xi}, t - \tau) = \delta(\mathbf{x} - \boldsymbol{\xi})$$

Theorem 8.12(i) shows that, as a function of $\mathbf{x}$ and t, K satisfies

$$K_t - \kappa \nabla_{\mathbf{x}}^2 K = \delta(\mathbf{x} - \boldsymbol{\xi})\,\delta(t - \tau) \tag{8.19}$$

In the context of time-dependent heat flow, (*8.19*) permits the following interpretation: K is the temperature distribution in $\mathbf{x}$ at time t due to the release of a unit heat pulse at position $\boldsymbol{\xi}$ at time τ. Theorem 8.12(ii) implies that, in $\boldsymbol{\xi}$ and τ, K satisfies

$$-K_\tau - \kappa \nabla_{\boldsymbol{\xi}}^2 K = \delta(\mathbf{x} - \boldsymbol{\xi})\,\delta(t - \tau) \tag{8.20}$$

Given a bounded region Ω to which the divergence theorem applies, a *fundamental solution* of the diffusion equation in Ω, $\phi(\mathbf{x}, t; \boldsymbol{\xi}, \tau)$, is defined in much the same way as a fundamental solution of Laplace's equation (Section 8.2). Specifically, we have:

$$\phi(\mathbf{x}, t; \boldsymbol{\xi}, \tau) = K(\mathbf{x} - \boldsymbol{\xi}, t - \tau) + J(\mathbf{x}, t; \boldsymbol{\xi}, \tau)$$

where J is any solution, in Ω, of the dual problems

$$\begin{aligned} J_t - \kappa \nabla_{\mathbf{x}}^2 J &= 0 \quad t > \tau \\ J &\equiv 0 \quad t < \tau \end{aligned} \qquad\qquad \begin{aligned} -J_\tau - \kappa \nabla_{\boldsymbol{\xi}}^2 J &= 0 \quad \tau < t \\ J &\equiv 0 \quad \tau > t \end{aligned}$$

EXAMPLE 8.4 Let Ω be the unit interval $(0, 1)$ of $\mathbf{R}^1$. Then, for $0 < x < 1$ and $0 < \xi < 1$, one fundamental solution of the one-dimensional diffusion equation is

$$\phi_1(x, t; \xi, \tau) = K(x-\xi, t-\tau) + H(t-\tau)\, e^{-\kappa(t-\tau)} \cos(x-\xi)$$

where K is given by (*8.18*) with $n = 1$.

Another possibility would be

$$\phi_2(x, t; \xi, \tau) = K(x-\xi, t-\tau) + \sum_{i=1}^{\infty} C_i K(x-\xi_i, t-\tau)$$

where the points $\xi_1, \xi_2, \ldots$ are *outside* (0, 1) and where the constants C_i are such that the series is convergent. Such a fundamental solution might arise when the method of images is used to satisfy boundary conditions.

Theorem 8.13: Let Ω be a bounded region to which the divergence theorem applies, and let $u(\mathbf{x}, t)$ solve

$$u_t - \kappa \nabla^2 u = f(\mathbf{x}, t) \qquad \mathbf{x} \text{ in } \Omega,\ t > 0$$
$$u(\mathbf{x}, 0) = u_0(\mathbf{x}) \qquad \mathbf{x} \text{ in } \Omega$$

Then, for any fundamental solution ϕ of the (homogeneous) diffusion equation,

$$u(\mathbf{x}, t) = \int_0^t \int_\Omega \phi(\mathbf{x}, t; \boldsymbol{\xi}, \tau) f(\boldsymbol{\xi}, \tau)\, d_\xi\Omega\, d\tau + \int_\Omega \phi(\mathbf{x}, t; \boldsymbol{\xi}, 0)\, u_0(\boldsymbol{\xi})\, d_\xi\Omega$$
$$+ \kappa \int_0^t \int_S \left(\phi \frac{\partial u}{\partial n} - u \frac{\partial \phi}{\partial n} \right) d_\xi S\, d\tau \tag{8.21}$$

If the integrands in (*8.21*) decay rapidly enough at infinity, Theorem 8.13 is valid for unbounded regions.

Assume now that the problem

$$u_t - \kappa \nabla^2 u = f(\mathbf{x}, t) \qquad \mathbf{x} \text{ in } \Omega,\ t > 0$$
$$u(\mathbf{x}, 0) = u_0(\mathbf{x}) \qquad \mathbf{x} \text{ in } \Omega$$
$$\alpha u + \beta \frac{\partial u}{\partial n} = g(\mathbf{x}, t) \qquad \mathbf{x} \text{ on } S,\ t > 0$$

has a unique solution for any continuous f, u_0, and g. The Green's function, $G(\mathbf{x}, t; \boldsymbol{\xi}, \tau)$, for this problem is a fundamental solution of the diffusion equation which satisfies

$$\alpha G + \beta \frac{\partial G}{\partial n} = 0 \qquad \mathbf{x} \text{ on } S,\ t > 0$$

For $\phi = G$, Theorem 8.13 yields $u(\mathbf{x}, t)$ as the sum of integrals of known functions (cf. Theorem 8.2 and Laplace's equation). Methods for constructing the Green's function include images, eigenfunction expansions, and integral transforms.

8.5 WAVE EQUATION

The singularity solution for the wave equation in n space-dimensions, $u_{tt} - a^2 \nabla^2 u = 0$, is

$$k(\mathbf{x}, t; \boldsymbol{\xi}, \tau) = \begin{cases} H(a(t-\tau) - |x-\xi|)\dfrac{1}{2a} & (n=1) \\[2ex] H(a(t-\tau) - |\mathbf{x}-\boldsymbol{\xi}|)\dfrac{1}{2\pi a\sqrt{[a(t-\tau)]^2 - |\mathbf{x}-\boldsymbol{\xi}|^2}} & (n=2) \\[2ex] \delta(a(t-\tau) - |\mathbf{x}-\boldsymbol{\xi}|)\dfrac{1}{4\pi a\, |\mathbf{x}-\boldsymbol{\xi}|} & (n=3) \end{cases} \tag{8.22}$$

As with the diffusion equation, the singularity solution is a point-source solution in that k satisfies

$$k_{tt} - a^2\nabla^2 k = \delta(\mathbf{x} - \boldsymbol{\xi})\,\delta(t-\tau)$$

In the variables $\mathbf{x}$, t, the function k represents the *causal Green's function* for the wave equation; in $\boldsymbol{\xi}$, τ, it is the *free-space Green's function.* Figure 8-1 suggests the difference in interpretation, for $n = 1$.

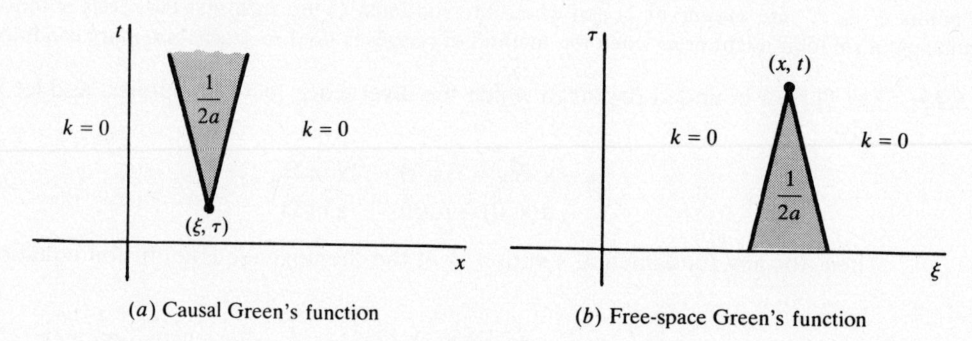

(*a*) Causal Green's function (*b*) Free-space Green's function

Fig. 8-1

In contrast to the diffusion equation, however, the singularity solution for the wave equation depends essentially on the dimension n. For $n = 3$, the delta function in (*8.22*) implies that, at time t, the disturbance due to a local impulse at $\boldsymbol{\xi} = (\xi, \eta, \zeta)$ at time $\tau < t$ is concentrated *on the surface of* a sphere of radius $a(t-\tau)$ with center at (ξ, η, ζ). However, for $n = 2$, because of the Heaviside function, the analogous disturbance is distributed *over the interior of* a circle of radius $a(t-\tau)$ centered on the source. Both disturbances are traveling radially with speed a; but, while in three dimensions there is a sharp wave front that leaves no wake, in two dimensions a wake exists that decays like $1/a(t-\tau)$ after the wave front passes. The foregoing observations form the basis of *Huygens' principle.*

Fundamental solutions of the wave equation and Green's functions for initial–boundary value problems are obtained from the singularity solution as in the parabolic case, Section 8.4. In particular, consider the problem

$$u_{tt} - a^2\nabla^2 u = f(\mathbf{x}, t) \qquad \mathbf{x} \text{ in } \Omega,\ t > 0 \tag{8.23}$$

$$u(\mathbf{x}, 0) = u_0(\mathbf{x}),\ \ u_t(\mathbf{x}, 0) = u_1(\mathbf{x}) \qquad \mathbf{x} \text{ in } \Omega \tag{8.24}$$

$$\alpha u + \beta\frac{\partial u}{\partial n} = g(\mathbf{x}, t) \qquad \mathbf{x} \text{ on } S,\ t > 0 \tag{8.25}$$

Theorem 8.14: If a (causal) Green's function, $G(\mathbf{x}, t; \boldsymbol{\xi}, \tau)$, for (*8.23*)–(*8.25*) exists, it is determined as the solution of either of the following problems:

(i)

$$G_{tt} - a^2\nabla^2 G = \delta(\mathbf{x} - \boldsymbol{\xi})\,\delta(t-\tau) \qquad \mathbf{x}, \boldsymbol{\xi} \text{ in } \Omega$$

$$G \equiv 0 \qquad t < \tau$$

$$\alpha G + \beta\frac{\partial G}{\partial n} = 0 \qquad \mathbf{x} \text{ on } S$$

(ii)

$$G_{tt} - a^2\nabla^2 G = 0 \qquad \mathbf{x}, \boldsymbol{\xi} \text{ in } \Omega,\ t > \tau$$

$$\lim_{t\to\tau+} G = 0,\ \ \lim_{t\to\tau+} G_t = \delta(\mathbf{x} - \boldsymbol{\xi}) \qquad \mathbf{x}, \boldsymbol{\xi} \text{ in } \Omega$$

$$\alpha G + \beta\frac{\partial G}{\partial n} = 0 \qquad \mathbf{x} \text{ on } S$$

Theorem 8.15: With G as in Theorem 8.14, the solution of (*8.23*)–(*8.25*) is, for $\alpha \neq 0$,

$$u(\mathbf{x}, t) = \int_0^t \int_\Omega G(\mathbf{x}, t; \boldsymbol{\xi}, \tau) f(\boldsymbol{\xi}, \tau)\, d_\xi\Omega + \int_\Omega [G(\mathbf{x}, t; \boldsymbol{\xi}, 0)\, u_1(\boldsymbol{\xi}) - G_\tau(\mathbf{x}, t; \boldsymbol{\xi}, 0)\, u_0(\boldsymbol{\xi})]\, d_\xi\Omega$$

$$-a^2 \int_0^t \int_S \frac{1}{\alpha(\boldsymbol{\xi})} \frac{\partial G}{\partial n}(\mathbf{x}, t; \boldsymbol{\xi}, \tau)\, g(\boldsymbol{\xi}, \tau)\, d_\xi S\, d\tau$$

where the normal derivative involves the ξ-gradient. If $\alpha = 0$, the last term is replaced by

$$+a^2 \int_0^t \int_S \frac{1}{\beta(\boldsymbol{\xi})} G(\mathbf{x}, t; \boldsymbol{\xi}, \tau)\, g(\boldsymbol{\xi}, \tau)\, d_\xi S\, d\tau$$

and, for the pure initial-value problem (*8.23*)–(*8.24*), the last term is dropped.

Solved Problems

8.1 Let Ω be an open region and let **T** denote the set of all infinitely differentiable (C^∞) functions on Ω having the property that each θ in **T** is identically zero outside some closed bounded subset of Ω. **T** is called the set of *test functions on* Ω (cf. Problem 5.15). A sequence $\{\theta_n\}$ of test functions is said to *converge to zero in* **T** if all the θ_n are zero outside some common bounded set and if $\{\theta_n\}$ and all its derived sequences converge uniformly to zero as $n \to \infty$. Any rule d which assigns to each test function θ a real number, $\langle d, \theta\rangle$, which satisfies the linearity condition (note that **T** is a vector space over the reals)

$$\langle d, \alpha\theta_1 + \beta\theta_2\rangle = \alpha\langle d, \theta_1\rangle + \beta\langle d, \theta_2\rangle \qquad (\alpha, \beta \text{ in } \mathbf{R};\ \theta_1, \theta_2 \text{ in } \mathbf{T}) \tag{1}$$

and the continuity condition

$$\lim_{n\to\infty} \langle d, \theta_n\rangle = 0 \quad \text{whenever} \quad \{\theta_n\} \text{ converges to zero in } \mathbf{T} \tag{2}$$

is called a *distribution* or *generalized function.* (*a*) If $f(\mathbf{x})$ is a continuous function in Ω, show that the rule

$$\langle f, \theta\rangle \equiv \int_\Omega f(\mathbf{x})\theta(\mathbf{x})\, d\Omega \tag{3}$$

defines a distribution. (*b*) If $\mathbf{x}_0$ is any point in Ω, show that the rule

$$\langle \delta, \theta\rangle \equiv \theta(\mathbf{x}_0) \tag{4}$$

defines a distribution.

(*a*) Since f is continuous and θ is C^∞ and vanishes outside a bounded subset of Ω, the integral on the right of (*3*) exists (i.e., is a real number). The linearity condition (*1*) follows immediately from the linearity in θ of the integral (*3*). To establish the continuity condition (*2*), let $\{\theta_n\}$ converge to zero in **T**. Since the θ_n all vanish outside some common set, Q, in Ω and $\{\theta_n\}$ converges *uniformly* to zero on Q,

$$\lim_{n\to\infty} \langle f, \theta_n\rangle = \lim_{n\to\infty} \int_\Omega f\theta_n\, d\Omega = \lim_{n\to\infty} \int_Q f\theta_n\, d\Omega = \int_Q (\lim_{n\to\infty} \theta_n) f\, d\Omega = 0$$

Thus, (*3*) defines a distribution. Because this distribution is determined exclusively by the continuous function f, it too has been given the symbol f.

(*b*) Since $\mathbf{x}_0$ is in Ω and θ is a test function on Ω, the right side of (*4*) is well defined. The rule δ assigns to the linear combination of test functions $\alpha\theta_1 + \beta\theta_2$ the number

$$\alpha\theta_1(\mathbf{x}_0) + \beta\theta_2(\mathbf{x}_0) = \alpha\langle\delta, \theta_1\rangle + \beta\langle\delta, \theta_2\rangle$$

so (*4*) satisfies (*1*). For any sequence $\{\theta_n\}$ which converges to zero in $\mathbf{T}$, it is clear that $\langle\delta, \theta_n\rangle = \theta_n(\mathbf{x}_0) \to 0$ as $n \to \infty$; therefore, (*2*) is also satisfied.

The distribution or generalized function (*4*) is known as the *Dirac delta distribution* or, more commonly, as the *Dirac delta function.* By a formal analogy with (*3*), it is common practice to write

$$\langle\delta, \theta\rangle = \int_\Omega \delta(\mathbf{x} - \mathbf{x}_0)\,\theta(\mathbf{x})\,d\Omega = \theta(\mathbf{x}_0) \tag{5}$$

even though there is no continuous function $f(\mathbf{x})$ that can be identified with $\delta(\mathbf{x} - \mathbf{x}_0)$.

A *linear combination* of two distributions, d_1 and d_2, is defined by

$$\langle\alpha d_1 + \beta d_2, \theta\rangle \equiv \alpha\langle d_1, \theta\rangle + \beta\langle d_2, \theta\rangle \qquad (\alpha, \beta \text{ in } \mathbf{R})$$

Products and quotients of distributions are not well-defined in general. However, if $\mathbf{x} = (x, y)$ and $\boldsymbol{\xi} = (\xi, \eta)$, the two-dimensional δ-function $\delta(\mathbf{x} - \boldsymbol{\xi})$ can be represented as a direct product of two one-dimensional δ-functions:

$$\delta(\mathbf{x} - \boldsymbol{\xi}) = \delta(x - \xi)\,\delta(y - \eta)$$

A distribution d is said to be *zero on an open set* Ω' in Ω if $\langle d, \theta\rangle = 0$ for every θ in $\mathbf{T}$ which is identically zero outside Ω'. In this sense, $\delta(\mathbf{x} - \mathbf{x}_0)$ is zero on any region that does not contain $\mathbf{x}_0$.

8.2 Establish Theorem 8.2 for $\mathbf{R}^2$.

Let $\boldsymbol{\xi} = (\xi, \eta)$ be any point in the two-dimensional region Ω and let $r = |\mathbf{x} - \boldsymbol{\xi}|$ be the distance from $\boldsymbol{\xi}$ to any other point $\mathbf{x} = (x, y)$ in Ω. Let s be a circle of radius ϵ centered at $\boldsymbol{\xi}$, and let Ω' be the portion of Ω exterior to s. From Green's second identity, (*1.8*),

$$\int_{\Omega'} (u\nabla^2\phi - \phi\nabla^2 u)\,d\Omega' = \int_S \left(u\frac{\partial\phi}{\partial n} - \phi\frac{\partial u}{\partial n}\right) dS + \int_s \left(u\frac{\partial\phi}{\partial n} - \phi\frac{\partial u}{\partial n}\right) ds \tag{1}$$

In both boundary integrals of (*1*) the contour is described so that Ω' is kept to the left (see Fig. 8-2).

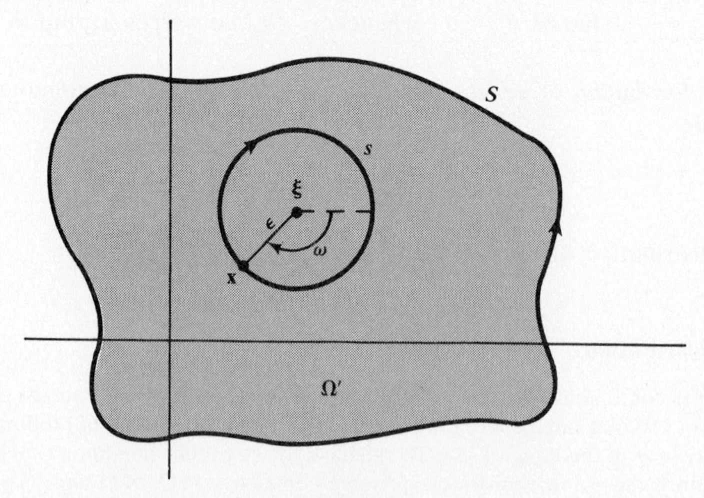

Fig. 8-2

Since ϕ is a fundamental solution with singularity at $\boldsymbol{\xi}$, which is not in Ω',

$$\nabla^2\phi = 0 = \nabla^2 u - f \qquad \text{in } \Omega' \tag{2}$$

On s the outward normal derivative of ϕ relative to Ω' is

$$\frac{\partial \phi}{\partial n} = -\frac{\partial \phi}{\partial r}\bigg|_{r=\epsilon} = -\frac{1}{2\pi\epsilon} - \frac{\partial h}{\partial r}\bigg|_{r=\epsilon}$$

wherefore (consult Fig. 8-2; $ds > 0$)

$$\int_s \left(u\frac{\partial \phi}{\partial n} - \phi\frac{\partial u}{\partial n}\right) ds = \int_0^{2\pi}\left(\frac{-u}{2\pi\epsilon} - u\frac{\partial h}{\partial r} + \phi\frac{\partial u}{\partial r}\right)\epsilon\, d\omega$$

$$= \frac{1}{2\pi}\int_0^{2\pi}(-u)\, d\omega - \epsilon\int_0^{2\pi}\left(u\frac{\partial h}{\partial r} - \phi\frac{\partial u}{\partial r}\right) d\omega \tag{3}$$

Now, the first term on the right in *(3)* is just the mean value of $-u$ over s; so, it approaches $-u(\boldsymbol{\xi})$ as $\epsilon \to 0$. As for the second term, since

$$\phi|_{r=\epsilon} = \frac{1}{2\pi}\log \epsilon + h|_{r=\epsilon}$$

it vanishes as $\epsilon \log \epsilon$, as $\epsilon \to 0$. Consequently, if we let $\epsilon \to 0$ in *(1)*, we obtain

$$-\int_\Omega \phi f\, d\Omega = \int_S \left(u\frac{\partial \phi}{\partial n} - \phi\frac{\partial u}{\partial n}\right) dS - u(\boldsymbol{\xi})$$

which rearranges to *(8.4)*.

An alternate (but purely formal) derivation of *(8.4)* can be given using the Dirac delta function. The fundamental solution ϕ satisfies

$$\nabla^2\phi = \delta(\mathbf{x} - \boldsymbol{\xi}) \qquad \text{in } \Omega \tag{4}$$

Substitute *(4)* into Green's identity

$$\int_\Omega (u\nabla^2\phi - \phi\nabla^2 u)\, d\Omega = \int_S \left(u\frac{\partial \phi}{\partial n} - \phi\frac{\partial u}{\partial n}\right) dS$$

and use the sifting property of the δ-function, *(5)* of Problem 8.1, to obtain *(8.4)*.

8.3 Assume that a Green's function, $G(\mathbf{x}; \boldsymbol{\xi})$, exists for the mixed boundary value problem

$$\nabla^2 u = f(\mathbf{x}) \qquad \text{in } \Omega \tag{1}$$

$$\alpha u + \beta\frac{\partial u}{\partial n} = 0 \qquad \text{on } S \tag{2}$$

with $\alpha \neq 0$. Formally show that *(a)* $G(\mathbf{x}; \boldsymbol{\xi})$ as a function of $\mathbf{x}$ satisfies

$$\nabla_{\mathbf{x}}^2 G = \delta(\mathbf{x} - \boldsymbol{\xi}) \qquad \text{in } \Omega \tag{3}$$

$$\alpha G + \beta\frac{\partial G}{\partial n} = 0 \qquad \text{on } S \tag{4}$$

(b) $G(\mathbf{x}; \boldsymbol{\xi})$ is symmetric, $G(\mathbf{x}; \boldsymbol{\xi}) = G(\boldsymbol{\xi}; \mathbf{x})$; *(c)* the solution to *(1)* subject to the nonhomogeneous boundary condition

$$\alpha u + \beta\frac{\partial u}{\partial n} = g(\mathbf{x}) \qquad \text{on } S \tag{5}$$

is given by

$$u(\mathbf{x}) = \int_\Omega G(\mathbf{x}; \boldsymbol{\xi}) f(\boldsymbol{\xi})\, d_\xi\Omega + \int_S \frac{1}{\alpha(\boldsymbol{\xi})} g(\boldsymbol{\xi}) \frac{\partial G}{\partial n}(\mathbf{x}; \boldsymbol{\xi})\, d_\xi S$$

where $\partial/\partial n \equiv \mathbf{n} \cdot \nabla_\xi$.

(*a*) If $G(\mathbf{x};\boldsymbol{\xi})$ is the Green's function for (*1*)–(*2*), then by definition

$$u(\mathbf{x})=\int_\Omega G(\mathbf{x};\boldsymbol{\xi})f(\boldsymbol{\xi})\,d_{\boldsymbol{\xi}}\Omega$$

From

$$f(\mathbf{x})=\nabla_{\mathbf{x}}^2 u=\int_\Omega \nabla_{\mathbf{x}}^2 G(\mathbf{x};\boldsymbol{\xi})f(\boldsymbol{\xi})\,d_{\boldsymbol{\xi}}\Omega$$

it follows that $\nabla_{\mathbf{x}}^2 G(\mathbf{x};\boldsymbol{\xi})$ has the sifting property of the Dirac delta function, and so (*3*) holds. From

$$0=\alpha u+\beta\frac{\partial u}{\partial n}=\int_\Omega\left[\alpha G+\beta\frac{\partial G}{\partial n}\right]f(\boldsymbol{\xi})\,d_{\boldsymbol{\xi}}\Omega \qquad (\mathbf{x}\text{ on }S)$$

and the arbitrariness of f, (*4*) follows.

(*b*) In Green's second identity,

$$\int_\Omega (u\nabla^2 v-v\nabla^2 u)\,d\Omega=\int_S\left(u\frac{\partial v}{\partial n}-v\frac{\partial u}{\partial n}\right)dS$$

adding and subtracting

$$\frac{\beta}{\alpha}\frac{\partial u}{\partial n}\frac{\partial v}{\partial n}$$

on the right gives

$$\int_\Omega (u\nabla^2 v-v\nabla^2 u)\,d\Omega=\int_S\frac{1}{\alpha}\left[\left(\alpha u+\beta\frac{\partial u}{\partial n}\right)\frac{\partial v}{\partial n}-\left(\alpha v+\beta\frac{\partial v}{\partial n}\right)\frac{\partial u}{\partial n}\right]dS \tag{6}$$

If u and v both satisfy (*2*), by (*6*),

$$\int_\Omega u\nabla^2 v\,d\Omega=\int_\Omega v\nabla^2 u\,d\Omega \tag{7}$$

Now, if $\mathbf{x}_1$ and $\mathbf{x}_2$ are distinct points in Ω, $u\equiv G(\mathbf{x};\mathbf{x}_1)$ and $v\equiv G(\mathbf{x};\mathbf{x}_2)$ both satisfy (*2*), by part (*a*); furthermore, $\nabla^2 u$ and $\nabla^2 v$ both have the sifting property. Thus, (*7*) gives:

$$\int_\Omega G(\mathbf{x};\mathbf{x}_1)\,\delta(\mathbf{x}-\mathbf{x}_2)\,d\Omega=\int_\Omega G(\mathbf{x};\mathbf{x}_2)\,\delta(\mathbf{x}-\mathbf{x}_1)\,d\Omega$$

or $G(\mathbf{x}_2;\mathbf{x}_1)=G(\mathbf{x}_1;\mathbf{x}_2)$.

(*c*) In (*6*) let $v=G(\mathbf{x};\boldsymbol{\xi})$ and let u be the solution of (*1*) and (*5*). Then, since v obeys (*3*)–(*4*), (*6*) yields

$$u(\boldsymbol{\xi})=\int_\Omega G(\mathbf{x};\boldsymbol{\xi})f(\mathbf{x})\,d_{\mathbf{x}}\Omega+\int_S\frac{1}{\alpha(\mathbf{x})}g(\mathbf{x})\frac{\partial G}{\partial n}(\mathbf{x};\boldsymbol{\xi})\,d_{\mathbf{x}}S \tag{8}$$

Interchanging the roles of $\mathbf{x}$ and $\boldsymbol{\xi}$ in (*8*) and using $G(\mathbf{x};\boldsymbol{\xi})=G(\boldsymbol{\xi};\mathbf{x})$ leads to the desired formula.

8.4 If Ω is the rectangle $0<x<a, 0<y<b$, find the Green's function for the boundary value problem,

$$\nabla^2 u=f \qquad \text{in }\Omega \tag{1}$$

$$u=g \qquad \text{on }S \tag{2}$$

The required Green's function is symmetric, $G(x,y;\xi,\eta)=G(\xi,\eta;x,y)$, and is defined by

$$G_{xx}+G_{yy}=\delta(x-\xi)\,\delta(y-\eta) \qquad (x,y)\text{ in }\Omega \tag{3}$$

$$G(x,y;\xi,\eta)=0 \qquad (x,y)\text{ on }S \tag{4}$$

Method 1 (*Eigenfunction Expansion*)

The eigenfunctions for the Laplacian ∇^2 on Ω subject to zero Dirichlet boundary conditions are the nonzero solutions of

$$\nabla^2 u = \lambda u \quad \text{in } \Omega \qquad u = 0 \quad \text{on } S \tag{5}$$

Using separation of variables to solve (*5*) gives the eigenvalues, λ_{mn}, and corresponding normalized eigenfunctions, u_{mn}, as

$$\lambda_{mn} = -\left[\left(\frac{m\pi}{a}\right)^2 + \left(\frac{n\pi}{b}\right)^2\right] \qquad u_{mn} = \frac{2}{\sqrt{ab}} \sin\frac{m\pi x}{a} \sin\frac{n\pi y}{b}$$

The expansion

$$G(x, y; \xi, \eta) = \sum_{m,n=1}^{\infty} C_{mn}(\xi, \eta)\, u_{mn}(x, y)$$

satisfies the boundary condition (*4*). Since $\nabla^2 u_{mn} = \lambda_{mn} u_{mn}$, it will satisfy (*3*) if

$$\sum_{m,n=1}^{\infty} C_{mn}(\xi, \eta)\lambda_{mn} u_{mn}(x, y) = \delta(x-\xi)\delta(y-\eta) \tag{6}$$

To find the C_{mn} multiply (*6*) by $u_{pq}(x, y)$ and integrate over Ω, using the orthonormality of the eigenfunctions (the weight function is unity) and the shifting property of the delta function:

$$\sum_{m,n=1} C_{mn}\lambda_{mn} \int_\Omega u_{mn}(x, y)u_{pq}(x, y)\,dx\,dy = \int_\Omega \delta(x-\xi)\delta(y-\eta)u_{pq}(x, y)\,dx\,dy$$

$$C_{pq}\lambda_{pq} = u_{pq}(\xi, \eta) \tag{7}$$

Therefore,

$$G(x, y; \xi, \eta) = \sum_{m,n=1}^{\infty} \frac{u_{mn}(x, y)\,u_{mn}(\xi, \eta)}{\lambda_{mn}}$$

$$= \frac{-4}{ab} \sum_{m,n=1}^{\infty} \frac{\sin\dfrac{m\pi x}{a} \sin\dfrac{n\pi y}{b} \sin\dfrac{m\pi\xi}{a} \sin\dfrac{n\pi\eta}{b}}{\left(\dfrac{m\pi}{a}\right)^2 + \left(\dfrac{n\pi}{b}\right)^2} \tag{8}$$

in agreement with Theorem 8.6, wherein r counts pairs (m, n).

Method 2 (*Partial Eigenfunction Expansion*)

In this method, G is expressed as an eigenfunction expansion in only one of the variables, say y. Let μ_n and $v_n(y)$ be the eigenvalues and the normalized eigenfunctions of the eigenvalue problem for the y-part of the Laplacian, subject to zero Dirichlet boundary conditions; i.e., $v'' = \mu_n v$, $v(0) = 0 = v(b)$. We have

$$\mu_n = -\left(\frac{n\pi}{b}\right)^2 \qquad v_n(y) = \sqrt{\frac{2}{b}} \sin\frac{n\pi y}{b}$$

An expansion for G of the form

$$G(x, y; \xi, \eta) = \sum_{n=1}^{\infty} A_n(x, \xi, \eta)\, v_n(y) \tag{9}$$

satisfies the boundary conditions on the horizontal boundaries, $y = 0$ and $y = b$. Substitution in (*3*) gives

$$\sum_{n=1}^{\infty} (A_n'' + \mu_n A_n)v_n(y) = \delta(x-\xi)\delta(y-\eta) \tag{10}$$

where $'$ denotes differentiation on x. Multiply (*10*) by $v_k(y)$, integrate both sides from $y = 0$ to $y = b$, and use orthonormality to obtain

$$A_k'' + \mu_k A_k = v_k(\eta)\delta(x-\xi)$$

To ensure that $G = 0$ on $x = 0$ and $x = a$, A_k must vanish on $x = 0$ and $x = a$. This means that the function

$$g(x;\xi) \equiv \frac{A_k(x,\xi,\eta)}{v_k(\eta)}$$

must be the Green's function for the ordinary two-point boundary value problem

$$u'' + \mu_k u = f \qquad 0 < x < a$$
$$u(0) = u(a) = 0$$

Now, from the theory of ordinary differential equations, we have

Theorem 8.16: Suppose that when $f = 0$, $u = 0$ is the only solution to the two-point boundary value problem

$$a(x)u''(x) + b(x)u'(x) + c(x)u(x) = f(x) \qquad x_1 < x < x_2$$
$$\alpha_1 u(x_1) + \beta_1 u'(x_1) = 0$$
$$\alpha_2 u(x_2) + \beta_2 u'(x_2) = 0$$

in which $a(x) \neq 0$. Then the Green's function for the problem is given by

$$g(x;\xi) = \begin{cases} \dfrac{u_1(x)u_2(\xi)}{a(\xi)W(\xi)} & x_1 < x < \xi \\[2ex] \dfrac{u_1(\xi)u_2(x)}{a(\xi)W(\xi)} & \xi < x < x_2 \end{cases}$$

where, for $i = 1, 2$, u_i satisfies

$$au_i'' + bu_i' + cu = 0 \qquad x_1 < x < x_2$$
$$\alpha_i u_i(x_i) + \beta_i u_i'(x_i) = 0$$

and where $W(\xi) \equiv u_1(\xi)u_2'(\xi) - u_2(\xi)u_1'(\xi)$ is the Wronskian of u_1 and u_2 evaluated at ξ.

By Theorem 8.16, with $u_1 = \sinh(k\pi x/b)$ and $u_2 = \sinh(k\pi(x-a)/b)$,

$$A_k(x,\xi,\eta) = \begin{cases} \dfrac{\sinh \dfrac{k\pi x}{b} \sinh \dfrac{k\pi(\xi - a)}{b}}{\dfrac{k\pi}{b} \sinh \dfrac{k\pi a}{b}} v_k(\eta) & 0 < x < \xi \\[4ex] \dfrac{\sinh \dfrac{k\pi \xi}{b} \sinh \dfrac{k\pi(x - a)}{b}}{\dfrac{k\pi}{b} \sinh \dfrac{k\pi a}{b}} v_k(\eta) & \xi < x < a \end{cases} \tag{11}$$

Although the series (*9*), with coefficients (*11*), is more complicated in appearance than (*8*), it is the more rapidly convergent, because

$$|A_k(x,\xi,\eta)| = O(e^{-k\pi|x-\xi|/b})$$

as $k \to \infty$.

8.5 Use the method of images to find the Green's function for

$$\nabla^2 u = f \qquad \text{in } \Omega$$
$$u = g \qquad \text{on } S$$

if Ω is (*a*) $\{-\infty < x < \infty, y > 0\}$, (*b*) $\{0 < x < \infty, y > 0\}$, (*c*) $\{-\infty < x < \infty, 0 < y < b\}$.

When applying the method of images to Laplace's equation in $\mathbf{R}^2$, it is often helpful to interpret the free-space Green's function, $\sigma(x, y; \xi, \eta)$, as the steady xy-temperature distribution due to a unit line sink at (ξ, η). ("Sink," not "source," because of Fourier's law and Theorem 8.1(iii).)

(*a*) In $-\infty < x < \infty$, $y > 0$, if $\sigma(x, y; \xi, \eta)$ represents a sink at (ξ, η), $\eta > 0$, then a symmetrically placed source at $(\xi, -\eta)$, $-\sigma(x, y; \xi, -\eta)$, will ensure a zero temperature on $y = 0$. Because $h(x, y; \xi, \eta) \equiv -\sigma(x, y; \xi, -\eta)$ is harmonic in $y > 0$, we have from Theorem 8.3:

$$G(x, y; \xi, \eta) = \sigma(x, y; \xi, \eta) - \sigma(x, y; \xi, -\eta) = \frac{1}{4\pi} \log \frac{(x-\xi)^2 + (y-\eta)^2}{(x-\xi)^2 + (y+\eta)^2}$$

(*b*) Placing sinks, $+\sigma$, and sources, $-\sigma$, as indicated in Fig. 8-3(*a*) shows the Green's function for the first quadrant to be

$$G(x, y; \xi, \eta) = \sigma(x, y; \xi, \eta) - \sigma(x, y; -\xi, \eta) + \sigma(x, y; -\xi, -\eta) - \sigma(x, y; \xi, -\eta)$$
$$= \frac{1}{4\pi} \log \left[\frac{(x-\xi)^2 + (y-\eta)^2}{(x+\xi)^2 + (y-\eta)^2} \frac{(x+\xi)^2 + (y+\eta)^2}{(x-\xi)^2 + (y+\eta)^2} \right]$$

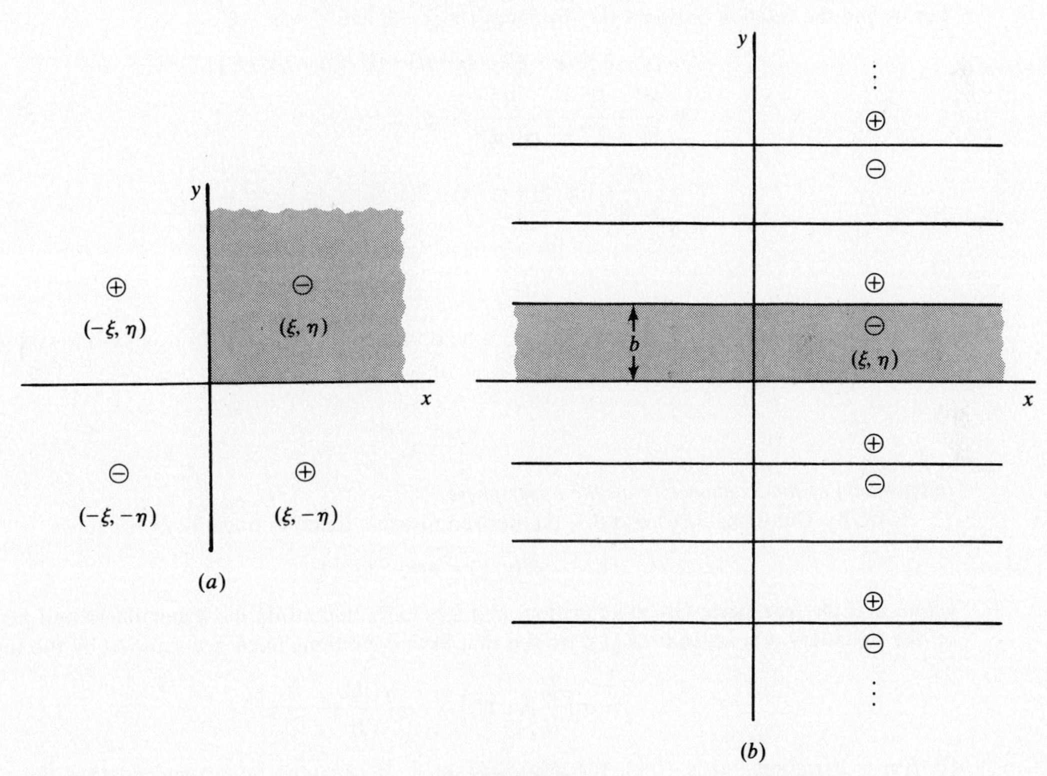

Fig. 8-3

(*c*) In this case, an infinite series of sources and sinks is required to make $G = 0$ on both $y = 0$ and $y = b$. For $0 < \eta < b$, Fig. 8-3(*b*) shows that sinks must be placed at η, $\eta \pm 2b$, $\eta \pm 4b, \ldots$ and sources at $-\eta$, $-\eta \pm 2b$, $-\eta \pm 4b, \ldots$. Thus,

$$G(x, y; \xi, \eta) = \sum_{k=-\infty}^{\infty} [\sigma(x, y; \xi, 2kb + \eta) - \sigma(x, y; \xi, 2kb - \eta)]$$

8.6 Find the steady xy-temperature distribution in the half-plane $y > 0$ due to a line source of strength 2 at $(x, y) = (1, 3)$ and unit line sinks at $(-5, 6)$ and $(4, 7)$. The boundary, $y = 0$, is held at temperature zero.

The Green's function, $G(x, y; \xi, \eta)$, which was constructed in Problem 8.5(*a*), gives the *xy*-temperature distribution due to a unit sink at (ξ, η) and zero temperature on $y = 0$. Therefore, the solution to the present problem is

$$u(x, y) = -2G(x, y; 1, 3) + G(x, y; -5, 6) + G(x, y; 4, 7)$$

8.7 Use the method of images to find the Green's function for the *n*-dimensional Dirichlet problem ($n \geq 2$)

$$\nabla^2 u = f(\mathbf{x}) \qquad |\mathbf{x}| < R$$
$$u = g(\mathbf{x}) \qquad |\mathbf{x}| = R$$

Let $\mathbf{x}$ and $\boldsymbol{\xi}$ be two points inside the hypersphere of radius R; their inverse points (Problem 3.29), outside the hypersphere, are

$$\mathbf{x}' = \frac{R^2}{|\mathbf{x}|^2}\mathbf{x} \qquad \boldsymbol{\xi}' = \frac{R^2}{|\boldsymbol{\xi}|^2}\boldsymbol{\xi}$$

Let us find the relation between the distances $r \equiv |\mathbf{x} - \boldsymbol{\xi}|$ and $r' \equiv |\mathbf{x}' - \boldsymbol{\xi}'|$.

$$\begin{aligned} r'^2 &= (\mathbf{x}' - \boldsymbol{\xi}') \cdot (\mathbf{x}' - \boldsymbol{\xi}') = (\mathbf{x}' \cdot \mathbf{x}') + (\boldsymbol{\xi}' \cdot \boldsymbol{\xi}') - 2(\mathbf{x}' \cdot \boldsymbol{\xi}') \\ &= \frac{R^4}{|\mathbf{x}|^2} + \frac{R^4}{|\boldsymbol{\xi}|^2} - 2\frac{R^4}{|\mathbf{x}|^2|\boldsymbol{\xi}|^2}(\mathbf{x} \cdot \boldsymbol{\xi}) \\ &= \frac{R^4}{|\mathbf{x}|^2|\boldsymbol{\xi}|^2}[(\boldsymbol{\xi} \cdot \boldsymbol{\xi}) + (\mathbf{x} \cdot \mathbf{x}) - 2(\mathbf{x} \cdot \boldsymbol{\xi})] = \frac{R^4}{|\mathbf{x}|^2|\boldsymbol{\xi}|^2} r^2 \end{aligned}$$

or

$$r' = \frac{R^2}{|\mathbf{x}|\,|\boldsymbol{\xi}|} r$$

From this it follows that if $\boldsymbol{\xi}$ (and with it, $\boldsymbol{\xi}'$) is held fixed while $\mathbf{x}$ (and with it, $\mathbf{x}'$) is allowed to approach the boundary point $\mathbf{s}$,

$$r'|_{\mathbf{s}} = \frac{R}{|\boldsymbol{\xi}|} r|_{\mathbf{s}} \tag{1}$$

independent of the location of $\mathbf{s}$ *on the hypersphere.*

Now, by Theorems 8.1(i) and 8.3, the desired Green's function must be of the form

$$G(\mathbf{x}; \boldsymbol{\xi}) = \sigma(r) + h(\mathbf{x}; \boldsymbol{\xi})$$

where σ is the free-space Green's function, and h is harmonic inside the hypersphere and just cancels σ on the boundary. On account of (*1*), we see that both conditions for h are satisfied by the function

$$-\sigma\left(\frac{|\boldsymbol{\xi}|}{R}|\mathbf{x} - \boldsymbol{\xi}'|\right) = -\sigma\left(\left|\frac{|\boldsymbol{\xi}|}{R}\mathbf{x} - \frac{R}{|\boldsymbol{\xi}|}\boldsymbol{\xi}\right|\right)$$

(If $f(\mathbf{x})$ is harmonic, so is $f(c\mathbf{x})$, for any constant c. It can now be recognized that the geometrical property of the sphere expressed in (*1*) is crucial to the method of images.)

8.8 For the linear differential operator

$$L[u] = u_{xx} + u_{yy} + 2u_x - u_y + u \qquad \text{in } \Omega$$

find (*a*) the adjoint operator; (*b*) the adjoint boundary conditions corresponding to

$$\text{(i)} \quad u = 0 \qquad \text{(ii)} \quad u + \frac{\partial u}{\partial n} = 0$$

on S.

(*a*) From (*8.8*), $L^*[v] = v_{xx} + v_{yy} - 2v_x + v_y + v$, and M_1 and M_2 in Lagrange's identity, (*8.9*), are

$$M_1 = vu_x - uv_x + 2uv \qquad M_2 = vu_y - uv_y - uv$$

Consequently,

$$\int_S (M_1n_1 + M_2n_2)\,dS = \int_S \left[v\frac{\partial u}{\partial n} - u\frac{\partial v}{\partial n} + (2n_1 - n_2)uv\right] dS \tag{1}$$

(*b*) (i) To make the integral (*1*) vanish for all u such that $B[u] = u = 0$ on S, we must require that $B^*[v] = v = 0$ on S.

(ii) For any u satisfying $B[u] = u + (\partial u/\partial n) = 0$ on S, the integrand in (*1*) equals

$$u\left[(2n_1 - n_2 - 1)v - \frac{\partial v}{\partial n}\right] \equiv u\,B^*[v]$$

which defines the adjoint boundary condition, $B^*[v] = 0$ on S.

8.9 If Ω is the rectangle $0 < x < a$, $0 < y < b$, find the Green's function for the boundary value problem

$$u_{xx} + u_{yy} + 2u_x = f(x, y) \qquad \text{in } \Omega \tag{1}$$

$$u = g(x, y) \qquad \text{on } S \tag{2}$$

(Compare Problem 8.4.)

The Green's function, $G = G(x, y; \xi, \eta)$, is defined by

$$G_{xx} + G_{yy} + 2G_x = \delta(x - \xi)\delta(y - \eta) \qquad \text{in } \Omega \tag{3}$$

$$G = 0 \qquad \text{on } S \tag{4}$$

An eigenfunction expansion is most simply obtained if the left side of (*3*) is put in self-adjoint form. From Problem 2.14 we see that the change of dependent variable $H = e^x G$ eliminates the first-order x-derivative, giving

$$L[H] \equiv H_{xx} + H_{yy} - H = e^x\delta(x - \xi)\delta(y - \eta) \qquad \text{in } \Omega \tag{5}$$

$$H = 0 \qquad \text{on } S \tag{6}$$

We now proceed to look for an expansion of H in terms of the eigenfunctions of problem (*5*)–(*6*); that is, we set

$$H = \sum_{m,n=1}^{\infty} c_{mn}w_{mn} \tag{7}$$

where $L[w_{mn}] = \lambda_{mn}w_{mn}$ in Ω and $w_{mn} = 0$ on S. Using separation of variables, we find for the eigenvalues and normalized eigenfunctions

$$\lambda_{mn} = -1 - \left(\frac{m\pi}{a}\right)^2 - \left(\frac{n\pi}{b}\right)^2 \qquad w_{mn} = \frac{2}{\sqrt{ab}}\sin\frac{m\pi x}{a}\sin\frac{n\pi y}{b}$$

Substituting (*7*) into (*5*) gives

$$\sum_{m,n=1}^{\infty} c_{mn}\lambda_{mn}w_{mn} = e^x\delta(x - \xi)\delta(y - \eta) \tag{8}$$

If (*8*) is multiplied by w_{pq} and integrated over Ω, then the orthonormality of the eigenfunctions implies

$$c_{pq}\lambda_{pq} = \frac{2}{\sqrt{ab}} e^{\xi} \sin\frac{p\pi\xi}{a} \sin\frac{q\pi\eta}{b}$$

Therefore,

$$G(x, y; \xi, \eta) = -\frac{4}{ab} e^{\xi - x} \sum_{m,n=1}^{\infty} \frac{\sin\frac{m\pi x}{a} \sin\frac{n\pi y}{b} \sin\frac{m\pi\xi}{a} \sin\frac{n\pi\eta}{b}}{1 + \left(\frac{m\pi}{a}\right)^2 + \left(\frac{n\pi}{b}\right)^2}$$

8.10 (*a*) Show that the Neumann problem

$$\nabla^2 u = f \quad \text{in } \Omega \tag{1}$$

$$\frac{\partial u}{\partial n} = g \quad \text{on } S \tag{2}$$

does not have a Green's function. (*b*) Define a modified Green's function that will give the solution of (*1*)–(*2*) up to an additive constant.

(*a*) Recall (Example 3.8) that a necessary condition for the existence of a solution of (*1*)–(*2*) is

$$\int_\Omega f\, d\Omega = \int_S g\, dS \tag{3}$$

Assume that (*3*) holds, so that (*1*)–(*2*) has a solution which is unique up to an additive constant. By analogy with the case of a Dirichlet boundary condition, the Green's function for (*1*)–(*2*) must satisfy

$$\nabla^2 G = \delta(\mathbf{x} - \boldsymbol{\xi}) \quad \text{in } \Omega \tag{4}$$

$$\frac{\partial G}{\partial n} = 0 \quad \text{on } S \tag{5}$$

But (*4*)–(*5*) has no solution, since (*3*) does not hold for it.

(*b*) Let a modified Green's function, or *Neumann function*, $N(\mathbf{x}; \boldsymbol{\xi})$, for (*1*)–(*2*) be defined, up to an additive constant, by

$$\nabla^2 N = \delta(\mathbf{x} - \boldsymbol{\xi}) - \frac{1}{V(\Omega)} \quad \text{in } \Omega \tag{6}$$

$$\frac{\partial N}{\partial n} = 0 \quad \text{on } S \tag{7}$$

where $V(\Omega)$ is the volume of Ω. Now condition (*3*) is met. Apply Green's second identity to solutions of (*1*)–(*2*) and (*6*)–(*7*):

$$\int_\Omega (u\nabla^2 N - N\nabla^2 u)\, d\Omega = \int_S \left(u\frac{\partial N}{\partial n} - N\frac{\partial u}{\partial n}\right) dS$$

$$\int_\Omega \left\{u\left[\delta(\mathbf{x} - \boldsymbol{\xi}) - \frac{1}{V(\Omega)}\right] - Nf\right\} d\Omega = \int_S (0 - Ng)\, dS$$

$$u(\boldsymbol{\xi}) - \bar{u} - \int_\Omega N(\mathbf{x}; \boldsymbol{\xi}) f(\mathbf{x})\, d\Omega = -\int_S N(\mathbf{x}; \boldsymbol{\xi}) g(\mathbf{x})\, dS$$

or, interchanging $\mathbf{x}$ and $\boldsymbol{\xi}$,

$$u(\mathbf{x}) = \int_\Omega N(\mathbf{x}; \boldsymbol{\xi}) f(\boldsymbol{\xi})\, d_\xi\Omega - \int_S N(\mathbf{x}; \boldsymbol{\xi}) g(\boldsymbol{\xi})\, d_\xi S + \bar{u} \tag{8}$$

In view of (*3*), the right side of (*8*) has the same value for all solutions N of (*6*)–(*7*). Hence, (*8*) determines $u(\mathbf{x})$ up to the additive constant $\bar{u}$, the mean value of u over Ω.

8.11 Show that the singularity solution for $u_t - \kappa u_{xx} = 0$,

$$K(x-\xi, t-\tau) = \frac{1}{\sqrt{4\pi\kappa(t-\tau)}} \exp\left[\frac{-(x-\xi)^2}{4\kappa(t-\tau)}\right] \qquad (t > \tau)$$

satisfies $\int_{-\infty}^{\infty} K(x-\xi, t-\tau)\, dx = 1$ for all $t > \tau$.

Let us exploit the fact that if $f(x)$ and $F(\alpha)$ are Fourier transform pairs (Section 6.4),

$$\int_{-\infty}^{\infty} f(x)\, dx = 2\pi F(0)$$

Now, from Table 6-2, line 1, and Table 6-1, lines 7 and 3,

$$f(x) = \frac{1}{\sqrt{4\pi\kappa(t-\tau)}} e^{-(x-\xi)^2/4\kappa(t-\tau)} \qquad \text{and} \qquad F(\alpha) = \frac{1}{2\pi} e^{-\kappa(t-\tau)\alpha^2} e^{-i\xi\alpha}$$

are Fourier transform pairs, and we see that $2\pi F(0) = 1$.

8.12 If f is a bounded continuous function, show that

$$\lim_{t\to 0+} \int_{-\infty}^{\infty} K(x-\xi, t-0) f(\xi)\, d\xi = f(x) \tag{1}$$

Since K is symmetric in x and ξ, Problem 8.11 implies

$$\int_{-\infty}^{\infty} K(x-\xi, t)\, d\xi = 1 \qquad (t > 0)$$

From

$$\int_{-\infty}^{\infty} K(x-\xi, t) f(\xi)\, d\xi = \int_{-\infty}^{\infty} K(x-\xi, t) f(x)\, d\xi + \int_{-\infty}^{\infty} K(x-\xi, t)[f(\xi) - f(x)]\, d\xi$$

$$= f(x) + \int_{-\infty}^{\infty} K(x-\xi, t)[f(\xi) - f(x)]\, d\xi$$

we see the last integral must be shown to approach zero as $t \to 0+$. Let $s \equiv (x-\xi)/\sqrt{4t}$; then, using the explicit expression for K,

$$\lim_{t\to 0+} \int_{-\infty}^{\infty} K(x-\xi, t)[f(\xi) - f(x)]\, d\xi = \lim_{t\to 0+} \frac{1}{\sqrt{\pi}} \int_{-\infty}^{\infty} e^{-s^2}[f(x + s\sqrt{4t}) - f(x)]\, ds$$

$$= \frac{1}{\sqrt{\pi}} \int_{-\infty}^{\infty} e^{-s^2} \lim_{t\to 0+} [f(x + s\sqrt{4t}) - f(x)]\, ds = 0$$

The boundedness of f ensures that the improper integral in s is uniformly convergent, which allows the limit to be taken under the integral. Then the continuity of f implies that

$$[f(x + s\sqrt{4t}) - f(x)] \to 0$$

as $t \to 0+$.

A similar, but more complicated argument establishes (*1*) if f is continuous and

$$|f(x)| < Ae^{Bx^2}$$

for constants A and B. Equation (*1*) shows that as t approaches τ from above, the singularity solution $K(x-\xi, t-\tau)$ approaches the delta function $\delta(x-\xi)$.

8.13 If Ω is the first quadrant ($x>0$, $y>0$), (*a*) find the Green's function for

$$u_t - \kappa(u_{xx}+u_{yy}) = f(x, y, t) \qquad \text{in } \Omega,\ t>0 \tag{1}$$

$$u(x, y, 0) = u_0(x, y) \qquad \text{in } \Omega \tag{2}$$

$$u(x, 0, t) = u_1(x, t) \qquad x>0,\ t>0 \tag{3}$$

$$u_x(0, y, t) = u_2(y, t) \qquad y>0,\ t>0 \tag{4}$$

and (*b*) use (*8.21*) to express the solution of (*1*)–(*4*) in terms of G and the data f, u_0, u_1, u_2.

(*a*) The Green's function is a fundamental solution, $G = K + J$, with J chosen to make $G = 0$ on $y = 0$ and $G_x = 0$ on $x = 0$. The singular part, $K(x, y, t; \xi, \eta, \tau)$, of G represents a point source (e.g., a burst of heat) at (ξ, η) in Ω at time τ. To make G zero on $y = 0$ a sink, $-K(x, y, t; \xi, -\eta, \tau)$, at the image point $(\xi, -\eta)$ is required. In an attempt to zero G_x on $x = 0$, place a source at $(-\xi, \eta)$. Finally, balance these two images with a sink, $-K(x, y, t; -\xi, -\eta, \tau)$:

$$G(x, y, t; \xi, \eta, \tau) = K(x, y, t; \xi, \eta, \tau) - K(x, y, t; \xi, -\eta, \tau) + K(x, y, t; -\xi, \eta, \tau) - K(x, y, t; -\xi, -\eta, \tau)$$

(*b*)

$$u(x, y, t) = \int_0^t \int_0^\infty \int_0^\infty G(x, y, t; \xi, \eta, \tau) f(\xi, \eta, \tau)\, d\xi\, d\eta\, d\tau$$

$$+ \int_0^\infty \int_0^\infty G(x, y, t; \xi, \eta, 0)\, u_0(\xi, \eta)\, d\xi\, d\eta$$

$$+ \kappa \int_0^t \int_0^\infty u_2(\eta, \tau)\, G(x, y, t; 0, \eta, \tau)\, d\eta\, d\tau$$

$$+ \kappa \int_0^t \int_0^\infty u_1(\xi, \tau)\, G_\eta(x, y, t; \xi, 0, \tau)\, d\xi\, d\tau$$

where the boundary integration is in the positive sense; i.e., first in the direction of decreasing η, then in the direction of increasing ξ.

8.14 Find the Green's function for the initial–boundary value problem

$$u_t - \kappa u_{xx} = f(x, t) \qquad 0<x<\ell,\ t>0$$

$$u(x, 0) = g(x) \qquad 0<x<\ell$$

$$u(0, t) = h_1(t),\quad u(\ell, t) = h_2(t) \qquad t>0$$

Method 1 (*Reflection*)

Proceeding as in Problem 8.5(*c*), place sources $+K$ along the x-axis at $\xi + 2n\ell$, and sinks $-K$ at $-\xi + 2n\ell$, where $n = 0, \pm 1, \pm 2, \ldots$. This yields

$$G(x, t; \xi, \tau) = \sum_{n=-\infty}^{\infty} [K(x - \xi - 2n\ell, t-\tau) - K(x + \xi - 2n\ell, t-\tau)]$$

for $t>\tau$; for $t<\tau$, $G \equiv 0$.

Method 2 (*Partial Eigenfunction Expansion*)

The Green's function may also be characterized as the solution of

$$G_t - \kappa G_{xx} = \delta(x-\xi)\delta(t-\tau) \tag{1}$$

$$G = 0 \qquad x = 0 \text{ and } x = \ell \tag{2}$$

$$G \equiv 0 \qquad t<\tau \tag{3}$$

The space part, $\partial^2/\partial x^2$, of the linear differential operator has eigenvalues $\lambda_n = -(n\pi/\ell)^2$ and corresponding normalized eigenfunctions

$$v_n(x) = \sqrt{\frac{2}{\ell}} \sin \frac{n\pi x}{\ell} \qquad (n = 1, 2, 3, \ldots)$$

(cf. Problem 8.4). Thus, the expansion

$$G = \sum_{n=1}^{\infty} c_n(t, \xi, \tau) v_n(x) \tag{4}$$

satisfies (2). Putting (4) into (1) gives, with a prime ′ denoting time differentiation,

$$\sum_{n=1}^{\infty} (c_n' - \kappa \lambda_n c_n) v_n(x) = \delta(x - \xi)\delta(t - \tau)$$

which, multiplied by $v_m(x)$ and integrated from $x = 0$ to $x = \ell$, becomes

$$c_m' - \kappa \lambda_m c_m = v_m(\xi)\delta(t - \tau) \tag{5}$$

On account of (3), $c_m \equiv 0$ for $t < \tau$. For $t_1 > \tau > 0$, integrate (5) from $t = 0$ to $t = t_1$, obtaining

$$c_m(t_1) - \kappa \lambda_m \int_{\tau}^{t_1} c_m(t)\, dt = v_m(\xi) \tag{6}$$

The solution of (6) is (verify by substitution):

$$c_m(t_1) = v_m(\xi) e^{\kappa \lambda_m (t_1 - \tau)}$$

Hence,

$$G(x, t; \xi, \tau) = \begin{cases} 0 & t < \tau \\ \dfrac{2}{\ell} \displaystyle\sum_{n=1}^{\infty} \exp\left[-\left(\frac{n\pi}{\ell}\right)^2 \kappa(t - \tau)\right] \sin \frac{n\pi x}{\ell} \sin \frac{n\pi \xi}{\ell} & t > \tau \end{cases}$$

8.15 Solve the initial value problem

$$\begin{aligned} u_{tt} - a^2 u_{xx} &= f(x, t) & -\infty < x < \infty,\ t > 0 \\ u(x, 0) &= u_0(x) & -\infty < x < \infty \\ u_t(x, 0) &= u_1(x) & -\infty < x < \infty \end{aligned}$$

using a Green's function.

By (8.22) with $n = 1$, the free-space Green's function is

$$G(x, t; \xi, \tau) = H(a(t - \tau) - |x - \xi|) \frac{1}{2a}$$

To use Theorem 8.15 to write the solution, $u(x, t)$, we need $G_\tau(x, t; \xi, 0)$. Now, the derivative of the Heaviside function is the δ-function (see Problem 8.19(b)), so that

$$G_\tau(x, t; \xi, 0) = -\frac{1}{2} \delta(at - |x - \xi|)$$

and

$$u(x, t) = \frac{1}{2a} \int_0^t \int_{x - a(t-\tau)}^{x + a(t-\tau)} f(\xi, \tau)\, d\xi\, d\tau + \frac{1}{2a} \int_{x-at}^{x+at} u_1(\xi)\, d\xi + \frac{1}{2}[u_0(x - at) + u_0(x + at)]$$

which is the D'Alembert solution.

8.16 Solve by a Green's function:

$$\begin{aligned} u_{tt} - \nabla^2 u &= f(\mathbf{x}, t) & \mathbf{x} \text{ in } \mathbf{R}^3,\ t > 0 \\ u(\mathbf{x}, 0) = 0,\ u_t(\mathbf{x}, 0) &= 1 & \mathbf{x} \text{ in } \mathbf{R}^3 \end{aligned}$$

The Green's function is the free-space Green's function for the three-dimensional wave equation (with $a = 1$),

$$G(\mathbf{x}, t; \boldsymbol{\xi}, \tau) = \frac{1}{4\pi|\mathbf{x} - \boldsymbol{\xi}|}\delta(t - \tau - |\mathbf{x} - \boldsymbol{\xi}|)$$

and Theorem 8.15 gives the solution as

$$u(\mathbf{x}, t) = \int_0^t \int_{\mathbf{R}^3} \frac{1}{4\pi\,|\mathbf{x} - \boldsymbol{\xi}|}\delta(t - \tau - |\mathbf{x} - \boldsymbol{\xi}|) f(\boldsymbol{\xi}, \tau)\, d_{\boldsymbol{\xi}}\Omega\, d\tau + \int_{\mathbf{R}^3} \frac{1}{4\pi\,|\mathbf{x} - \boldsymbol{\xi}|}\delta(t - |\mathbf{x} - \boldsymbol{\xi}|)\, d_{\boldsymbol{\xi}}\Omega$$

When the order of integration is reversed, the double integral is seen to have the value

$$\int_{|\mathbf{x}-\boldsymbol{\xi}|<t} \frac{1}{4\pi\,|\mathbf{x} - \boldsymbol{\xi}|} f(\boldsymbol{\xi}, t - |\mathbf{x} - \boldsymbol{\xi}|)\, d_{\boldsymbol{\xi}}\Omega \tag{1}$$

When the origin of ξ-space is shifted to the point $\mathbf{x}$ and the polar coordinate $r = |\boldsymbol{\xi} - \mathbf{x}|$ is introduced, the other integral is seen to have the value

$$\int_0^\infty \frac{1}{4\pi r}\delta(t - r)\, 4\pi r^2\, dr = t \tag{2}$$

(as would have been found immediately if the original problem had been split into two subproblems).

The expression (*1*), the part of u that is independent of the initial conditions, may be interpreted as the superposition of disturbances which arose at points $\boldsymbol{\xi}$ at times previous to t and, traveling at speed 1, are just reaching point $\mathbf{x}$ at time t. For this reason, the integrand in (*1*) is called the *retarded potential.*

Supplementary Problems

8.17 (*a*) If Ω is the set of real numbers and $a > 0$, show that

$$\theta(x) = \begin{cases} \exp\left(\dfrac{1}{x^2 - a^2}\right) & |x| < a \\ 0 & |x| \geq a \end{cases}$$

is a test function on Ω. (*b*) Show that

$$\theta(x, y) = \begin{cases} \exp\left(\dfrac{1}{x^2 + y^2 - a^2}\right) & x^2 + y^2 < a^2 \\ 0 & x^2 + y^2 \geq a^2 \end{cases}$$

is a test function on $\bar{\Omega} = \{(x, y)\colon x^2 + y^2 < R^2, R > a > 0\}$.

8.18 Let $\theta_n(x)$ be the test function obtained by replacing x and a in the function of Problem 8.17(*a*) by x/n and a/n, where $n = 1, 2, 3, \ldots$. Show that $\{\theta_n(x)\}$ converges to zero in **T**. (See Problem 8.1.)

8.19 If Ω is the real line, define the *distributional derivative*, d', of a generalized function d by

$$\langle d', \theta\rangle = -\langle d, \theta'\rangle$$

for every test function θ on Ω. (*a*) If f is a continuously differentiable function, show that the ordinary and distributional derivatives are equivalent in the sense that the distribution defined by the continuous function f' [via (*3*) of Problem 8.1] is the distributional derivative of the distribution defined by f. (*b*) Show that the rule

$$\langle H, \theta\rangle = \int_{x_0}^\infty \theta(x)\, dx$$

where x_0 is a fixed real number, defines a distribution corresponding to the (discontinuous) Heaviside function $H(x-x_0)$; show further that $H' = \delta$, the distribution defined by (4) of Problem 8.1. (We usually express this result by saying that the derivative of the Heaviside *function* is the delta *function*.)

8.20 With Ω the x-axis, let d be a generalized function, $\theta(x)$ a test function on Ω, and $g(x)$ a C^∞ function. Define

$$\langle g(x)d, \theta(x)\rangle = \langle d, g(x)\theta(x)\rangle$$

Prove: (a) $g(x)\delta(x-y) = g(y)\delta(x-y)$ and (b) $x\delta(x) = 0$.

8.21 Find the generalized derivatives of (a) $|x|$, (b) $e^x\delta(x)$, (c) $x\delta(x)$.

8.22 Derive the formal sine and cosine series for the delta function, $\delta(x-\xi)$, on the interval $(0, \ell)$, with $0 < x, \xi < \ell$.

8.23 Given a bounded region Ω, let a Green's function be defined by

$$\begin{aligned} \nabla^2 G - c^2 G &= \delta(\mathbf{x}-\boldsymbol{\xi}) && \text{in } \Omega \\ G &= 0 && \text{on } S \end{aligned}$$

Expand this Green's function in terms of the eigenfunctions u_n, where $\nabla^2 u_n = -\lambda_n^2 u_n$ in Ω, $u_n = 0$ on S, $u_n \not\equiv 0$.

8.24 Represent the Green's function for

$$\begin{aligned} \nabla^2 u - c^2 u &= f(x, y) && -\infty < x < \infty,\ 0 < y < a \\ u(x, 0) = u(x, a) &= 0 && -\infty < x < \infty \end{aligned}$$

by an eigenvalue expansion.

8.25 Verify that the Green's function for the problem

$$\begin{aligned} \nabla^2 u &= f(x, y, z) && z > 0 \\ u_z - cu &= 0 && z = 0 \end{aligned}$$

where $c \geq 0$, is

$$G(x, y, z; \xi, \eta, \zeta) = -\frac{1}{4\pi[(x-\xi)^2 + (y-\eta)^2 + (z-\zeta)^2]^{1/2}} - \frac{1}{4\pi[(x-\xi)^2 + (y-\eta)^2 + (z+\zeta)^2]^{1/2}}$$
$$+ \frac{c}{2\pi}\int_{-\infty}^{-\zeta} \frac{e^{c(s+\zeta)}}{[(x-\xi)^2 + (y-\eta)^2 + (z-s)^2]^{1/2}}\, ds$$

8.26 Use the method of images to find $G(x, y; \xi, \eta)$ for

$$\nabla^2 u = f \qquad 0 < x < \infty,\ 0 < y < b$$

if (a) $u = 0$ for $x = 0$, for $y = 0$, and for $y = b$; (b) $u = 0$ for $y = 0$ and for $y = b$, and $u_x = 0$ for $x = 0$. [*Hint*: Make use of the Green's function, $G_0(x, y; \xi, \eta)$, of Problem 8.5(c).]

8.27 Use a Green's function to solve Problem 7.14.

8.28 Use Theorem 8.8(ii) and the mapping

$$w = \frac{R(z-\zeta)}{z\bar{\zeta} - R^2} \qquad (R > 0)$$

which carries $|z| < R$ onto $|w| < 1$, to derive the Green's function for the problem

$$\begin{aligned} u_{xx} + u_{yy} &= f && x^2 + y^2 < R^2 \\ u &= g && x^2 + y^2 = R^2 \end{aligned}$$

Check your answer against Problem 8.7 ($n = 2$), identifying $\mathbf{x}$ and $\boldsymbol{\xi}$ with the complex vectors z and ζ.

8.29 Show that $w = -e^{-i\pi z/b}$ $(b > 0)$ maps the infinite strip $0 < x < b, -\infty < y < \infty$ of the z-plane onto the upper half of the w-plane. Then use Theorem 8.8(i) to show that the Green's function for Laplace's equation which vanishes on $x = 0$ and $x = b$ is given by

$$G(x, y; \xi, \eta) = \frac{1}{2\pi} \log \left| \frac{\sin \dfrac{\pi(z - \zeta)}{2b}}{\sin \dfrac{\pi(z + \bar{\zeta})}{2b}} \right|$$

where $z = x + iy$, $\zeta = \xi + i\eta$. Finally, use the infinite product representation

$$\sin z = z \prod_{n=1}^{\infty} \left(1 - \frac{z^2}{n^2\pi^2}\right)$$

to compare this result with the one obtained by the method of images (interchange x and y in Problem 8.5(c)).

8.30 Show that in two dimensions the biharmonic equation, $\nabla^4 u = 0$, has the singularity solution $r^2(1 - \log r)/8$.

8.31 Find the adjoint of each of the following differential operators:

(a) $L[u] = u_{xx} + u_{yy} + u_x - u_y + 3u$

(b) $L[u] = u_{xx} - u_t$

(c) $L[u] = u_{xx} - u_{tt}$

(d) $L[u] = u_{xx} + u_{yy} + xu_x + yu_y$

(e) $L[u] = u_{xx} + u_{yy} + yu_x + xu_y$

(f) $L[u] = x^2 u_{xx} + y^2 u_{yy}$

8.32 If $L[u] = au_{xx} + 2bu_{xy} + cu_{yy} + du_x + eu_y + fu$ has constant coefficients, show that L is self-adjoint if and only if $d = e = 0$.

8.33 Show that the linear PDE $au_{xx} + 2bu_{xy} + cu_{yy} + du_x + eu_y + fu = g$ is self-adjoint if and only if it can be written in the form

$$(au_x + bu_y)_x + (bu_x + cu_y)_y + fu = g$$

8.34 Let $L[\ \ ]$ be the differential operator of Problem 8.33 and assume that the coefficient functions are in C^2. Show that L is self-adjoint if and only if $a_x + b_y = d$ and $b_x + c_y = e$.

8.35 If the PDE of Problem 8.33 is not self-adjoint, under what conditions will a *reducing factor*, $R(x, y)$, exist such that after multiplication of the equation by R a self-adjoint equation results?

8.36 Find a reducing factor for $u_{xx} + 2u_{xy} + 2u_{yy} + u_x + u_y + 3u = 0$.

8.37 Given the boundary value problem

$$L[u] \equiv u_{xx} + u_{yy} + 2u_x + 3u_y = f \qquad 0 < x < a,\ 0 < y < b$$
$$u = 0 \qquad \text{on } x = 0 \text{ and } x = a$$
$$u_y = 0 \qquad \text{on } y = 0 \text{ and } y = b$$

(a) write the adjoint operator and the adjoint boundary conditions; (b) write a boundary value problem in x and y for the Green's function.

8.38 Solve

$$u_{xx} + u_{yy} + 6u_y = \delta(x - \tfrac{1}{2})\delta(y) \qquad 0 < x < 1,\ -\infty < y < \infty$$
$$u(0, y) = u(1, y) = 0 \qquad -\infty < y < \infty$$

[*Hint*: Let $u = e^{-3y}v$, make a partial eigenfunction expansion of v, and use the Fourier transform to find the coefficient functions.]

8.39 Refer to Problem 8.7. Show that in ξ-space, the normal derivative of the Green's function on $|\boldsymbol{\xi}| = R$ is given by

$$\frac{\partial G}{\partial n} = \frac{R^2 - |\mathbf{x}|^2}{R} \frac{\sigma'(r)}{r}$$

From this, infer Poisson's integral formula, (*3.5*).

8.40 Using the free-space Green's function and the definition

$$\operatorname{erf} z \equiv \frac{2}{\sqrt{\pi}} \int_0^z e^{-s^2}\, ds = 1 - \operatorname{erfc} z$$

(cf. Problem 2.17 and Table 6-4, line 9), solve the initial value problem $u_t - \kappa u_{xx} = 0$ $(-\infty < x < \infty, t > 0)$, $u(x, 0) = x/|x|$ $(x \neq 0)$.

8.41 Using superposition arguments (and nothing else), infer from the solution of Problem 8.40 the solutions of the following problems:

(*a*)
$$\begin{aligned} u_t - \kappa u_{xx} &= 0 && x>0,\ t>0 \\ u(x,0) &= U && x>0 \\ u(0,t) &= 0 && t>0 \end{aligned}$$

(*b*)
$$\begin{aligned} u_t - \kappa u_{xx} &= 0 && x>0,\ t>0 \\ u(x,0) &= 0 && x>0 \\ u(0,t) &= U && t>0 \end{aligned}$$

(*c*)
$$\begin{aligned} u_t - \kappa u_{xx} &= 0 && -\infty<x<\infty,\ t>0 \\ u(x,0) &= UH(x) && x \neq 0 \end{aligned}$$

(*d*)
$$\begin{aligned} u_t - \kappa u_{xx} &= 0 && -\infty<x<\infty,\ t>0 \\ u(x,0) &= U && |x|<\ell \\ u(x,0) &= 0 && |x|>\ell \end{aligned}$$

(*e*)
$$\begin{aligned} u_t - \kappa u_{xx} &= 0 && -\infty<x<\infty,\ t>0 \\ u(x,0) &= 0 && |x|<\ell \\ u(x,0) &= U && |x|>\ell \end{aligned}$$

(*f*)
$$\begin{aligned} u_t - \kappa u_{xx} &= 0 && 0<x<\infty,\ t>0 \\ u(x,0) &= U && b<x<c \\ u(x,0) &= 0 && 0<x<b \text{ or } x>c \\ u(0,t) &= 0 && t>0 \end{aligned}$$

8.42 Prove the *product law*: If $v(x, t)$ satisfies $v_t - \kappa v_{xx} = 0$ and $w(y, t)$ satisfies $w_t - \kappa w_{yy} = 0$, then $u = vw$ satisfies $u_t - \kappa(u_{xx} + u_{yy}) = 0$.

8.43 Find the Green's function and the solution for the following problems:

(*a*)
$$\begin{aligned} &u_t - \kappa u_{xx} = f(x,t) \quad -\infty<x<\infty,\ t>0 \\ &u(x,0) = h(x) \end{aligned}$$

(*b*)
$$\begin{aligned} &u_t - \kappa u_{xx} = f(x,t) \quad 0<x<\infty,\ t>0 \\ &u(x,0) = h(x),\ \ u(0,t) = p(t) \end{aligned}$$

(*c*)
$$\begin{aligned} &u_t - \kappa u_{xx} = f(x,t) \quad 0<x<\infty,\ t>0 \\ &u(x,0) = h(x),\ \ u_x(0,t) = p(t) \end{aligned}$$

(*d*)
$$\begin{aligned} &u_t - \kappa u_{xx} = f(x,t) \quad 0<x<\ell,\ t>0 \\ &u(x,0) = h(x),\ \ u(0,t) = p(t),\ \ u(\ell,t) = q(t) \end{aligned}$$

(*e*)
$$\begin{aligned} &u_t - \kappa u_{xx} = f(x,t) \quad 0<x<\ell,\ t>0 \\ &u(x,0) = h(x),\ \ u_x(0,t) = p(t),\ \ u_x(\ell,t) = q(t) \end{aligned}$$

8.44 Given $\xi > 0$, find the Green's function satisfying

$$\begin{aligned} G_{tt} - a^2 G_{xx} &= \delta(x-\xi)\delta(t-\tau) && x>0,\ x \neq \xi \\ G &= 0 && x = 0 \\ G &\equiv 0 && 0<t<\tau \end{aligned}$$

8.45 Given $-\ell/2 < \xi < \ell/2$, construct the Green's function obeying

$$\begin{aligned} G_{tt} - a^2 G_{xx} &= \delta(x-\xi)\delta(t-\tau) && -\ell/2<x<\ell/2,\ x \neq \xi \\ G &= 0 && \text{for } x = \pm\ell/2 \\ G &\equiv 0 && 0<t<\tau \end{aligned}$$

(*a*) by a partial eigenfunction expansion, (*b*) by the method of images.

Chapter 9

Difference Methods for Parabolic Equations

9.1 DIFFERENCE EQUATIONS

The various partial derivatives of a function $u(x, t)$ can be expressed as a *difference quotient* plus a *truncation error* (T.E.).

Forward Difference for u_t

$$u_t(x, t) = \frac{u(x, t+k) - u(x, t)}{k} + \text{T.E.}$$
$$\text{T.E.} = -\frac{k}{2} u_{tt}(x, \bar{t}) \qquad (t < \bar{t} < t + k) \tag{9.1}$$

Centered Difference for u_x

$$u_x(x, t) = \frac{u(x+h, t) - u(x-h, t)}{2h} + \text{T.E.}$$
$$\text{T.E.} = -\frac{h^2}{6} u_{xxx}(\bar{x}, t) \qquad (x - h < \bar{x} < x + h) \tag{9.2}$$

Centered Difference for u_{xx}

$$u_{xx}(x, t) = \frac{u(x+h, t) - 2u(x, t) + u(x-h, t)}{h^2} + \text{T.E.}$$
$$\text{T.E.} = -\frac{h^2}{12} u_{xxxx}(\bar{x}, t) \qquad (x - h < \bar{x} < x + h) \tag{9.3}$$

Centered Difference for u_{xt}

$$u_{xt}(x, t) = \frac{u(x+h, t+k) - u(x+h, t-k) - u(x-h, t+k) + u(x-h, t-k)}{4hk} + \text{T.E.}$$
$$\text{T.E.} = -\frac{h^2}{6} u_{xxxt}(\bar{x}, \bar{t}) - \frac{k^2}{6} u_{xttt}(x', t') \qquad (x - h < \bar{x}, x' < x + h, t - k < \bar{t}, t' < t + k) \tag{9.4}$$

Usually it is only the order of magnitude of the truncation error which is of interest. A function $f(h)$ is said to be of the *order of magnitude* $g(h)$ *as* $h \to 0$, where g is a nonnegative function, if

$$\lim_{h \to 0} \frac{f(h)}{g(h)} = \text{constant}$$

In the *O-notation* we write $f(h) = O(g(h))$ $(h \to 0)$. It is easy to see that if, as $h_1, h_2 \to 0$, $f_1 = O(g_1)$ and $f_2 = O(g_2)$, then $f_1 + f_2 = O(g_1 + g_2)$.

EXAMPLE 9.1 For *(9.1)*, T.E. $= O(k)$ $(k \to 0)$, provided u_{tt} is bounded. For *(9.2)*, T.E. $= O(h^2)$ $(h \to 0)$, provided u_{xxx} is bounded. In *(9.4)*, T.E. $= O(h^2 + k^2)$, provided u_{xxxt} and u_{xttt} are bounded; we omit as understood the $(h, k \to 0)$.

A *grid* or *mesh* in the xt-plane is a set of points $(x_n, t_j) = (x_0 + nh, t_0 + jk)$, where n and j are integers and (x_0, t_0) is a reference point. The (x_n, t_j) are called *grid points, mesh points,* or *nodes.* The positive numbers h and k are respectively the x and t *grid spacings* or *grid sizes.* If h and k are constants, the grid is called *uniform*; if $h = k =$ constant, the grid is said to be *square.* The compact subscript notation

$$u_{nj} \equiv u(x_n, t_j)$$

is convenient and widely used.

EXAMPLE 9.2 The difference formulas (*9.1*) and (*9.3*) may be written

$$u_t(x_n, t_j) = \frac{u_{n,j+1} - u_{nj}}{k} + O(k)$$

$$u_{xx}(x_n, t_j) = \frac{u_{n+1,j} - 2u_{nj} + u_{n-1,j}}{h^2} + O(h^2) \equiv \frac{\delta_x^2 u_{nj}}{h^2} + O(h^2)$$

where the *difference operator* δ_x^2 is the analog of the differential operator $\delta^2/\delta x^2$. We say that (*9.1*) is *two-level* (in t) because it involves only two j-values, these being consecutive.

Let a region Ω in the xt-plane be covered by a grid, (x_n, t_j). If all the derivatives in the PDE

$$L[u] = f \qquad (x, t) \text{ in } \Omega \tag{9.5}$$

are replaced by difference quotients, the result is the *finite-difference equation*

$$D[U_{nj}] = f_{nj} \qquad (x_n, t_j) \text{ in } \Omega \tag{9.6}$$

The continuous problem (*9.5*) was *differenced* or *discretized* to produce the discrete problem (*9.6*), whose solution, U_{nj}, approximates $u(x, t)$ at the grid points.

9.2 CONSISTENCY AND CONVERGENCE

If discretization is to provide a useful approximation, the solution to (*9.5*) should very nearly satisfy (*9.6*), when h and k are taken sufficiently small. The amount by which the solution to $L[u] = f$ fails to satisfy the difference equation is called the *local truncation error*; it may be expressed as

$$T_{nj} \equiv D[u_{nj}] - f_{nj}$$

The difference equation (*9.6*) is said to be *consistent with* the PDE (*9.5*) if

$$\lim_{h,\,k \to 0} T_{nj} = 0 \tag{9.7}$$

With the exception of the DuFort–Frankel method (Problem 9.10), all difference methods to be treated are consistent with their corresponding PDEs.

In addition to consistency, we want the accuracy of the approximation to improve as $h, k \to 0$. If U_{nj} is the exact solution to (*9.6*) and u_{nj} is the solution of (*9.5*) evaluated at (x_n, t_j), the *discretization error* is defined as $U_{nj} - u_{nj}$. The difference method (*9.6*) is said to be *convergent* if

$$\lim_{h,\,k \to 0} |U_{nj} - u_{nj}| = 0 \qquad (x_n, t_j) \text{ in } \Omega \tag{9.8}$$

It is possible for a difference method to be consistent but not convergent.

9.3 STABILITY

Let U_{nj} satisfy (*9.6*), with initial values U_{n0} and possibly boundary values prescribed. Let V_{nj} be the solution to a perturbed difference system which differs only in the initial values, and write $V_{n0} \equiv U_{n0} + E_{n0}$. Then, assuming exact arithmetic, the initial perturbation, or "error," E_{n0}, can be shown to propagate, with increasing j, according to the homogeneous difference equation

$$D[E_{nj}] = 0$$

subject to homogeneous boundary conditions.

When applying (*9.6*) to approximate $u(x, T)$ for fixed $T = t_0 + jk$, it is clear that letting $h, k \to 0$ entails letting $j \to \infty$. Also, on a fixed grid, if we apply (*9.6*) to approximate $u(x_n, t_j)$ at successively larger t_j, then again we have the case $j \to \infty$ to consider. For a PDE with a bounded solution, the difference method (*9.6*) is said to be *stable* if the E_{nj} are uniformly bounded in n as $j \to \infty$; i.e., if for some constant M and some positive integer J

$$|E_{nj}| < M \qquad (j > J) \tag{9.9}$$

If h and k must be functionally related for (*9.9*) to hold, the difference method is *conditionally stable.* When the PDE has a solution that is unbounded in t, the stability condition (*9.9*) is relaxed to allow errors to grow with the solution (see Problem 9.7).

One of the concerns in applying a difference method is whether or not rounding errors in the calculation grow to such an extent that they dominate the numerical solution. When a stable method is used, rounding errors do not generally cause any difficulties.

It is usually easier to check a difference method for consistency and stability than for convergence. Fortunately, stability and convergence are equivalent for a large class of problems.

Theorem 9.1 (*Lax Equivalence Theorem*): Given a well-posed initial–boundary value problem and a finite-difference problem consistent with it, stability is both necessary and sufficient for convergence.

In certain cases (e.g., Problem 9.11) the boundary conditions imposed on (*9.5*) make themselves felt as modifications in the form of the operator $D[\ ]$ of (*9.6*), as it applies at grid points adjacent to the boundary. Let us agree to call U_{nj} an *extended solution* of (*9.6*) if it satisfies the equation for the *unmodified* operator; an extended solution would be an actual solution in the event that (*9.5*) held over the entire *xt*-plane.

von Neumann stability criterion. A difference method for an initial–boundary value problem with a bounded solution is *von Neumann stable* if every extended solution to $D[U_{nj}] = 0$ of the form

$$U_{nj} = \xi^j e^{i\beta n} \quad (\beta \text{ real}, \ \xi = \xi(\beta) \text{ complex})$$

has the property $|\xi| \le 1$. For a problem with an unbounded solution, the criterion becomes $|\xi| \le 1 + O(k)$.

For a rationale of the von Neumann criterion, see Problem 9.5. Stability in the von Neumann sense is a necessary condition for stability in the general sense of (*9.9*); moreover,

Theorem 9.2: For two-level difference methods, von Neumann stability is both necessary and sufficient for stability.

Consider an initial–boundary value problem with N nodes in the x-direction and define a column vector of errors at level j, $\mathbf{E}_j = (E_{1j}, E_{2j}, E_{3j}, \ldots, E_{Nj})^T$. For two-level difference methods, the errors at levels j and $j+1$ are related by

$$\mathbf{E}_{j+1} = \mathbf{C}\mathbf{E}_j$$

where $\mathbf{C}$ is an $N \times N$ matrix. Let $\rho(\mathbf{C})$, the *spectral radius* of $\mathbf{C}$, denote the maximum of the magnitudes of the eigenvalues of $\mathbf{C}$.

Matrix stability criterion. A two-level difference method for an initial–boundary value problem with a bounded solution is *matrix stable* if $\rho(\mathbf{C}) \le 1$. For a problem with an unbounded solution, the criterion becomes $\rho(\mathbf{C}) \le 1 + O(k)$.

Matrix stability is a necessary condition for the stability of a two-level method. Furthermore,

Theorem 9.3: Let $\mathbf{C}$ be symmetric or similar to a symmetric matrix, whereby all eigenvalues of $\mathbf{C}$ are real. Then matrix stability is necessary and sufficient for stability.

Although a matrix stability analysis incorporates the boundary conditions of the problem (as reflected in the form of **C**) whereas a von Neumann stability analysis neglects the boundary conditions, the conclusions reached regarding the stability of a difference method are nearly always the same. This indicates that the stability of a method is determined more by the character of the difference equations than by the way in which the boundary conditions are accounted for.

9.4 PARABOLIC EQUATIONS

The one-dimensional diffusion equation,

$$u_t = a^2 u_{xx} \tag{9.10}$$

(to avoid confusion between κ and k we write a^2 for the diffusivity) is used as a guide in developing finite-difference methods for parabolic PDEs in general. For the grid $(x_n, t_j) = (nh, jk)$, we shall state three commonly used difference equations for (*9.10*). All three are two-level equations whereby the solution, known at level j, is advanced to level $j+1$.

Explicit (Forward-Difference) Method

$$\frac{U_{n,j+1} - U_{nj}}{k} = a^2 \frac{U_{n+1,j} - 2U_{nj} + U_{n-1,j}}{h^2}$$

or

$$U_{n,j+1} = (1 + r\delta_x^2)U_{nj} \qquad (r \equiv a^2k/h^2) \tag{9.11}$$

Implicit (Backward-Difference) Method

$$\frac{U_{n,j+1} - U_{nj}}{k} = a^2 \frac{U_{n+1,j+1} - 2U_{n,j+1} + U_{n-1,j+1}}{h^2}$$

or

$$(1 - r\delta_x^2)U_{n,j+1} = U_{nj} \qquad (r \equiv a^2k/h^2) \tag{9.12}$$

Implicit (Crank–Nicolson) Method

$$\frac{U_{n,j+1} - U_{nj}}{k} = \frac{a^2}{2} \frac{\delta_x^2 U_{nj} + \delta_x^2 U_{n,j+1}}{h^2}$$

or

$$\left(1 - \frac{r}{2}\delta_x^2\right) U_{n,j+1} = \left(1 + \frac{r}{2}\delta_x^2\right) U_{nj} \qquad (r \equiv a^2k/h^2) \tag{9.13}$$

Theorem 9.4: The forward-difference method (*9.11*) has local truncation error $O(k + h^2)$; it is (conditionally) stable if and only if $r \leq 1/2$.

Theorem 9.5: The backward-difference method (*9.12*) has local truncation error $O(k + h^2)$; it is stable.

Theorem 9.6: The Crank–Nicolson method (*9.13*) has local truncation error $O(k^2 + h^2)$; it is stable.

For the two-dimensional diffusion equation,

$$u_t = a^2(u_{xx} + u_{yy}) \tag{9.14}$$

let $(x_m, y_n, t_j) = (mh, nh, jk)$ and $U_{mnj} \approx u_{mnj} = u(x_m, y_n, t_j)$. The above methods have as their counterparts:

Explicit (Forward-Difference) Method

$$U_{mn,j+1} = [1 + r(\delta_x^2 + \delta_y^2)]U_{mnj} \qquad (r \equiv a^2k/h^2) \tag{9.15}$$

Implicit (Backward-Difference) Method

$$[1 - r(\delta_x^2 + \delta_y^2)]U_{mn,j+1} = U_{mnj} \qquad (r \equiv a^2k/h^2) \tag{9.16}$$

Implicit (Crank–Nicolson) Method

$$\left[1-\frac{r}{2}(\delta_x^2+\delta_y^2)\right]U_{mn,j+1}=\left[1+\frac{r}{2}(\delta_x^2+\delta_y^2)\right]U_{mnj} \qquad (r\equiv a^2k/h^2) \tag{9.17}$$

Theorem 9.7: The forward-difference method (*9.15*) has local truncation error $O(k+h^2)$; it is (conditionally) stable if and only if $r\leq 1/4$.

Theorem 9.8: The backward-difference method (*9.16*) has local truncation error $O(k+h^2)$; it is stable.

Theorem 9.9: The Crank–Nicolson method (*9.17*) has local truncation error $O(k^2+h^2)$; it is stable.

As a class, the explicit methods enjoy the property of directly marching the solution forward in time, from one level to the next. They apply either to pure initial value problems or to initial–boundary value problems, but suffer the drawback of conditional stability. On the other hand, the implicit methods, which are stable, require (in effect) a matrix inversion at each step forward in time. Thus, these methods are applicable only to initial–boundary value problems with a finite number of spatial grid points.

For the one-dimensional implicit methods (*9.12*) and (*9.13*), the matrix to be inverted is tridiagonal; for the two-dimensional (*9.16*) and (*9.17*), pentadiagonal. *Alternating-direction implicit* (ADI) methods for parabolic problems in $x_1, x_2, \ldots$ preserve the tridiagonal feature by first solving a sequence of one-dimensional difference equations in x_1; then a sequence in x_2; and so on. Thus, if Dirichlet boundary conditions are specified for (*9.14*); if $m=0,1,2,\ldots,M$; and if $n=0,1,2,\ldots,N$; then we have $(r\equiv a^2k/h^2)$:

Peaceman–Rachford ADI Method

$$\begin{aligned}\left(1-\frac{r}{2}\delta_x^2\right)U^*_{mn,j+1}&=\left(1+\frac{r}{2}\delta_y^2\right)U_{mnj} \qquad (n=1,2,\ldots,N-1)\\ \left(1-\frac{r}{2}\delta_y^2\right)U_{mn,j+1}&=\left(1+\frac{r}{2}\delta_x^2\right)U^*_{mn,j+1} \qquad (m=1,2,\ldots,M-1)\end{aligned} \tag{9.18}$$

Theorem 9.10: The Peaceman–Rachford ADI method (*9.18*) has local truncation error $O(k^2+h^2)$; it is stable.

With only slight modifications, the above difference methods become applicable to the general linear parabolic PDE.

Solved Problems

9.1 Derive the difference formula (*9.2*).

By Taylor's theorem,

$$u(x+h,t)=u(x,t)+u_x(x,t)h+u_{xx}(x,t)\frac{h^2}{2}+u_{xxx}(\bar{x},t)\frac{h^3}{6} \tag{1}$$

$$u(x-h,t)=u(x,t)-u_x(x,t)h+u_{xx}(x,t)\frac{h^2}{2}-u_{xxx}(\hat{x},t)\frac{h^3}{6} \tag{2}$$

where $x<\bar{x}<x+h$ and $x-h<\hat{x}<x$. Subtracting (*2*) from (*1*) and solving for u_x yields

$$u_x(x,t)=\frac{u(x+h,t)-u(x-h,t)}{2h}-[u_{xxx}(\bar{x},t)+u_{xxx}(\hat{x},t)]\frac{h^2}{12}$$

If u_{xxx} is continuous, the mean-value theorem implies that

$$\frac{u_{xxx}(\bar{x}, t) + u_{xxx}(\hat{x}, t)}{2} = u_{xxx}(\tilde{x}, t) \qquad (\hat{x} < \tilde{x} < \bar{x})$$

and (*9.2*) results.

9.2 Derive a difference formula similar to (*9.2*) in the case of a nonuniform grid,

$$x_{i+1} = x_i + h_{i+1}$$

By Taylor's theorem,

$$u(x_{i+1}, t) = u(x_i, t) + u_x(x_i, t)h_{i+1} + \frac{1}{2}u_{xx}(x_i, t)h_{i+1}^2 + O(h_{i+1}^3) \tag{1}$$

$$u(x_{i-1}, t) = u(x_i, t) - u_x(x_i, t)h_i + \frac{1}{2}u_{xx}(x_i, t)h_i^2 - O(h_i^3) \tag{2}$$

Subtracting (*2*) from (*1*) gives

$$u_x(x_i, t) = \frac{u(x_{i+1}, t) - u(x_{i-1}, t)}{h_{i+1} + h_i} + \frac{1}{2}u(x_i, t)(h_{i+1} - h_i) + \frac{O(h_{i+1}^3) + O(h_i^3)}{h_{i+1} + h_i} \tag{3}$$

Note that in (*3*) the dominant term in the truncation error is $O(h_{i+1} - h_i)$. To maintain control over the truncation error, the grid spacings h_i should not be allowed to vary too rapidly with i.

Multiplying (*1*) by h_i and (*2*) by h_{i+1} and then adding yields

$$h_i u(x_{i+1}, t) + h_{i+1}u(x_{i-1}, t) = (h_i + h_{i+1})u(x_i, t) + u_{xx}(x_i, t)\frac{h_i h_{i+1}^2 + h_{i+1}h_i^2}{2} + h_i O(h_{i+1}^3) + h_{i+1}O(h_i^3)$$

which when solved for u_{xx} gives

$$u_{xx}(x_i, t) = \frac{h_{i+1}u(x_{i-1}, t) - (h_i + h_{i+1})u(x_i, t) + h_i u(x_{i+1}, t)}{0.5(h_i h_{i+1}^2 + h_{i+1}h_i^2)} + \text{T.E.}$$

9.3 Show that the explicit method (*9.11*) has local truncation error $O(k + h^2)$. Then show that if

$$\frac{k}{h^2} = \frac{1}{6a^2}$$

the local truncation error can be reduced to $O(k^2 + h^4)$.

With $(x_n, t_j) = (nh, jk)$, (*9.1*) and (*9.3*) give

$$(u_t - a^2u_{xx})_{nj} = \frac{u_{n,j+1} - u_{nj}}{k} - a^2\frac{\delta_x^2 u_{nj}}{h^2} - \frac{k}{2}u_{tt}(x_n, \bar{t}_j) + a^2\frac{h^2}{12}u_{xxxx}(\bar{x}_n, t_j)$$

where $t_j < \bar{t}_j < t_{j+1}$ and $x_{n-1} < \bar{x}_n < x_{n+1}$. The amount by which the solution of $u_t - a^2u_{xx} = 0$ fails to satisfy the difference equation (*9.11*) is

$$T_{nj} = \frac{k}{2}u_{tt}(x_n, \bar{t}_j) - a^2\frac{h^2}{12}u_{xxxx}(\bar{x}_n, t_j) = O(k + h^2)$$

provided u_{tt} and u_{xxxx} are bounded.

Now, by Taylor's theorem and $(u_t - a^2u_{xx})_{nj} = 0$,

$$\frac{u_{n,j+1} - u_{nj}}{k} - a^2\frac{\delta_x^2 u_{nj}}{h^2} = \left[\frac{k}{2}u_{tt} - a^2\frac{h^2}{12}u_{xxxx}\right]_{nj} + O(k^2) + O(h^4) \tag{1}$$

Since $u_t = a^2u_{xx}$, $u_{tt} = a^2u_{xxt} = a^2(u_t)_{xx}$; whence

$$u_{tt} = a^2(a^2u_{xx})_{xx} = a^4u_{xxxx}$$

This shows that the bracketed terms in (*1*) can be written as

$$\left(\frac{k}{2}a^4 - a^2\frac{h^2}{12}\right)u_{xxxx}(x_n, t_j)$$

which will be zero if we choose $k = h^2/6a^2$.

9.4 Show that if $r \leq 1/2$, then the explicit method (*9.11*) is convergent when applied to the problem

$$\begin{aligned} u_t - a^2u_{xx} &= 0 \qquad 0<x<1, t>0 \\ u(x, 0) &= f(x) \qquad 0<x<1 \\ u(0, t) = p(t),\ u(1, t) &= q(t) \qquad t>0 \end{aligned}$$

Let Ω be the region $0<x<1, 0<t<T$; take $(x_n, t_j) = (nh, jk)$ for $n = 0, 1, 2, \ldots, N$ and $j = 0, 1, 2, \ldots, J$, with $Nh = 1$ and $Jk = T$. Let U_{nj} satisfy the difference system

$$U_{n,j+1} = U_{nj} + r\delta_x^2 U_{nj} \qquad (r \equiv a^2k/h^2)$$

$$U_{n0} = f(x_n) \qquad U_{0j} = p(t_j) \qquad U_{Nj} = q(t_j)$$

and set $w_{nj} \equiv U_{nj} - u_{nj}$. Then w_{nj} satisfies

$$w_{n,j+1} = rw_{n-1,j} + (1-2r)w_{nj} + rw_{n+1,j} + \frac{k^2}{2}u_{tt}(x_n, \bar{t}_j) - \frac{kh^2a^2}{12}u_{xxxx}(\bar{x}_n, t_j) \tag{1}$$

$$w_{n0} = 0 \qquad w_{0j} = 0 \qquad w_{Nj} = 0$$

where $t_j < \bar{t}_j < t_{j+1}$ and $x_{n-1} < \bar{x}_n < x_{n+1}$.

If u_{tt} and u_{xxxx} are continuous and if we write

$$A \equiv \max\left|\frac{1}{2}u_{tt}(x, t)\right| \qquad B \equiv \max\left|\frac{a^2}{12}u_{xxxx}(x, t)\right|$$

for (x, t) in $\bar{\Omega}$, then, since $r \leq 1/2$, it follows from (*1*) that

$$\begin{aligned} |w_{n,j+1}| &\leq r|w_{n-1,j}| + (1-2r)|w_{nj}| + r|w_{n+1,j}| + Ak^2 + Bkh^2 \\ &\leq \|w_j\| + Ak^2 + Bkh^2 \quad (\|w_j\| \equiv \max_{0<n<N} |w_{nj}|) \end{aligned} \tag{2}$$

From (*2*) we have

$$\|w_{j+1}\| \leq \|w_j\| + Ak^2 + Bkh^2 \tag{3}$$

Because $\|w_0\| = 0$, (*3*) implies

$$\|w_j\| \leq j(Ak^2 + Bkh^2) \leq T(Ak + Bh^2)$$

which shows that $|w_{nj}| \to 0$ uniformly in Ω as $h, k \to 0$.

9.5 Use the von Neumann criterion to establish the condition $r \leq 1/2$ for the stability of the explicit method (*9.11*).

With (*9.11*) expressed in the form

$$U_{n,j+1} = rU_{n+1,j} + (1-2r)U_{nj} + rU_{n-1,j} \tag{1}$$

suppose that, at level j, an error is introduced at one or more of the x-nodes, perturbing the exact solution, U_{nj}, by an amount E_{nj}. If $U_{nj} + E_{nj}$ is used to advance the numerical solution to level $j+1$, the result is the exact solution, $U_{n,j+1}$, plus an error, $E_{n,j+1}$. Putting $U_{nj} + E_{nj}$ and $U_{n+1,j} + E_{n+1,j}$ into (*1*), we see that E_{nj} satisfies that equation.

Using separation of variables, we identify complex solutions of (*1*) of the form

$$a_j b_n = \xi^j e^{in\beta} \tag{2}$$

where ξ is some (possibly complex-valued) function of the real parameter β. From this, by superposition, we are led to the following expression for the error E_{nj}:

$$E_{nj} = \int_{-\infty}^{\infty} \xi(\beta)^j e^{in\beta} \, d\beta \tag{3}$$

(Strictly, the real part of the integral should be taken.) By a comparison with (*6.17*), (*3*) may be interpreted as a Fourier integral, representing the error at node n in terms of its frequency spectrum. Thus, at level j, the error amplitude corresponding to a given frequency $\beta/2\pi$ is $\xi(\beta)^j$. The Fourier–von Neumann rule, $|\xi| \leq 1$, amounts to the prescription that at no frequency should the error amplitude grow without limit as $j \to \infty$.

Substitution of (*2*) in (*1*) gives, after division by $\xi^j e^{in\beta}$,

$$\xi = re^{i\beta} + (1-2r) + re^{-i\beta} = 1 - 2r(1-\cos\beta) = 1 - 4r\sin^2\frac{\beta}{2}$$

and so $-1 \leq \xi \leq 1$ for all β—in particular, for $\beta = \pi$—if and only if $0 \leq r \leq 1/2$.

9.6 Show that the implicit method (*9.12*) is (von Neumann) stable.

The method can be written as

$$-rU_{n-1,j+1} + (1+2r)U_{n,j+1} - rU_{n+1,j+1} = U_{nj} \tag{1}$$

Substituting $\xi^j e^{i\beta n}$ into (*1*) and dividing by $\xi^j e^{i\beta n}$, we have

$$\xi[-re^{i\beta} + (1+2r) - re^{-i\beta}] = 1 \qquad \text{or} \qquad \xi = \left(1 + 4r\sin^2\frac{\beta}{2}\right)^{-1}$$

and we see that $|\xi| \leq 1$ for every β whatever the value of r.

9.7 (*a*) Modify the explicit method (*9.11*) to apply to the PDE

$$u_t = a^2 u_{xx} + cu \tag{1}$$

(*b*) Make a von Neumann stability analysis of the modified method.

(*a*) An explicit difference equation for (*1*) that reduces to (*9.11*) when $c = 0$ is

$$\frac{U_{n,j+1} - U_{nj}}{k} = a^2 \frac{\delta_x^2 U_{nj}}{h^2} + cU_{nj}$$

or

$$U_{n,j+1} = rU_{n-1,j} + (1 - 2r + ck)U_{nj} + rU_{n+1,j} \tag{2}$$

(*b*) Substituting $\xi^j e^{i\beta n}$ into (*2*), we find

$$\xi = 1 - 4r\sin^2\frac{\beta}{2} + ck \tag{3}$$

If $c < 0$, the solution of (*1*) is bounded, and the stability criterion is $|\xi| \leq 1$ for all β. This is satisfied if

$$r \leq \frac{1}{2} + \frac{ck}{4} \qquad (c < 0)$$

Note that for $c < 0$ and $r = 1/2$, (*2*) is not stable, but the asymptotic stability condition as $h, k \to 0$ is $r \leq 1/2$.

If $c > 0$, (*1*) can have exponentially growing solutions, which means that U_{nj} and its error must also be permitted to grow exponentially. Thus, the stability criterion is taken to be

$$|\xi| \leq 1 + O(k)$$

which is satisfied if again

$$r \leq \frac{1}{2} + \frac{ck}{4} \qquad (c > 0)$$

Now we have (conditional) stability for $r = 1/2$; once more, the asymptotic stability condition as $h, k \to 0$ is $r \leq 1/2$.

9.8 Show that the eigenvalues of the real, symmetric, tridiagonal, $N \times N$ matrix

$$\mathbf{C} = \begin{bmatrix} p & q & & & & 0 \\ q & p & q & & & \\ & q & p & q & & \\ & & \cdot & \cdot & \cdot & \\ & & & q & p & q \\ 0 & & & & q & p \end{bmatrix}$$

all lie in the interval $[p - |2q|, p + |2q|]$.

For **C** real and symmetric, we know that all eigenvalues are real, and that the largest and smallest eigenvalues are the absolute extrema of the normalized quadratic form

$$Q \equiv \frac{\boldsymbol{\zeta}^T \mathbf{C} \boldsymbol{\zeta}}{\boldsymbol{\zeta}^T \boldsymbol{\zeta}} = \frac{p(\zeta_1^2 + \zeta_2^2 + \cdots + \zeta_N^2) + 2q(\zeta_1\zeta_2 + \zeta_2\zeta_3 + \zeta_3\zeta_4 + \cdots + \zeta_{N-1}\zeta_N)}{\zeta_1^2 + \zeta_2^2 + \cdots + \zeta_N^2}$$

$$\equiv p + 2qR(\zeta_1, \zeta_2, \ldots, \zeta_N)$$

Now, by Cauchy's inequality,

$$|\zeta_1\zeta_2 + \zeta_2\zeta_3 + \zeta_3\zeta_4 + \cdots + \zeta_{N-1}\zeta_N|^2 \leq (\zeta_1^2 + \zeta_2^2 + \cdots + \zeta_{N-1}^2)(\zeta_2^2 + \zeta_3^2 + \cdots + \zeta_N^2)$$
$$\leq (\zeta_1^2 + \zeta_2^2 + \cdots + \zeta_N^2)^2$$

which implies that $|R| \leq 1$ and yields the desired interval.

From the fact that

$$R(\alpha, \alpha, \ldots, \alpha) = 1 - \frac{1}{N} \qquad R(\alpha, -\alpha, \alpha, \ldots, \pm\alpha) = -1 + \frac{1}{N}$$

we derive the inequalities

$$(p + |2q|) - \frac{|2q|}{N} \leq \lambda_{\max} \leq p + |2q|$$

$$p - |2q| \leq \lambda_{\min} \leq (p - |2q|) + \frac{|2q|}{N}$$

These show that the estimates

$$\lambda_{\max} \approx p + |2q| \qquad \lambda_{\min} \approx p - |2q| \tag{1}$$

are not necessarily sharp, even as $N \to \infty$, since q can—and, in applications, usually does—vary with N. In consequence, when *(1)* is used to investigate the stability of matrix **C**, it provides a *conservative* condition. Indeed, it is possible that the exact values of $\lambda_{\min}$ and $\lambda_{\max}$, from Problem 11.11, can be used to establish stability when *(1)* guarantees nothing. These same remarks apply to the Gerschgorin Circle Theorem (Problem 9.26).

9.9 Use the matrix stability criterion to show that the explicit method *(9.11)* is stable when applied to the initial–boundary value problem

$$u_t = a^2 u_{xx} \qquad 0 < x < 1, t > 0$$
$$u(x, 0) = f(x) \qquad 0 < x < 1$$
$$u(0, t) = u(1, t) = 0 \qquad t > 0$$

if and only if $r \leq 1/2$.

Let $(x_n, t_j) = (nh, jk)$ $(n = 0, 1, 2, \ldots, N; j = 0, 1, 2, \ldots)$, with $Nh = 1$, and define a column vector $\mathbf{U}_j$ by

$$\mathbf{U}_j = [U_{1j}, U_{2j}, U_{3j}, \ldots, U_{N-1,j}]^T$$

The explicit method (*9.11*) can be expressed in matrix form as

$$\mathbf{U}_{j+1} = \mathbf{C}\mathbf{U}_j \qquad (j = 0, 1, 2, 3, \ldots) \tag{1}$$

where $\mathbf{U}_0 = [f_1, f_2, f_3, \ldots, f_{N-1}]^T$ and $\mathbf{C}$ is the $(N-1) \times (N-1)$ tridiagonal matrix

$$\mathbf{C} = \begin{bmatrix} (1-2r) & r & & & & 0 \\ r & (1-2r) & r & & & \\ & r & (1-2r) & r & & \\ & & \cdot & \cdot & \cdot & \\ & & & r & (1-2r) & r \\ 0 & & & & r & (1-2r) \end{bmatrix}$$

Suppose that at time level j errors E_{nj} are introduced at x_n $(n = 1, 2, \ldots, N-1)$, perturbing the solution of (*1*) to $\mathbf{U}_j + \mathbf{E}_j$, where $\mathbf{E}_j$ is a column vector with nth component E_{nj}. Then, using (*1*) to advance the solution, we have

$$\mathbf{U}_{j+1} + \mathbf{E}_{j+1} = \mathbf{C}\mathbf{U}_j + \mathbf{C}\mathbf{E}_j \qquad \text{or} \qquad \mathbf{E}_{j+1} = \mathbf{C}\mathbf{E}_j$$

or, after m steps,

$$\mathbf{E}_{j+m} = \mathbf{C}^m \mathbf{E}_j \tag{2}$$

Let $\lambda_1, \lambda_2, \ldots, \lambda_{N-1}$ and $\mathbf{V}_1, \mathbf{V}_2, \ldots, \mathbf{V}_{N-1}$ be the eigenvalues and associated linearly independent eigenvectors of the symmetric matrix $\mathbf{C}$. Writing $\mathbf{E}_j$ as a linear combination of the $\mathbf{V}_k$,

$$\mathbf{E}_j = \sum_{k=1}^{N-1} a_k \mathbf{V}_k$$

and using (*2*) and $\mathbf{C}\mathbf{V}_k = \lambda_k \mathbf{V}_k$, we see that

$$\mathbf{E}_{j+m} = \sum_{k=1}^{N-1} \lambda_k^m a_k \mathbf{V}_k \tag{3}$$

Equation (*3*) shows that the errors E_{nj} remain bounded if and only if $|\lambda_k| \leq 1$ for $k = 1, 2, \ldots, N-1$. By Problem 9.8, with $p = (1-2r)$ and $q = r$,

$$\lambda_{\max} \approx (1-2r) + |2r| = 1 \qquad \lambda_{\min} \approx (1-2r) - |2r| = 1 - 4r$$

which yields the condition $1 - 4r \geq -1$, or $r \leq 1/2$. (This same condition is obtained when the exact expressions for the eigenvalues, from Problem 11.11, are employed.)

9.10 A stable (see Problem 9.20) overlapping-steps method is the *DuFort–Frankel method*,

$$\frac{U_{n,j+1} - U_{n,j-1}}{2k} = a^2 \frac{U_{n-1,j} - (U_{n,j+1} + U_{n,j-1}) + U_{n+1,j}}{h^2} \tag{9.19}$$

(*a*) Show that (*9.19*) is consistent with $u_t = a^2 u_{xx}$ only if

$$\lim_{h,\, k \to 0} \frac{k}{h} = 0$$

(*b*) Show that if k/h is held constant as $h, k \to 0$, then (*9.19*) is consistent with an equation of hyperbolic type.

If the more natural central term $-2U_{nj}$ were taken in (*9.19*), the method would be unstable for every positive $r \equiv a^2 k/h^2$.

(*a*) From (*9.2*) (for u_t) and (*9.3*),

$$\frac{u_{n,j+1}-u_{n,j-1}}{2k}=u_t(x_n,t_j)+O(k^2) \tag{1}$$

$$\frac{u_{n-1,j}-2u_{nj}+u_{n+1,j}}{h^2}=u_{xx}(x_n,t_j)+O(h^2) \tag{2}$$

Further, $u_{n,j+1}+u_{n,j-1}=2u_{nj}+k^2u_{tt}(x_n,t_j)+O(k^4)$; so that (*2*) gives

$$\frac{u_{n-1,j}-u_{n,j+1}-u_{n,j-1}+u_{n+1,j}}{h^2}=u_{xx}(x_n,t_j)-\frac{k^2}{h^2}u_{tt}(x_n,t_j)+O(h^2)+O\left(\frac{k^4}{h^2}\right) \tag{3}$$

Equations (*1*) and (*3*), together with $u_t=a^2u_{xx}$, show that u fails to satisfy (*9.19*) by an amount

$$T_{nj}=a^2\frac{k^2}{h^2}u_{tt}(x_n,t_j)+O\left(k^2+h^2+\frac{k^4}{h^2}\right) \tag{4}$$

It follows that (*9.19*) is consistent with the diffusion equation only if k/h tends to zero along with k and h.

(*b*) If $k/h\equiv e$, then it is obvious from (*4*) that (*9.19*) will be consistent with the PDE

$$u_t-a^2u_{xx}+a^2e^2u_{tt}=0$$

9.11 Using centered differences to approximate all x-derivatives and the implicit method (*9.12*), derive difference equations for the Neumann initial–boundary value problem

$$\begin{aligned} u_t&=a^2u_{xx} && 0<x<1,\ t>0\\ u(x,0)&=f(x) && 0<x<1\\ u_x(0,t)=p(t),\quad u_x(1,t)&=q(t) && t>0 \end{aligned}$$

Let $(x_n,t_j)=(nh,jk)$ $(n=-1,0,1,2,\ldots,N-1,N,N+1;\ j=0,1,2,\ldots)$, with $Nh=1$. The *ghost points* x_{-1} and x_{N+1} are introduced so that the boundary conditions can be approximated via the centered differences

$$\frac{U_{1j}-U_{-1j}}{2h}=p(t_j)\qquad \frac{U_{N+1,j}-U_{N-1,j}}{2h}=q(t_j) \tag{1}$$

By (*9.12*), the linear equations for the unknowns $U_{n,j+1}$ are

$$-rU_{n-1,j+1}+(1+2r)U_{n,j+1}-rU_{n+1,j+1}=U_{nj}\qquad (n=0,1,\ldots,N)$$

From these, $U_{-1,j+1}$ and $U_{N+1,j+1}$ may be eliminated by using (*1*) to write

$$U_{-1,j+1}=U_{1,j+1}-2hp_{j+1}\qquad U_{N+1,j+1}=U_{N-1,j+1}+2hq_{j+1}$$

The resultant system is expressed in matrix form as

$$\begin{bmatrix} (1+2r) & -2r & & & & 0\\ -r & (1+2r) & -r & & & \\ & -r & (1+2r) & -r & & \\ & & \cdot & \cdot & \cdot & \\ & & & -r & (1+2r) & -r\\ 0 & & & & -2r & (1+2r) \end{bmatrix}\begin{bmatrix} U_{0,j+1}\\ U_{1,j+1}\\ U_{2,j+1}\\ \cdots\\ U_{N-1,j+1}\\ U_{N,j+1}\end{bmatrix}=\begin{bmatrix} U_{0j}-2hrp_{j+1}\\ U_{1j}\\ U_{2j}\\ \cdots\\ U_{N-1,j}\\ U_{Nj}+2hrq_{j+1}\end{bmatrix}$$

Observe the two anomalous entries, $-2r$, in the transition matrix, which arise from the Neumann boundary conditions.

9.12 Write a computer program that uses the explicit method (*9.11*) to approximate the solution to

$$u_t = u_{xx} \qquad 0<x<1, t>0$$
$$u(x,0) = 100 \sin \pi x \qquad 0<x<1$$
$$u(0,t) = u(1,t) = 0 \qquad t>0$$

At $t = 0.5$ compare the numerical results with those from the exact solution,

$$u = 100\, e^{-\pi^2 t} \sin \pi x$$

A FORTRAN-77 program, EHEAT, is listed in Fig. 9-1. Two stable runs are given in Fig. 9-2. The excellent agreement between the numerical and exact solutions in the case $r \approx 1/6$ is explained in Problem 9.3.

```
      PROGRAM EHEAT
C     TITLE:  DEMO PROGRAM FOR EXPLICIT METHOD
C             FOR HEAT EQUATION, UT = KAPPA*UXX
C     INPUT:  N, NUMBER OF X-SUBINTERVALS
C             K, TIME STEP
C             TMAX, MAXIMUM COMPUTATION TIME
C             KAPPA, DIFFUSIVITY VALUE
C             [X1,X2], X-INTERVAL
C             P(T), LEFT BOUNDARY CONDITION
C             Q(T), RIGHT BOUNDARY CONDITION
C             F(X), INITIAL CONDITION
C             E(X,T), EXACT SOLUTION
C     OUTPUT: NUMERICAL AND EXACT SOLUTION AT T=TMAX
      COMMON U(0:51),V(0:51)
      REAL K,KAPPA
      DATA T,X1,X2,KAPPA/0,0,1,1/
      P(T) = 0
      Q(T) = 0
      F(X) = 100*SIN(PI*X)
      E(X,T) = 100*EXP(-PI*PI*T)*SIN(PI*X)
      PRINT*,'ENTER TMAX,NUMBER OF X-SUBINTERVALS AND TIME STEP'
      READ*,TMAX,N,K
      H = (X2-X1)/N
      R = KAPPA*K/H/H
      PI = 4*ATAN(1.)
C     SET INITIAL CONDITION
      DO 10 I = 0,N
         X = X1 + I*H
         V(I) = F(X)
10    CONTINUE
15    DO 20 I = 1,N-1
         U(I) = V(I) + R*(V(I+1) -2*V(I) + V(I-1))
20    CONTINUE
      T = T + K
      U(0) = P(T)
      U(N) = Q(T)
C     WRITE U OVER V TO PREPARE FOR NEXT TIME STEP
      DO 30 I = 0,N
         V(I) = U(I)
30    CONTINUE
C     IF T IS LESS THAN TMAX, TAKE A TIME STEP
      IF(ABS(TMAX-T).GT.K/2) GOTO 15
C     OTHERWISE, PRINT RESULT
      WRITE(6,100)
      WRITE(6,110) N,K,TMAX
      WRITE(6,120) T
      DO 40 I = 0,N
         X = X1 + I*H
         EXACT = E(X,T)
         WRITE(6,130) X,U(I),EXACT
40    CONTINUE
100   FORMAT(///,T9,'RESULTS FROM PROGRAM EHEAT',/)
110   FORMAT('N =',I4,T15,'K = ',F8.6,T30,'TMAX =',F5.2,/)
120   FORMAT('T = ',F5.2,T18,'NUMERICAL',T35,'EXACT',/)
130   FORMAT( 'X = ',F4.1,T13,F13.6,T30,F13.6)
      END
```

Fig. 9-1

```
N =  10          K = 0.001667    TMAX = 0.50

T =  0.50          NUMERICAL          EXACT

X =  0.            0.                 0.
X =  0.1           0.222262           0.222242
X =  0.2           0.422767           0.422730
X =  0.3           0.581889           0.581838
X =  0.4           0.684051           0.683992
X =  0.5           0.719254           0.719192
X =  0.6           0.684051           0.683992
X =  0.7           0.581889           0.581838
X =  0.8           0.422767           0.422730
X =  0.9           0.222262           0.222243
X =  1.0           0.                 0.000000

N =  10          K = 0.005000    TMAX = 0.50

T =  0.50          NUMERICAL          EXACT

X =  0.            0.                 0.
X =  0.1           0.204463           0.222242
X =  0.2           0.388912           0.422730
X =  0.3           0.535292           0.581838
X =  0.4           0.629273           0.683991
X =  0.5           0.661657           0.719191
X =  0.6           0.629273           0.683991
X =  0.7           0.535292           0.581838
X =  0.8           0.388912           0.422730
X =  0.9           0.204463           0.222242
X =  1.0           0.                 0.000000
```

Fig. 9-2

9.13 For the initial–boundary value problem

$$\begin{aligned} u_t &= a^2 u_{xx} && 0<x<1,\ t>0 \\ u(x,0) &= f(x) && 0<x<1 \\ u(0,t)=p(t),\ u(1,t) &= q(t) && t>0 \end{aligned}$$

show how to imbed the implicit method (*9.12*) and the Crank–Nicolson method (*9.13*) in a single algorithm.

With $(x_n, t_j) = (nh, jk)$ $(n = 0, 1, \ldots, N; j = 0, 1, 2, \ldots; Nh = 1)$, let $U_{n0} = f(x_n)$ $(n = 1, 2, \ldots, N-1)$, $U_{00} = [f(0)+p(0)]/2$, $U_{N0} = [f(1)+q(0)]/2$; further, for $j = 1, 2, \ldots$, let $U_{0j} = p(t_j)$, $U_{Nj} = q(t_j)$. In the implicit and Crank–Nicolson methods a system of linear equations must be solved to advance the solution from t_j to t_{j+1}: it is not possible simply to march the solution forward as in the explicit method.

The *weighted-difference method*,

$$U_{n,j+1} - U_{nj} = r[(1-w)\delta_x^2 U_{nj} + w\delta_x^2 U_{n,j+1}] \qquad (r \equiv a^2k/h^2) \tag{9.20}$$

reduces to (*9.12*) when $w = 1$ and to (*9.13*) when $w = 0.5$. Incorporating the boundary and initial conditions into (*9.20*), we find that the unknowns $U_{1,j+1}, U_{2,j+1}, \ldots, U_{N-1,j+1}$ satisfy the following tridiagonal system:

$$\begin{bmatrix} (1+2wr) & -wr & & & & 0 \\ -wr & (1+2wr) & -wr & & & \\ & -wr & (1+2wr) & -wr & & \\ & & \cdot & \cdot & \cdot & \\ & & -wr & (1+2wr) & -wr & \\ 0 & & & -wr & (1+2wr) & \end{bmatrix} \begin{bmatrix} U_{1,j+1} \\ U_{2,j+1} \\ U_{3,j+1} \\ \cdots \\ U_{N-2,j+1} \\ U_{N-1,j+1} \end{bmatrix} = \begin{bmatrix} D_1 \\ D_2 \\ D_3 \\ \cdots \\ D_{N-2} \\ D_{N-1} \end{bmatrix}$$

where $D_n \equiv U_{nj} + (1-w)r\delta_x^2 U_{nj}$ $(n = 2, 3, \ldots, N-2)$ and

$$D_1 \equiv U_{1j} + (1-w)r\delta_x^2 U_{1j} + wrU_{0,j+1} \qquad D_{N-1} \equiv U_{N-1,j} + (1-w)r\delta_x^2 U_{N-1,j} + wrU_{N,j+1}$$

The weighted-difference method program is given in Fig. 9-3. Two runs are shown in Fig. 9-4. Compare these with the first run in Fig. 9-2.

```
      PROGRAM IHEAT
C     TITLE:  DEMO PROGRAM FOR IMPLICT AND CRANK-
C             NICOLCON METHODS FOR UT = KAPPA*UXX
C     INPUT:  N, NUMBER OF X-SUBINTERVALS
C             K, TIME STEP
C             TMAX, MAXIMUM COMPUTATION TIME
C             KAPPA, DIFFUSIVITY VALUE
C             (X1,X2), X-INTERVAL
C             P(T), LEFT BOUNDARY CONDITION
C             Q(T), RIGHT BOUNDARY CONDITION
C             F(X), INITIAL CONDITION
C             E(X,T), EXACT SOLUTION
C             W, W=1 FOR IMPLICIT-W=.5 FOR CRANK-NICOLSON
C     OUTPUT: NUMERICAL AND EXACT SOLUTION AT T=TMAX
      COMMON/BLOCK1/A(51),B(51),C(51),D(51),L
      COMMON/BLOCK2/U(0:51)
      REAL K,KAPPA
      DATA T,X1,X2,KAPPA/0,0,1,1/
      P(T) = 0
      Q(T) = 0
      F(X) = 100*SIN(PI*X)
      E(X,T) = 100*EXP(-PI*PI*T)*SIN(PI*X)
      PRINT*,'ENTER TMAX,NUMBER OF X-SUBINTERVALS AND TIME STEP'
      READ*,TMAX,N,K
      PRINT*,'ENTER 1 FOR IMPLICIT, .5 FOR CRANK-NICOLSON METHOD'
      READ*, W
      H = (X2-X1)/N
      R = KAPPA*K/H/H
      PI = 4*ATAN(1.)
C     SET INITIAL CONDITION
      DO 10 I = 0,N
         X = X1 + I*H
         U(I) = F(X)
10    CONTINUE
C     DEFINE TRIDIAGONAL LINEAR SYSTEM
      L = N-1
15    DO 20 I = 1,L
         A(I) = -W*R
         B(I) = 1 + 2*W*R
         C(I) = -W*R
         D(I) = U(I) + (1-W)*R*(U(I-1) - 2*U(I) + U(I+1))
20    CONTINUE
C     CALL TRIDIAGONAL LINEAR EQUATION SOLVER
      CALL TRIDI
C     WRITE SOLUTION AT TIME T+K INTO THE U-ARRAY
      T = T + K
      DO 30 I = 1,N-1
         U(I) = D(I)
30    CONTINUE
      U(0) = P(T)
      U(N) = Q(T)
C     IF T IS LESS THAN TMAX, TAKE A TIME STEP
      IF(ABS(TMAX-T).GT.K/2) GOTO 15
C     OTHERWISE, PRINT RESULT
      WRITE(6,100) W
      WRITE(6,110) N,K,TMAX
      WRITE(6,120) T
      DO 40 I = 0,N
```

Fig. 9-3 *(Program continues on next page)*

```
      X = X1 + I*H
      EXACT = E(X,T)
      WRITE(6,130) X,U(I),EXACT
40    CONTINUE
100   FORMAT(///,T4,'RESULTS FROM PROGRAM IHEAT  W=',F5.2,/)
110   FORMAT('N =',I4,T15,'K = ',F8.6,T30,'TMAX =',F5.2,/)
120   FORMAT('T = ',F5.2,T18,'NUMERICAL',T35,'EXACT',/)
130   FORMAT( 'X = ',F4.1,T13,F13.6,T30,F13.6)
      END
      SUBROUTINE TRIDI
      COMMON/BLOCK1/A(51),B(51),C(51),D(51),L
C     TITLE:  TRIDIAGONAL LINEAR EQUATION SOLVER FOR
C             A SYSTEM WITH A NONZERO DETERMINANT
C     INPUT:  A, SUBDIAGONAL OF COEFFICIENT MATRIX
C             B, DIAGONAL OF COEFFICIENT MATRIX
C             C, SUPERDIAGONAL OF COEFFICIENT MATRIX
C             D, RIGHT HAND SIDE OF LINEAR SYSTEM
C             L, NUMBER OF LINEAR EQUATIONS
C     OUTPUT: SOLUTION OF LINEAR SYSTEM STORED IN D-ARRAY
C     FORWARD SUBSTITUTE TO ELIMINATE THE SUBDIAGONAL ELEMENTS
      DO 1 I = 2,L
        RT = -A(I)/B(I-1)
        B(I) = B(I) + RT*C(I-1)
        D(I) = D(I) + RT*D(I-1)
1     CONTINUE
C     BACK SUBSTITUTE AND STORE THE SOLUTION IN D-ARRAY
      D(L) = D(L)/B(L)
      DO 2 I = L-1,1,-1
        D(I) = (D(I) - C(I)*D(I+1))/B(I)
2     CONTINUE
      RETURN
      END
```

Fig. 9-3 (*Continued*)

```
W = 0.5 -- CRANK-NICOLSON METHOD
N =  10          K = 0.005000   TMAX = 0.50

T =  0.50          NUMERICAL         EXACT

X =  0.            0.                0.
X =  0.1           0.231190          0.222242
X =  0.2           0.439749          0.422730
X =  0.3           0.605262          0.581838
X =  0.4           0.711528          0.683991
X =  0.5           0.748145          0.719191
X =  0.6           0.711528          0.683991
X =  0.7           0.605262          0.581838
X =  0.8           0.439749          0.422730
X =  0.9           0.231189          0.222242
X =  1.0           0.                0.000000
```

```
W = 1.00 -- IMPLICIT METHOD
N =  10          K = 0.005000     TMAX = 0.50

T =  0.50          NUMERICAL           EXACT

X =  0.            0.                  0.
X =  0.1           0.259879            0.222242
X =  0.2           0.494319            0.422730
X =  0.3           0.680372            0.581838
X =  0.4           0.799825            0.683991
X =  0.5           0.840986            0.719191
X =  0.6           0.799825            0.683991
X =  0.7           0.680372            0.581838
X =  0.8           0.494319            0.422730
X =  0.9           0.259879            0.222242
X =  1.0           0.                  0.000000
```

Fig. 9-4

9.14 Write a computer program that uses the Peaceman–Rachford ADI method (*9.18*) to approximate the solution of

$$\begin{aligned} u_t &= u_{xx} + u_{yy} && 0<x,\ y<1,\ t>0 \\ u(x, y, 0) &= 100 \sin \pi x \sin \pi y && 0<x,\ y<1 \\ u(0, y, t) &= u(1, y, t) = 0 && 0<y<1,\ t>0 \\ u(x, 0, t) &= u(x, 1, t) = 0 && 0<x<1,\ t>0 \end{aligned}$$

Compare the numerical solution with the exact solution, $u = 100e^{-2\pi^2 t} \sin \pi x \sin \pi y$, at $t = 0.1$.

For a program, see Fig. 9-5. Though the symmetry of the solution was not exploited in constructing the program, the numerical results do display the expected symmetries. It therefore suffices to compare the numerical and exact solutions on $0<y\le x$, $0<x\le 1/2$. See Fig. 9-6.

```
      PROGRAM ADI
C     TITLE:  DEMO PROGRAM FOR ADI METHOD FOR
C             UT = KAPPA*(UXX + UYY)
C     INPUT:  MMAX & NMAX, NUMBER OF X & Y-SUBINTERVALS
C             K, TIME STEP
C             TMAX, MAXIMUM COMPUTATION TIME
C             KAPPA, DIFFUSIVITY VALUE
C             (X1,X2) & (Y1,Y2), X & Y-INTERVALS
C             P1(Y,T) & Q1(Y,T), LEFT & RIGHT BOUNDARY CONDITIONS
C             P2(X,T) & Q2(X,T),UPPER & LOWER BOUNDARY CONDITIONS
C             F(X,Y), INITIAL CONDITION
C             E(X,Y,T), EXACT SOLUTION
C     OUTPUT: NUMERICAL AND EXACT SOLUTION AT T=TMAX
      COMMON/BLOCK1/A(51),B(51),C(51),D(51),L
      COMMON/BLOCK2/U(0:51,0:51),V(0:51,0:51)
      REAL K,KAPPA
      DATA T,X1,X2,Y1,Y2,KAPPA/0,0,1,0,1,1/
      P1(Y,T) = 0
      Q1(Y,T) = 0
      P2(X,T) = 0
      Q2(X,T) = 0
      F(X,Y) = 100*SIN(PI*X)*SIN(PI*Y)
      E(X,Y,T) = 100*EXP(-2*PI*PI*T)*SIN(PI*X)*SIN(PI*Y)
      PRINT*,'ENTER TMAX AND TIME STEP'
      READ*,TMAX,K
      PRINT*,'ENTER NUMBER OF X-SUBINTERVALS, NUMBER OF Y-SUBINTERVALS'
      READ*,MMAX,NMAX
C     SET INITIAL CONDITION
      PI = 4*ATAN(1.)
      HX = (X2-X1)/MMAX
      HY = (Y2-Y1)/NMAX
      DO 10 M = 0,MMAX
      DO 10 N = 0,NMAX
         X = X1 + M*HX
         Y = Y1 + N*HY
         U(M,N) = 100*SIN(PI*X)*SIN(PI*Y)
10    CONTINUE
C     CALCULATE INTERMEDIATE VALUES SWEEPING VERTICALLY
      RX = KAPPA*K/HX/HX
15    DO 20 N = 1,NMAX-1
         Y = Y1 + N*HY
      DO 30 M = 1,MMAX-1
         A(M) = -.5*RX
         B(M) = 1 + RX
         C(M) = -.5*RX
30       D(M) = .5*RX*(U(M-1,N)+U(M+1,N)) + (1-RX)*U(M,N)
C     SOLVE TRIDIAGONAL SYSTEM FOR VALUES ON N-TH HORIZONTAL LINE
      L = MMAX -1
      CALL TRIDI
C     WRITE INTERMEDIATE VALUES INTO THE V-ARRAY
      DO 40 M = 1,MMAX-1
         V(M,N) = D(M)
40    CONTINUE
      V(0,N) = P1(Y,T)
      V(MMAX,N) = Q1(Y,T)
20    CONTINUE
C     CALCULATION OF INTERMEDIATE VALUES IS COMPLETE
C     BEGIN HORIZONTAL SWEEP TO COMPLETE THE TIME STEP
      RY = K/HY/HY
      DO 50 M = 1,MMAX-1
          X = X1 + M*HX
      DO 60 N = 1,NMAX-1
         A(N) = -.5*RY
         B(N) = 1 + RY
         C(N) = -.5*RY
         D(N) = .5*RY*(V(M,N-1)+V(M,N+1)) + (1-RY)*V(M,N)
60    CONTINUE
C     SOLVE TRIDIAGONAL SYSTEM FOR VALUES ON M-TH VERTICAL LINE
      L = NMAX - 1
      CALL TRIDI
```

Fig. 9-5 *(Program continues on next page)*

```
C       WRITE T+K VALUES INTO THE U-ARRAY
        DO 70 N = 1,NMAX-1
           U(M,N) = D(N)
70      CONTINUE
        U(M,0) = P2(X,T)
        U(M,NMAX) = Q2(X,T)
50      CONTINUE
C       TIME STEP IS COMPLETE
        T = T+K
C       IF T IS LESS THAN TMAX, TAKE ANOTHER TIME STEP
        IF(ABS(TMAX-T).GT.K/2) GOTO 15
C       IF T EQUALS TMAX, PRINT RESULT
        WRITE(6,100)
        WRITE(6,110) MMAX,NMAX,K,TMAX
        WRITE(6,120) T
        DO 80 M = 1,MMAX/2
        DO 80 N = 1,M
           X = X1 + M*HX
           Y = Y1 + N*HY
           EXACT = E(X,Y,T)
           WRITE(6,130) M,N,U(M,N),EXACT
80      CONTINUE
100     FORMAT(///,T9,'RESULTS FROM PROGRAM ADI',/)
110     FORMAT('MMAX=',I2,' NMAX=',I2,T18,'K = ',F5.2,T30,'TMAX =',F5.2,/)
120     FORMAT('T = ',F5.2,T18,'NUMERICAL',T35,'EXACT',/)
130     FORMAT( 'M,N = ',I1,',',I1,T13,F13.6,T30,F13.6)
        END
        SUBROUTINE TRIDI
        COMMON/BLOCK1/A(51),B(51),C(51),D(51),L
C       TITLE:  TRIDIAGONAL LINEAR EQUATION SOLVER FOR
C               A SYSTEM WITH A NONZERO DETERMINANT
C       INPUT:  A, SUBDIAGONAL OF COEFFICIENT MATRIX
C               B, DIAGONAL OF COEFFICIENT MATRIX
C               C, SUPERDIAGONAL OF COEFFICIENT MATRIX
C               D, RIGHT HAND SIDE OF LINEAR SYSTEM
C               L, NUMBER OF LINEAR EQUATIONS
C       OUTPUT: SOLUTION OF LINEAR SYSTEM STORED IN D-ARRAY
C       FORWARD SUBSTITUTE TO ELIMINATE THE SUBDIAGONAL ELEMENTS
        DO 1 I = 2,L
           RT = -A(I)/B(I-1)
           B(I) = B(I) + RT*C(I-1)
           D(I) = D(I) + RT*D(I-1)
1       CONTINUE
C       BACK SUBSTITUTE AND STORE THE SOLUTION IN D-ARRAY
        D(L) = D(L)/B(L)
        DO 2 I = L-1,1,-1
           D(I) = (D(I) - C(I)*D(I+1))/B(I)
2       CONTINUE
        RETURN
        END
```

Fig. 9-5 *(Continued)*

```
MMAX=10 NMAX=10  K =  0.01    TMAX = 0.10
T =  0.10        NUMERICAL       EXACT

M,N = 1,1         1.346013        1.326484
M,N = 2,1         2.560268        2.523122
M,N = 2,2         4.869920        4.799263
M,N = 3,1         3.523907        3.472779
M,N = 3,2         6.702869        6.605618
M,N = 3,3         9.225708        9.091854
M,N = 4,1         4.142601        4.082497
M,N = 4,2         7.879695        7.765370
M,N = 4,3        10.845469       10.688116
M,N = 4,4        12.749614       12.564634
M,N = 5,1         4.355789        4.292591
M,N = 5,2         8.285203        8.164993
M,N = 5,3        11.403603       11.238150
M,N = 5,4        13.405738       13.211238
M,N = 5,5        14.095628       13.891117
```

Fig. 9-6

9.15 For a parabolic initial–boundary value problem, the *method of lines* consists in discretizing only the spatial variables to obtain a system of ordinary differential equations in t. Illustrate the method of lines by applying it to

$$u_t = u_{xx} \qquad 0<x<1,\ t>0 \tag{1}$$
$$u(x,0) = f(x) \qquad 0<x<1 \tag{2}$$
$$u(0,t) = u(1,t) = 0 \qquad t>0 \tag{3}$$

Take $h = 0.25$.

Let $x_n = nh$ $(n = 0, 1, 2, 3, 4)$ and let $U_n(t)$ be an approximation to $u(x_n, t)$. In (*1*) approximate u_t by $U_n'(t)$ and u_{xx} by $\delta_x^2 U_n(t)$ to obtain the following system of ordinary differential equations.

$$\begin{aligned} U_1' &= \frac{1}{h^2}(-2U_1 + U_2) \\ U_2' &= \frac{1}{h^2}(U_1 - 2U_2 + U_3) \\ U_3' &= \frac{1}{h^2}(U_2 - 2U_3) \end{aligned} \tag{4}$$

If we look for solutions to the system (*4*) of the form $U_n = a_n e^{\lambda t}$, we are led to the following eigenvalue problem in λ:

$$h^{-2}\begin{bmatrix} -2 & 1 & 0 \\ 1 & -2 & 1 \\ 0 & 1 & -2 \end{bmatrix}\begin{bmatrix} a_1 \\ a_2 \\ a_3 \end{bmatrix} = \lambda \begin{bmatrix} a_1 \\ a_2 \\ a_3 \end{bmatrix} \tag{5}$$

By Problem 11.11, the eigenvalues λ of (*5*) are

$$\lambda_k = h^{-2}\left(-2 + 2\cos\frac{k\pi}{4}\right) \qquad (k = 1, 2, 3)$$

or $\lambda_1 = -8 + 4\sqrt{2}$, $\lambda_2 = -8$, $\lambda_3 = -8 - 4\sqrt{2}$, with corresponding eigenvectors

$$\mathbf{V}_1 = [\sqrt{2}, 2, \sqrt{2}]^T \qquad \mathbf{V}_2 = [1, 0, -1]^T \qquad \mathbf{V}_3 = [-\sqrt{2}, -1, \sqrt{2}]^T$$

The solution of (*4*) can be expressed as

$$[U_1(t), U_2(t), U_3(t)]^T = c_1\mathbf{V}_1 e^{\lambda_1 t} + c_2\mathbf{V}_2 e^{\lambda_2 t} + c_3\mathbf{V}_3 e^{\lambda_3 t} \tag{6}$$

and all that remains is to set $t = 0$ and $U_n(0) = f(x_n)$ in (*6*) to obtain three linear equations for c_1, c_2, c_3.

In practice, the number of x-nodes is usually much larger than five, and the PDE may have variable coefficients or may be nonlinear. In these circumstances it is desirable, or necessary, to obtain an approximate solution to the system of ordinary differential equations by a numerical method.

Supplementary Problems

9.16 In the centered-difference formula (*9.2*), suppose that the computed values of $u(x \pm h, t)$ are $\hat{u}(x \pm h, t)$ plus rounding errors of magnitude at most ϵ. Also, suppose that M is an upper bound for $u_{xxx}(x, t)$. (*a*) Show that

$$\left| u_x(x,t) - \frac{\hat{u}(x+h,t) - \hat{u}(x-h,t)}{2h} \right| \le \frac{\epsilon}{h} + \frac{h^2}{6}M$$

(*b*) What is the effect of rounding errors as $h \to 0$?

9.17 *Lagrange's interpolation formula* for three points gives

$$y = u_0\frac{(x-x_1)(x-x_2)}{(x_0-x_1)(x_0-x_2)} + u_1\frac{(x-x_0)(x-x_2)}{(x_1-x_0)(x_1-x_2)} + u_2\frac{(x-x_0)(x-x_1)}{(x_2-x_0)(x_2-x_1)}$$

as the quadratic function which assumes the values u_0, u_1, u_2 for the arguments x_0, x_1, x_2. Choosing $u_i = u(x_i, t)$ $(i = 0, 1, 2)$, use $y'(x_1)$ and y'' as finite-difference approximations of $u_x(x_1, t)$ and $u_{xx}(x_1, t)$, respectively. Verify that your formulas agree with those of Problem 9.2.

9.18 Prove that the eigenvalues μ of $\mathbf{C} \equiv (\mathbf{I}+\mathbf{B})^{-1}(\mathbf{I}-\mathbf{B})$ are given by

$$\mu = \frac{1-\lambda}{1+\lambda}$$

where λ is an eigenvalue of $\mathbf{B}$.

9.19 (*a*) Show that errors in the Crank–Nicolson method are governed by

$$(\mathbf{I}+\mathbf{B})\mathbf{E}_{j+1} = (\mathbf{I}-\mathbf{B})\mathbf{E}_j$$

where $\mathbf{B}$ is a symmetric, tridiagonal matrix with diagonal entries r and sub- and superdiagonal entries $-r/2$. (*b*) From (*a*), Problem 9.18, and Problem 9.8, infer the stability of the Crank–Nicolson method.

9.20 Show that the DuFort–Frankel method (*9.19*) is (von Neumann) stable. [*Hint*: Establish that

$$\xi = \frac{2r\cos\beta \pm \sqrt{1-(2r\sin\beta)^2}}{1+2r}$$

and consider the cases $r \le 1/2$, $r > 1/2$ and $|2r\sin\beta| \le 1$, $r > 1/2$ and $|2r\sin\beta| > 1$.]

9.21 Show that the Peaceman–Rachford ADI method (*9.18*) is (von Neumann) stable. [*Hint*: For separable solutions of the form $\xi^j e^{i(\beta m+\gamma n)}$, show that

$$\xi = \frac{1-2r\sin^2\dfrac{\beta}{2}}{1+2r\sin^2\dfrac{\beta}{2}} \cdot \frac{1-2r\sin^2\dfrac{\gamma}{2}}{1+2r\sin^2\dfrac{\gamma}{2}}$$

whence $|\xi| \le 1$.

9.22 (*a*) Exhibit in matrix form a backward-difference method for the problem

$$\begin{aligned} u_t &= u_{xx} - u && 0<x<1,\ t>0 \\ u(x,0) &= f(x) && 0<x<1 \\ u(0,t) &= u(1,t) = 0 && t>0 \end{aligned}$$

(*b*) Perform a matrix stability analysis, utilizing Problem 11.11 and the fact that the eigenvalues of $\mathbf{C}^{-1}$ are the reciprocals of the eigenvalues of $\mathbf{C}$.

9.23 For $j = 1, 2$, exhibit the solution U_{nj} to (*9.11*) that assumes the initial values $U_{n0} = 0$ $(n = \pm1, \pm2, \ldots)$, $U_{00} = 100$. Choose grids with (*a*) $r = 1/4$, (*b*) $r = 1$ (unstable).

9.24 Show that the explicit method (*9.11*), when applied to the problem

$$\begin{aligned} u_t &= a^2 u_{xx} && 0<x<1,\ t>0 \\ u(x,0) &= f(x) && 0<x<1 \\ u_x(0,t) &= u_x(1,t) = 0 && t>0 \end{aligned}$$

on the x-grid $0 = x_0 < x_1 < \cdots < x_N = 1$, has the matrix formulation

$$\begin{bmatrix} U_{0,j+1} \\ U_{1,j+1} \\ \cdots \\ U_{N-1,j+1} \\ U_{N,j+1} \end{bmatrix} = \begin{bmatrix} 1-2r & 2r & & & 0 \\ r & 1-2r & r & & \\ & \cdot & \cdot & \cdot & \\ & & r & 1-2r & r \\ 0 & & & 2r & 1-2r \end{bmatrix} \begin{bmatrix} U_{0j} \\ U_{1j} \\ \cdots \\ U_{N-1,j} \\ U_{Nj} \end{bmatrix}$$

9.25 In Problem 9.24, prove that

$$\int_0^1 u(x, t)\, dx = \int_0^1 f(x)\, dx$$

(conservation of diffused material between impermeable walls at $x = 0$ and $x = 1$) and that, correspondingly,

$$\sum_{n=0}^{N} U_{nj} = \sum_{n=0}^{N} U_{n0}$$

9.26 The *Gerschgorin Circle Theorem* states that if $\mathbf{A} = [a_{ij}]$ is an $N \times N$ matrix and C_i is the circle in the complex plane with center a_{ii} and radius

$$\sum_{\substack{j=1 \\ j \neq i}}^{N} |a_{ij}|$$

then all the eigenvalues of $\mathbf{A}$ are contained in the union of $C_1, C_2, \ldots, C_N$. Use this theorem to show that (*a*) the difference method of Problem 9.24 is stable provided $r \leq 1/2$, (*b*) the difference method of Problem 9.11 is unconditionally stable.

9.27 For the problem

$$\begin{aligned} u_t &= a^2 u_{xx} && 0 < x < \ell, t > 0 \\ u(x, 0) &= f(x) && 0 < x < \ell \\ \alpha u(0, t) + \beta u_x(0, t) &= p(t) && t > 0 \\ u(\ell, t) &= q(t) && t > 0 \end{aligned}$$

where α and $\beta \neq 0$ are constants, use the explicit method (*9.11*) and a ghost point, x_{-1}, to derive a difference system for U_{nj} $(n = 0, 1, \ldots, N-1; j = 0, 1, 2, \ldots)$.

9.28 For the problem

$$\begin{aligned} u_t + cu_x - a^2 u_{xx} &= 0 && 0 < x < 1, t > 0 \\ u(x, 0) &= 0 && 0 < x < 1 \\ u(0, t) = 1, \quad u(1, t) &= 0 && t > 0 \end{aligned}$$

use the backward-difference method (*9.12*), together with a centered difference for u_x, to formulate a difference system for U_{nj} $(n = 1, 2, \ldots, N-1; j = 0, 1, 2, \ldots)$.

Chapter 10

Difference Methods for Hyperbolic Equations

10.1 ONE-DIMENSIONAL WAVE EQUATION

Methods similar to those given in Section 9.4 may be used to approximate smooth solutions to

$$u_{tt} = c^2 u_{xx} \tag{10.1}$$

Let $(x_n, t_j) = (nh, jk)$ $(n, j = 0, 1, 2, \ldots)$ and write $s \equiv k/h$; we have as representatives of the two sorts of methods:

Explicit Method

$$\frac{U_{n,j+1} - 2U_{nj} + U_{n,j-1}}{k^2} = c^2 \frac{U_{n+1,j} - 2U_{nj} + U_{n-1,j}}{h^2}$$

or

$$\delta_t^2 U_{nj} = c^2 s^2 \delta_x^2 U_{nj} \tag{10.2}$$

Implicit Method

$$\delta_t^2 U_{nj} = c^2 s^2 \frac{\delta_x^2 U_{n,j+1} + \delta_x^2 U_{n,j-1}}{2}$$

or

$$-c^2 s^2 U_{n-1,j+1} + (2 + 2c^2 s^2) U_{n,j+1} - c^2 s^2 U_{n+1,j+1} = 4U_{nj} - 2U_{n,j-1} + c^2 s^2 \delta_x^2 U_{nj} \tag{10.3}$$

The local truncation errors given in Theorems 10.1 and 10.2 assume that u is four times continuously differentiable in x and t.

Theorem 10.1: The explicit method (*10.2*) has local truncation error $O(k^2 + h^2)$; it is stable if and only if $c^2 s^2 \leq 1$.

Theorem 10.2: The implicit method (*10.3*) has local truncation error $O(k^2 + h^2)$; it is stable.

In Problem 10.1 it is shown how initial conditions are used to evaluate U_{n0} and U_{n1}, which are needed to start the calculations in either method. The sorts of problems to which the two methods properly apply are as in the parabolic case, assuming smooth solutions. For problems to which the solution is not smooth, the method of characteristics usually provides a more accurate numerical solution (see Section 10.2).

The stability condition of Theorem 10.1 is often referred to as a *Courant–Friedrichs–Lewy* (CFL) *condition.* The CFL condition for the stability of an explicit finite-difference method is *that the numerical domain of dependence must contain the analytical domain of dependence.* Thus, for (*10.2*) to be stable, the backward characteristics through (x_n, t_{j+1}) must pass between (x_{n-1}, t_j) and (x_{n+1}, t_j).

10.2 NUMERICAL METHOD OF CHARACTERISTICS FOR A SECOND-ORDER PDE

The Cauchy problem for the quasilinear hyperbolic equation

$$au_{xx} + 2bu_{xy} + cu_{yy} = f \qquad (b^2 > ac) \tag{10.4}$$

wherein u, u_x, and u_y are prescribed along some initial curve Γ that is nowhere tangent to a characteristic, becomes a Cauchy problem for a first-order system of the same type when u_x and u_y are taken as new dependent variables. Writing

$$\lambda_+ \equiv \frac{b + \sqrt{b^2 - ac}}{a} \qquad \lambda_- \equiv \frac{b - \sqrt{b^2 - ac}}{a}$$

we obtain from the theory of Chapter 5 (see also Section 10.4) the following basic results for *(10.4)*:

Theorem 10.3: The level curves of the surfaces $z = F(x, y)$ and $z = G(x, y)$ are respectively the α- and β-characteristics of *(10.4)* if

$$\frac{dy}{dx} = -\frac{F_x}{F_y} = \lambda_+ \quad \text{along} \quad F(x, y) = \beta \tag{10.5}$$

$$\frac{dy}{dx} = -\frac{G_x}{G_y} = \lambda_- \quad \text{along} \quad G(x, y) = \alpha \tag{10.6}$$

The introduction of α and β as new coordinates in the vicinity of Γ leads to the replacement of *(10.4)* by the system of characteristic equations

$$y_\alpha = \lambda_+ x_\alpha \tag{10.7}$$

$$y_\beta = \lambda_- x_\beta \tag{10.8}$$

$$\lambda_+ a(u_x)_\alpha + c(u_y)_\alpha = f y_\alpha \tag{10.9}$$

$$\lambda_- a(u_x)_\beta + c(u_y)_\beta = f y_\beta \tag{10.10}$$

$$u_\alpha = u_x x_\alpha + u_y y_\alpha \tag{10.11a}$$

$$u_\beta = u_x x_\beta + u_y y_\beta \tag{10.11b}$$

The *numerical method of characteristics* begins with the selection of grid points P_i on Γ (Fig. 10-1); $u(P_i)$, $u_x(P_i)$, and $u_y(P_i)$ are therefore known. Next, all α- and β-derivatives in *(10.7)*–*(10.11)* are replaced by difference quotients; e.g.,

$$(u_x)_\alpha \approx \frac{u_x(Q_i) - u_x(P_i)}{\Delta\alpha} \qquad y_\beta \approx \frac{y(P_{i+1}) - y(Q_i)}{\Delta\beta}$$

The result, after cancellation of $\Delta\alpha$ and $\Delta\beta$, is a system of five algebraic equations in the five unknowns $x(Q_i)$, $y(Q_i)$, $u(Q_i)$, $u_x(Q_i)$, and $u_y(Q_i)$. In general, the system is nonlinear and must be solved by an iterative technique (see Problem 10.5). With the new grid points Q_i located and with new starting values at hand, the transition can be made to the R_i; and so forth.

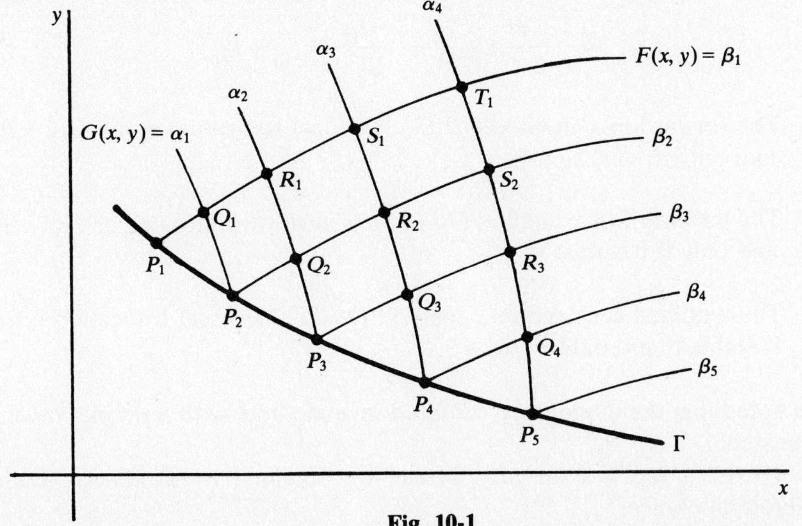

Fig. 10-1

In the event that a, b, and c in (*10.4*) are independent of u, an a priori integration of the ordinary differential equations (*10.5*) and (*10.6*) may be possible, yielding the characteristic curves and their points of intersection, the grid points. In that case, (*10.7*) and (*10.8*) may be dropped from the numerical algorithm.

Only one of the equations (*10.11*) need figure in the numerical method of characteristics. In the case of a pure initial value problem, it is a good idea to check the solution obtained when (*10.11a*) is used against that involving (*10.11b*). For an initial–boundary value problem, in calculations adjacent to a boundary, the choice of equations (*10.11*) is dictated by the relative orientations of the boundary and the α- and β-characteristics. See Problem 10.20.

10.3 FIRST-ORDER EQUATIONS

We start with the simple equation

$$au_x + u_t = f \tag{10.12}$$

where a is a constant and $f = f(x, t)$, because difference methods for (*10.12*) carry over directly to a hyperbolic system of m linear first-order equations in m functions of x and t. Letting $(x_n, t_j) = (nh, jk)$ and $s \equiv k/h$, we have

Explicit (Forward-in-x) Method

$$a\frac{U_{n+1,j} - U_{nj}}{h} + \frac{U_{n,j+1} - U_{nj}}{k} = f_{nj}$$

or

$$U_{n,j+1} = (1 + sa)U_{nj} - sa\,U_{n+1,j} + kf_{nj} \tag{10.13}$$

Explicit (Backward-in-x) Method

$$a\frac{U_{nj} - U_{n-1,j}}{h} + \frac{U_{n,j+1} - U_{nj}}{k} = f_{nj}$$

or

$$U_{n,j+1} = saU_{n-1,j} + (1 - sa)U_{nj} + kf_{nj} \tag{10.14}$$

Explicit (Modified Centered-in-x) Method

$$a\frac{U_{n+1,j} - U_{n-1,j}}{2h} + \frac{U_{n,j+1} - \frac{1}{2}(U_{n+1,j} + U_{n-1,j})}{k} = f_{nj}$$

or

$$U_{n,j+1} = \frac{1 + sa}{2}U_{n-1,j} + \frac{1 - sa}{2}U_{n+1,j} + kf_{nj} \tag{10.15}$$

Theorem 10.4: The forward-in-x method (*10.13*) has local truncation error $O(k + h)$; it is stable if and only if $-1 \le sa \le 0$.

Theorem 10.5: The backward-in-x method (*10.14*) has local truncation error $O(k + h)$; it is stable if and only if $0 \le sa \le 1$.

Theorem 10.6: The modified centered-in-x method (*10.15*) has local truncation error $O(k + h^2)$; it is stable if and only if $|sa| \le 1$.

It should be noted that the unmodified centered-in-x method, with a simple forward difference in time, is always unstable.

A three-level explicit method for (*10.12*) can be obtained by estimating both $u_x(x_n, t_j)$ and $u_t(x_n, t_j)$ by centered differences:

Leapfrog Method

$$a\frac{U_{n+1,j}-U_{n-1,j}}{2h}+\frac{U_{n,j+1}-U_{n,j-1}}{2k}=f_{nj}$$

or

$$U_{n,j+1}=U_{n,j-1}-saU_{n-1,j}-saU_{n+1,j}+2kf_{nj} \tag{10.16}$$

Theorem 10.7: The leapfrog method (*10.16*) has local truncation error $O(k^2+h^2)$; it is stable if and only if $|sa|\leq 1$.

A two-level implicit method results from approximating both partial derivatives as an average of forward differences:

Wendroff's Implicit Method

$$a\frac{(U_{n+1,j}-U_{nj})+(U_{n+1,j+1}-U_{n,j+1})}{2h}+\frac{(U_{n,j+1}-U_{nj})+(U_{n+1,j+1}-U_{n+1,j})}{2k}=f(x_n+(h/2),\,t_j+(k/2))$$

or

$$(1+sa)U_{n+1,j+1}+(1-sa)U_{n,j+1}-(1-sa)U_{n+1,j}-(1+sa)U_{nj}=2kf_{n+(1/2),j+(1/2)} \tag{10.17}$$

Theorem 10.8: Wendroff's implicit method (*10.17*) has local truncation error $O(k^2+h^2)$; it is stable.

Wendroff's implicit method cannot be applied to a pure initial-value problem. However, for an initial–boundary value problem, (*10.17*) can be used in an explicit manner (see Problem 10.8). Each of the methods (*10.13*)–(*10.17*) can be modified to apply to the general quasilinear first-order PDE in two independent variables.

As is shown in Problem 10.4, the scalar conservation-law equation

$$[F(u)]_x+u_t=0 \tag{10.18}$$

admits the

Lax–Wendroff Method (Scalar)

$$U_{n,j+1}=U_{nj}-\frac{s}{2}(F_{n+1,j}-F_{n-1,j})+\frac{s^2}{4}[(F'_{n+1,j}+F'_{nj})(F_{n+1,j}-F_{nj})-(F'_{nj}+F'_{n-1,j})(F_{nj}-F_{n-1,j})] \tag{10.19}$$

Here, $F_{nj}\equiv F(U_{nj})$, $F'_{nj}\equiv F'(U_{nj})$.

Theorem 10.9: The Lax–Wendroff method (*10.19*) has local truncation error $O(k^2+h^2)$.

The stability criterion for (*10.19*) will depend on the function F; if $F(u)=au$ ($a=\text{const.}$), the method is stable if and only if $|sa|\leq 1$.

To avoid the calculation of F', several two-step modifications of (*10.19*) have been devised. For example,

$$U^*_{nj}=\frac{1}{2}(U_{n+1,j}+U_{nj})-\frac{s}{2}(F_{n+1,j}-F_{nj})$$
$$U_{n,j+1}=U_{nj}-s(F^*_{nj}-F^*_{n-1,j}) \tag{10.20}$$

has the same local truncation error, $O(k^2+h^2)$, as (*10.19*) and reduces to (*10.19*) in the linear case $F(u)=au$.

Next, consider the hyperbolic system of M linear first-order equations

$$\mathbf{A}\mathbf{u}_x+\mathbf{u}_t=\mathbf{f} \tag{10.21}$$

where $\mathbf{A}$ is a constant $M\times M$ matrix and

$$\mathbf{u}=[u_1(x,t),u_2(x,t),\ldots,u_M(x,t)]^T \qquad \mathbf{f}=[f_1(x,t),f_2(x,t),\ldots,f_M(x,t)]^T$$

By Section 5.2, all M eigenvalues of $\mathbf{A}$ are real. The scalar numerical methods (*10.13*)–(*10.17*) become vector numerical methods for *(10.21)* when a is replaced by $\mathbf{A}$ (and $1 \pm sa$ by $\mathbf{I} \pm s\mathbf{A}$), and U_{nj} and f_{nj} are replaced by

$$\mathbf{U}_{nj} = [U_{1,nj}, U_{2,nj}, \ldots, U_{M,nj}]^T \qquad \mathbf{f}_{nj} = [f_{1,nj}, f_{2,nj}, \ldots, f_{M,nj}]^T \tag{10.22}$$

Theorems 10.4–10.8 hold for these vector methods if, in the statements, a is replaced by λ, any eigenvalue of $\mathbf{A}$.

Similarly, the Lax–Wendroff method may be extended to handle the conservation-law system

$$\frac{\partial}{\partial x} F_1(u_1, u_2, \ldots, u_M) + \frac{\partial u_1}{\partial t} = 0$$

$$\frac{\partial}{\partial x} F_2(u_1, u_2, \ldots, u_M) + \frac{\partial u_2}{\partial t} = 0$$

$$\cdots\cdots\cdots\cdots\cdots\cdots\cdots\cdots$$

$$\frac{\partial}{\partial x} F_M(u_1, u_2, \ldots, u_M) + \frac{\partial u_M}{\partial t} = 0$$

i.e., the vector conservation-law equation

$$[\mathbf{F}(\mathbf{u})]_x + \mathbf{u}_t = \mathbf{0} \tag{10.23}$$

Define the vectors $\mathbf{U}_{nj}$ as in *(10.22)* and write

$$\mathbf{F}_{nj} \equiv \begin{bmatrix} F_1(U_{1,nj}, U_{2,nj}, \ldots, U_{M,nj}) \\ F_2(U_{1,nj}, U_{2,nj}, \ldots, U_{M,nj}) \\ \cdots \\ F_M(U_{1,nj}, U_{2,nj}, \ldots, U_{M,nj}) \end{bmatrix}$$

Let $\mathbf{J}(u_1, u_2, \ldots, u_M) \equiv [\partial F_p / \partial u_q]$ $(p, q = 1, 2, \ldots, M)$ be the Jacobian matrix of the functions $F_1, \ldots, F_M$, and write

$$\mathbf{J}_{nj} \equiv \mathbf{J}(U_{1,nj}, U_{2,nj}, \ldots, U_{M,nj})$$

Then, for *(10.23)*, we have the

Lax–Wendroff Method (Vector)

$$\mathbf{U}_{n,j+1} = \mathbf{U}_{nj} - \frac{s}{2}(\mathbf{F}_{n+1,j} - \mathbf{F}_{n-1,j}) + \frac{s^2}{4}[(\mathbf{J}_{n+1,j} + \mathbf{J}_{nj})(\mathbf{F}_{n+1,j} - \mathbf{F}_{nj}) - (\mathbf{J}_{nj} + \mathbf{J}_{n-1,j})(\mathbf{F}_{nj} - \mathbf{F}_{n-1,j})] \tag{10.24}$$

The vector version of the two-step modification *(10.20)* avoids calculation of the Jacobian matrix.

EXAMPLE 10.1 For $\mathbf{f} = \mathbf{0}$, the linear system *(10.21)* becomes the special case $\mathbf{F}(\mathbf{u}) = \mathbf{A}\mathbf{u}$ of *(10.23)*. In this case,

$$\mathbf{F}_{nj} = \mathbf{A}\mathbf{U}_{nj} \qquad \mathbf{J}_{nj} = \mathbf{A} = \text{const.}$$

and we obtain for the homogeneous *(10.21)* the *linear Lax–Wendroff method*

$$\mathbf{U}_{n,j+1} = \mathbf{U}_{nj} - \frac{s}{2}\mathbf{A}(\mathbf{U}_{n+1,j} - \mathbf{U}_{n-1,j}) + \frac{s^2}{2}\mathbf{A}^2(\mathbf{U}_{n+1,j} - 2\mathbf{U}_{nj} + \mathbf{U}_{n-1,j})$$

$$= \frac{s}{2}\mathbf{A}(s\mathbf{A} - \mathbf{I})\mathbf{U}_{n+1,j} - (s\mathbf{A} - \mathbf{I})(s\mathbf{A} + \mathbf{I})\mathbf{U}_{nj} + \frac{s}{2}\mathbf{A}(s\mathbf{A} + \mathbf{I})U_{n-1,j} \tag{10.25}$$

Problem 10.10 treats the stability of *(10.25)*.

The difference methods presented above, like those of Section 10.1, work best if the exact solution is smooth. If discontinuities are present, greater accuracy will be furnished by a numerical method of characteristics.

10.4 NUMERICAL METHOD OF CHARACTERISTICS FOR FIRST-ORDER SYSTEMS

The numerical method of characteristics is applicable to initial value problems for either a single first-order equation or a hyperbolic system of first-order equations.

First, consider the quasilinear PDE

$$au_x + bu_y = c \qquad (b \neq 0) \tag{10.26}$$

and suppose that u is given on the noncharacteristic initial curve Γ. Let Q be any fixed point on Γ and $\mathscr{C}$ be the characteristic of (*10.26*) passing through Q [Fig. 10-2(*a*)]. By (*4b*) of Problem 5.3,

$$b\,dx - a\,dy = 0 \qquad \text{and} \qquad b\,du - c\,dy = 0$$

along $\mathscr{C}$. Approximating dx, dy, and du by $x(P)-x(Q)$, $y(P)-y(Q)$, and $u(P)-u(Q)$, we obtain a pair of algebraic equations,

$$b[x(P)-x(Q)] - a[y(P)-y(Q)] = 0 \tag{10.27}$$

$$b[u(P)-u(Q)] - c[y(P)-y(Q)] = 0 \tag{10.28}$$

After one of the coordinates, $x(P)$ or $y(P)$, of P has been selected, this system determines the other coordinate of P and the value of u at P. The system (*10.27*)–(*10.28*) is linear only if (*10.26*) is linear with constant coefficients.

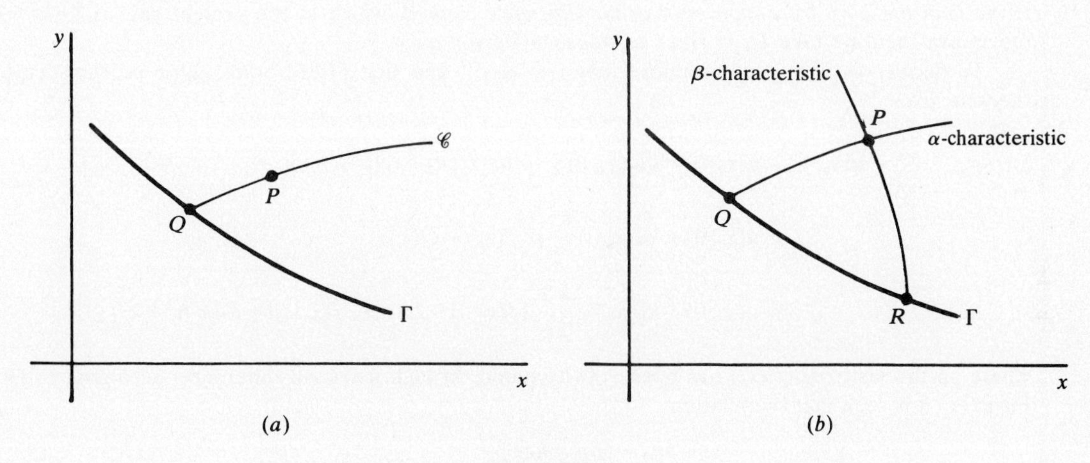

Fig. 10-2

Next, consider the 2×2 quasilinear hyperbolic system

$$\mathbf{A}\mathbf{u}_x + \mathbf{B}\mathbf{u}_y = \mathbf{c} \qquad (\det(\mathbf{B}) \neq 0) \tag{10.29}$$

with $\mathbf{u} = [u, v]^T$ given on Γ. Using the theory of Chapter 5 (see especially Problem 5.12) to transform from the variables x, y to characteristic coordinates α, β, we obtain the canonical equations for (*10.29*):

$$\begin{aligned} x_\alpha &= \lambda_1 y_\alpha \\ x_\beta &= \lambda_2 y_\beta \\ b_{11}^* u_\alpha + b_{12}^* v_\alpha &= c_1^* y_\alpha \\ b_{21} u_\beta + b_{22}^* v_\beta &= c_2^* y_\beta \end{aligned}$$

where λ_1 and λ_2 are the (by assumption, real) zeros of $\det(\mathbf{A} - \lambda\mathbf{B})$ and where the starred coefficients are known functions of x, y, u, v. Replacing the α- and β-derivatives by difference quotients (per Fig. 10-2(*b*)) yields the following numerical method:

$$x(P)-x(Q)=\lambda_1[y(P)-y(Q)] \tag{10.30}$$

$$x(P)-x(R)=\lambda_2[y(P)-y(R)] \tag{10.31}$$

$$b_{11}^*[u(P)-u(Q)]+b_{12}^*[v(P)-v(Q)]=c_1^*[y(P)-y(Q)] \tag{10.32}$$

$$b_{21}^*[u(P)-u(R)]+b_{22}^*[v(P)-v(R)]=c_2^*[y(P)-y(R)] \tag{10.33}$$

In general, the algebraic system (*10.30*)–(*10.33*) must be solved for the unknown $x(P)$, $y(P)$, $u(P)$, and $v(P)$ by an iterative procedure.

Solved Problems

10.1 Given the initial conditions $u(x,0)=f(x)$ and $u_t(x,0)=g(x)$ for the wave equation (*10.1*), show how to obtain starting values U_{n0} and U_{n1} for the difference methods (*10.2*) and (*10.3*).

The guiding principle here is that *the starting values should represent the initial data with an error no worse than the local truncation error of the difference method,* which in the present case is $O(k^2+h^2)$. Obviously, then, we take $U_{n0}=f(x_n)$, as this incurs error zero.

To decide on U_{n1}, let us suppose that f is in C^2 and that (*10.1*) holds at $t=0$. Then Taylor's theorem gives

$$\begin{aligned} u(x_n,t_1) &= u(x_n,0)+ku_t(x_n,0)+\frac{k^2}{2}u_{tt}(x_n,0)+O(k^3) \\ &= u(x_n,0)+kg(x_n)+\frac{k^2}{2}c^2f''(x_n)+O(k^3) \\ &= u(x_n,0)+kg(x_n)+\frac{k^2c^2}{2h^2}[f(x_{n-1})-2f(x_n)+f(x_{n+1})]+O(k^2h^2+k^3) \end{aligned} \tag{1}$$

where, in the last step, $f''(x_n)$ has been approximated through a second difference, according to (*9.3*). From (*1*) it is seen that the relation

$$g(x_n)=\frac{U_{n1}-U_{n0}}{k}-\frac{kc^2}{2h^2}[f(x_{n-1})-2f(x_n)+f(x_{n+1})] \tag{2}$$

is satisfied by the exact solution u to within $O(kh^2+k^2)$; i.e., letting (*2*) determine U_{n1} results in an error of higher order than $O(k^2+h^2)$.

10.2 Write a computer program that uses the explicit method (*10.2*), with starting values as in Problem 10.1, to approximate the solution to the initial–boundary value problem

$$\begin{aligned} u_{tt}-4u_{xx} &= 0 && 0<x<1,\ t>0 \\ u(x,0) &= \sin 2\pi x && 0<x<1 \\ u_t(x,0) &= 0 && 0<x<1 \\ u(0,t)=u(1,t) &= 0 && t>0 \end{aligned}$$

At $t=1$ compare the numerical results with the exact solution, $u=\cos 4\pi t \sin 2\pi x$.

Figure 10-3 gives a program listing, and Fig. 10-4 shows two runs, one stable and one unstable.

```
      PROGRAM EWAVE
C     TITLE:  DEMO PROGRAM FOR EXPLICIT METHOD
C             FOR WAVE EQUATION, UTT = C*C*UXX
C     INPUT:  N, NUMBER OF X-SUBINTERVALS
C             K, TIME STEP
C             TMAX, MAXIMUM COMPUTATION TIME
C             [X1,X2], X-INTERVAL
C             P1(T), LEFT BOUNDARY CONDITION
C             P2(T), RIGHT BOUNDARY CONDITION
C             F(X), INITIAL CONDITION ON U
C             G(X), INITIAL CONDITION ON UT
C             E(X,T), EXACT SOLUTION
C     OUTPUT: NUMERICAL AND EXACT SOLUTION AT T=TMAX
      COMMON U(0:51),V(0:51),W(0:51)
      REAL K
      DATA T,X1,X2,C/0,0,1,2/
      PI = 4*ATAN(1.)
      P1(T) = 0
      P2(T) = 0
      F(X) = SIN(2*PI*X)
      G(X) = 0
      E(X,T) = COS(4*PI*T)*SIN(2*PI*X)
      PRINT*,'ENTER TMAX,NUMBER OF X-SUBINTERVALS AND TIME STEP'
      READ*,TMAX,N,K
      H = (X2-X1)/N
      S = K/H
      Q = C*C*S*S
C     SET T = 0 VALUES
      DO 10 I = 0,N
         X = X1 + I*H
         W(I) = F(X)
10    CONTINUE
C     SET T = K VALUES
      T = K
      DO 20 I = 1,N-1
         X = X1 + I*H
         V(I) = W(I) + K*G(X) + .5*Q*(W(I+1) -2*W(I) + W(I-1))
20    CONTINUE
      V(0) = P1(T)
      V(N) = P2(T)
C     ADVANCE SOLUTION TO TIME T+K
15    DO 30 I = 1,N-1
         U(I) = 2*V(I) - W(I) + Q*(V(I+1) -2*V(I) + V(I-1))
30    CONTINUE
      T = T + K
      U(0) = P1(T)
      U(N) = P2(T)
C     WRITE V OVER W AND U OVER V TO PREPARE FOR NEXT TIME STEP
      DO 40 I = 0,N
         W(I) = V(I)
         V(I) = U(I)
40    CONTINUE
C     IF T IS LESS THAN TMAX, TAKE A TIME STEP
      IF(ABS(TMAX-T).GT.K/2) GOTO 15
C     OTHERWISE, PRINT RESULT
      WRITE(6,100)
      WRITE(6,110) N,K,TMAX
      WRITE(6,120) T
      ISTEP = .1/H
      DO 50 I = 0,N,ISTEP
         X = X1 + I*H
         EXACT = E(X,T)
         WRITE(6,130) X,U(I),EXACT
50    CONTINUE
100   FORMAT(///,T9,'RESULTS FROM PROGRAM EWAVE',/)
110   FORMAT('N =',I4,T15,'K = ',F8.6,T30,'C*C*S*S =',F5.2,/)
120   FORMAT('T = ',F5.2,T18,'NUMERICAL',T35,'EXACT',/)
130   FORMAT( 'X = ',F4.1,T13,F13.6,T30,F13.6)
      END
```

Fig. 10-3

```
N =  10        K = 0.100000   C*C*S*S = 1.00

T =  1.00          NUMERICAL          EXACT

X =  0.            0.                 0.
X =  0.1        -908.209473           0.587785
X =  0.2        1556.156250           0.951056
X =  0.3       -1734.383179           0.951056
X =  0.4        1350.902466           0.587785
X =  0.5        -465.308990           0.000000
X =  0.6        -654.874756          -0.587785
X =  0.7        1566.169800          -0.951056
X =  0.8       -1818.750488          -0.951057
X =  0.9        1211.699341          -0.587785
X =  1.0           0.                -0.000000

N =  20        K = 0.010000   C*C*S*S = 1.00

T =  1.00          NUMERICAL          EXACT

X =  0.            0.                 0.
X =  0.1           0.587232           0.587785
X =  0.2           0.950161           0.951056
X =  0.3           0.950161           0.951056
X =  0.4           0.587232           0.587785
X =  0.5           0.000000           0.000000
X =  0.6          -0.587232          -0.587785
X =  0.7          -0.950161          -0.951056
X =  0.8          -0.950161          -0.951057
X =  0.9          -0.587231          -0.587785
X =  1.0           0.                -0.000000
```

Fig. 10-4

10.3 Rework Problem 10.2 using the implicit method (*10.3*).

See Fig. 10-5 for a program listing, and Fig. 10-6 for a (stable) run.

```
      PROGRAM IWAVE
C     TITLE:  DEMO PROGRAM FOR IMPLICT METHOD
C             FOR UTT = C1*C1*UXX
C     INPUT:  N, NUMBER OF X-SUBINTERVALS
C             K, TIME STEP
C             TMAX, MAXIMUM COMPUTATION TIME
C             C1, CELERITY VALUE
C             (X1,X2), X-INTERVAL
C             P1(T), LEFT BOUNDARY CONDITION
C             P2(T), RIGHT BOUNDARY CONDITION
C             F(X), INITIAL CONDITION FOR U
C             G(X), INITIAL CONDITION FOR UT
C             E(X,T), EXACT SOLUTION
C     OUTPUT: NUMERICAL AND EXACT SOLUTION AT T=TMAX
      COMMON/BLOCK1/A(51),B(51),C(51),D(51),L
      COMMON/BLOCK2/U(0:51),V(0:51),W(0:51)
      REAL K
      DATA T,X1,X2,C1/0,0,1,2/
      PI = 4*ATAN(1.)
      P1(T) = 0
      P2(T) = 0
      F(X) = SIN(2*PI*X)
      G(X) = 0
      E(X,T) = COS(4*PI*T)*SIN(2*PI*X)
      PRINT*,'ENTER TMAX,NUMBER OF X-SUBINTERVALS AND TIME STEP'
      READ*, TMAX,N,K
      H = (X2-X1)/N
      S = K/H
      P = C1*C1*S*S
C     SET T = 0 VALUES
      DO 10 I = 0,N
        X = X1 + I*H
        W(I) = F(X)
10    CONTINUE
```

Fig. 10-5 (*Program continues on next page*)

```
C       SET T = K VALUES
        T = K
        DO 20 I = 1,N-1
           X = X1 + I*H
           V(I) = W(I) + K*G(X) + .5*P*(W(I+1) - 2*W(I) + W(I-1))
20      CONTINUE
        V(0) = P1(T)
        V(N) = P2(T)
C       DEFINE TRIDIAGONAL LINEAR SYSTEM
        L = N-1
15      DO 30 I = 1,L
           A(I) = -P
           B(I) = 2 + 2*P
           C(I) = -P
           D(I) = 4*V(I) - 2*W(I) + P*(W(I-1) - 2*W(I) + W(I+1))
30      CONTINUE
C       CALL TRIDIAGONAL LINEAR EQUATION SOLVER
        CALL TRIDI
C       WRITE SOLUTION AT TIME T+K INTO THE U-ARRAY
        DO 40 I = 1,N-1
           U(I) = D(I)
40      CONTINUE
        T = T + K
        U(0) = P1(T)
        U(N) = P2(T)
C       WRITE V OVER W AND U OVER V TO PREPARE FOR NEXT TIME STEP
        DO 50 I = 0,N
           W(I) = V(I)
           V(I) = U(I)
50      CONTINUE
C       IF T IS LESS THAN TMAX, TAKE A TIME STEP
        IF(ABS(TMAX-T).GT.K/2) GOTO 15
C       OTHERWISE, PRINT RESULT
        WRITE(6,100)
        WRITE(6,110) N,K,P
        WRITE(6,120) T
        ISTEP = .1/H
        DO 60 I = 0,N,ISTEP
           X = X1 + I*H
           EXACT = E(X,T)
           WRITE(6,130) X,U(I),EXACT
60      CONTINUE
100     FORMAT(///,T9,'RESULTS FROM PROGRAM IWAVE',/)
110     FORMAT('N =',I4,T15,'K = ',F8.6,T30,'C1*C1*S*S =',F5.2,/)
120     FORMAT('T = ',F5.2,T18,'NUMERICAL',T35,'EXACT',/)
130     FORMAT( 'X = ',F4.1,T13,F13.6,T30,F13.6)
        END
        SUBROUTINE TRIDI
        COMMON/BLOCK1/A(51),B(51),C(51),D(51),L
C       TITLE:  TRIDIAGONAL LINEAR EQUATION SOLVER FOR
C               A SYSTEM WITH A NONZERO DETERMINANT
C       INPUT:  A, SUBDIAGONAL OF COEFFICIENT MATRIX
C               B, DIAGONAL OF COEFFICIENT MATRIX
C               C, SUPERDIAGONAL OF COEFFICIENT MATRIX
C               D, RIGHT HAND SIDE OF LINEAR SYSTEM
C               L, NUMBER OF LINEAR EQUATIONS
C       OUTPUT: SOLUTION OF LINEAR SYSTEM STORED IN D-ARRAY
C       FORWARD SUBSTITUTE TO ELIMINATE THE SUBDIAGONAL ELEMENTS
        DO 1 I = 2,L
           RT = -A(I)/B(I-1)
           B(I) = B(I) + RT*C(I-1)
           D(I) = D(I) + RT*D(I-1)
1       CONTINUE
C       BACK SUBSTITUTE AND STORE THE SOLUTION IN D-ARRAY
        D(L) = D(L)/B(L)
        DO 2 I = L-1,1,-1
           D(I) = (D(I) - C(I)*D(I+1))/B(I)
2       CONTINUE
        RETURN
        END
```

Fig. 10-5 *(Continued)*

N = 20	K = 0.010000	C*C*S*S = 0.16
T = 1.00	NUMERICAL	EXACT
X = 0.	0.	0.
X = 0.1	0.585313	0.587785
X = 0.2	0.947056	0.951056
X = 0.3	0.947057	0.951056
X = 0.4	0.585313	0.587785
X = 0.5	-0.000001	0.000000
X = 0.6	-0.585314	-0.587785
X = 0.7	-0.947057	-0.951056
X = 0.8	-0.947058	-0.951057
X = 0.9	-0.585314	-0.587785
X = 1.0	0.	-0.000000

N = 20	K = 0.050000	C*C*S*S = 4.00
T = 1.00	NUMERICAL	EXACT
X = 0.	0.	0.
X = 0.1	0.365416	0.587785
X = 0.2	0.591255	0.951056
X = 0.3	0.591255	0.951056
X = 0.4	0.365415	0.587785
X = 0.5	0.000000	0.000000
X = 0.6	-0.365415	-0.587785
X = 0.7	-0.591254	-0.951056
X = 0.8	-0.591254	-0.951057
X = 0.9	-0.365415	-0.587785
X = 1.0	0.	-0.000000

Fig. 10-6

10.4 Derive the Lax–Wendroff method *(10.19)*.

A Taylor expansion in t gives

$$u(x_n, t_{j+1}) = u(x_n, t_j) + ku_t(x_n, t_j) + \frac{k^2}{2} u_{tt}(x_n, t_j) + \cdots \tag{1}$$

By *(10.18)*, $u_t = -[F(u)]_x$, and so, using a centered x-difference,

$$u_t(x_n, t_j) \approx -\frac{F(u_{n+1,j}) - F(u_{n-1,j})}{2h} \tag{2}$$

Furthermore, $u_{tt} = [F'(u)\,[F(u)]_x]_x$. Now, the usual centered second difference is the forward difference of a backward difference; that is,

$$\delta^2\phi_n = (\phi_{n+1} - \phi_n) - (\phi_n - \phi_{n-1})$$

Hence we approximate the "inside" x-derivative above as

$$[F(u)]_x \approx \frac{F(u_{nj}) - F(u_{n-1,j})}{h}$$

and represent its multiplier by a mean value:

$$F'(u) \approx \frac{F'(u_{nj}) + F'(u_{n-1,j})}{2}$$

The forward differencing corresponding to the "outside" derivative then gives

$$u_{tt}(x_n, t_j) \approx \frac{1}{h}\left[\frac{F'(u_{n+1,j}) + F'(u_{nj})}{2}\,\frac{F(u_{n+1,j}) - F(u_{nj})}{h} - \frac{F'(u_{nj}) + F'(u_{n-1,j})}{2}\,\frac{F(u_{nj}) - F(u_{n-1,j})}{h}\right] \tag{3}$$

Substitution of *(2)* and *(3)* in *(1)*, and replacement of u by U, yields *(10.19)*.

10.5 In terms of Fig. 10-2(*b*), the difference equations corresponding to *(10.7)*–*(10.11a)* are

$$y(P) - y(Q) = \lambda_+[x(P) - x(Q)] \tag{1}$$

$$y(P) - y(R) = \lambda_-[x(P) - x(R)] \tag{2}$$

$$\lambda_+ a[u_x(P) - u_x(Q)] + c[u_y(P) - u_y(Q)] = f[y(P) - y(Q)] \tag{3}$$

$$\lambda_- a[u_x(P) - u_x(R)] + c[u_y(P) - u_y(R)] = f[y(P) - y(R)] \tag{4}$$

$$u(P) - u(Q) = \frac{u_x(P) + u_x(Q)}{2}[x(P) - x(Q)] + \frac{u_y(P) + u_y(Q)}{2}[y(P) - y(Q)] \tag{5}$$

Give an iterative method for the solution of this nonlinear system.

One possibility is as follows. Calculate a first estimate, $x(P^1)$, $y(P^1)$, of $x(P)$, $y(P)$, by solving (*1*)–(*2*) with $\lambda_+ = \lambda_+(Q)$ and $\lambda_- = \lambda_-(R)$:

$$x(P^1) = \frac{y(R) - y(Q) + \lambda_+(Q)x(Q) - \lambda_-(R)x(R)}{\lambda_+(Q) - \lambda_-(R)} \tag{6}$$

$$y(P^1) = \frac{\lambda_+(Q)y(R) - \lambda_-(R)y(Q) + \lambda_+(Q)\lambda_-(R)[x(Q) - x(R)]}{\lambda_+(Q) - \lambda_-(R)} \tag{7}$$

Next, calculate a first approximation, $u_x(P^1)$, $u_y(P^1)$, to $u_x(P)$, $u_y(P)$, by solving (*3*)–(*4*) with $\lambda_+ = \lambda_+(Q)$, $a = a(Q)$, $c = c(Q)$, $f = f(Q)$, $y(P) = y(P^1)$, in (*3*); and $\lambda_- = \lambda_-(R)$, $a = a(R)$, $c = c(R)$, $f = f(R)$, $y(P) = y(P^1)$, in (*4*):

$$u_x(P^1) = \frac{c(R)B_1(Q) - c(Q)B_2(R)}{\lambda_+(Q)a(Q)c(R) - \lambda_-(R)a(R)c(Q)} \tag{8}$$

$$u_y(P^1) = \frac{\lambda_+(Q)a(Q)B_2(R) - \lambda_-(R)a(R)B_1(Q)}{\lambda_+(Q)a(Q)c(R) - \lambda_-(R)a(R)c(Q)} \tag{9}$$

where

$$B_1(Q) \equiv \lambda_+(Q)a(Q)u_x(Q) + c(Q)u_y(Q) + f(Q)[y(P^1) - y(Q)]$$
$$B_2(R) \equiv \lambda_-(R)a(R)u_x(R) + c(R)u_y(R) + f(R)[y(P^1) - y(R)]$$

Now $u(P^1)$ can be calculated from (*5*) as

$$u(P^1) = u(Q) + \frac{u_x(P^1) + u_x(Q)}{2}[x(P^1) - x(Q)] + \frac{u_y(P^1) + u_y(Q)}{2}[y(P^1) - y(Q)] \tag{10}$$

Upon the introduction of the averaged coefficients

$$\bar{\lambda}_+^j = [\lambda_+(Q) + \lambda_+(P^j)]/2 \qquad \hat{\lambda}_-^j = [\lambda_-(R) + \lambda_-(P^j)]/2$$
$$\bar{a}^j = [a(Q) + a(P^j)]/2 \qquad \hat{a}^j = [a(R) + a(P^j)]/2$$
$$\bar{c}^j = [c(Q) + c(P^j)]/2 \qquad \hat{c}^j = [c(R) + c(P^j)]/2$$
$$\bar{f}^j = [f(Q) + f(P^j)]/2 \qquad \hat{f}^j = [f(R) + f(P^j)]/2$$
$$\bar{B}_1^j = \lambda_+\bar{a}^j u_x(Q) + \bar{c}^j u_y(Q) + \bar{f}^j[y(P^j) - y(Q)]$$
$$\hat{B}_2^j = \lambda_-\hat{a}^j u_x(R) + \hat{c}^j u_y(R) + \hat{f}^j[y(P^j) - y(R)]$$

successive approximations can be calculated for $j = 1, 2, \ldots,$ as follows:

$$x(P^{j+1}) = \frac{y(R) - y(Q) + \bar{\lambda}_+^j x(Q) - \hat{\lambda}_-^j x(R)}{\bar{\lambda}_+^j - \hat{\lambda}_-^j} \tag{11}$$

$$y(P^{j+1}) = \frac{\bar{\lambda}_+^j y(R) - \hat{\lambda}_-^j y(Q) + \bar{\lambda}_+^j \hat{\lambda}_-^j [x(Q) - x(R)]}{\bar{\lambda}_+^j - \hat{\lambda}_-^j} \tag{12}$$

$$u_x(P^{j+1}) = \frac{\hat{c}^j \bar{B}_1^j - \bar{c}^j \hat{B}_2^j}{\bar{\lambda}_+^j \bar{a}^j \hat{c}^j - \hat{\lambda}_-^j \hat{a}^j \bar{c}^j} \tag{13}$$

$$u_y(P^{j+1}) = \frac{\bar{\lambda}_+^j \bar{a}^j \hat{B}_2^j - \hat{\lambda}_-^j \hat{a}^j \bar{B}_1^j}{\bar{\lambda}_+^j \bar{a}^j \hat{c}^j - \hat{\lambda}_-^j \hat{a}^j \bar{c}^j} \tag{14}$$

$$u(P^{j+1}) = u(Q) + \frac{u_x(P^{j+1}) + u_x(Q)}{2}[x(P^{j+1}) - x(Q)] + \frac{u_y(P^{j+1}) + u_y(Q)}{2}[y(P^{j+1}) - y(Q)] \tag{15}$$

The iterations using (*11*)–(*15*) are continued until two successive estimates agree to within some set tolerance.

10.6 Use the numerical method of characteristics to approximate the solution to

$$u_{xx} - u^2 u_{yy} = 0 \qquad u(x, 0) = x \qquad u_y(x, 0) = 2$$

at the first characteristic grid point P between $Q = (1, 0)$ and $R = (1.2, 0)$.

Using the notation of Problem 10.5, we have:

$$x(Q)=1 \qquad y(Q)=0 \qquad x(R)=1.2 \qquad y(R)=0$$
$$u_x(Q)=1 \qquad u_y(Q)=2 \qquad u_x(R)=1 \qquad u_y(R)=2$$
$$u(Q)=1 \qquad u(R)=1.2 \qquad \lambda_+(Q)=u(Q) \qquad \lambda_-(R)=-u(R)$$
$$a(Q)=1=a(R) \qquad c(Q)=-u(Q)^2 \qquad c(R)=-u(R)^2 \qquad f(Q)=0=f(R)$$

Putting the above values in (*6*)–(*10*) of Problem 10.5, we obtain

$$x(P^1)=1.109 \qquad y(P^1)=0.109 \qquad u_x(P^1)=1 \qquad u_y(P^1)=2 \qquad u(P^1)=1.3273$$

Using these values to initiate the successive approximations defined by (*11*)–(*15*) of Problem 10.5, we obtain the values displayed in Table 10-1. (The exact solution is $u = x + 2y$.)

Table 10-1

j	$x(P^j)$	$y(P^j)$	$u_x(P^j)$	$u_y(P^j)$	$u(P^j)$
1	1.10909	0.10909	1	2	1.32727
2	1.10412	0.12116	1	2	1.34644
3	1.10409	0.12212	1	2	1.34832
4	1.10408	0.12221	1	2	1.34851
5	1.10408	0.12221	1	2	1.34853
6	1.10408	0.12221	1	2	1.34853

10.7 Use the linear Lax–Wendroff method (*10.25*) (1×1 version) to approximate the solution to

$$u_x + u_t = 0 \qquad x>0,\ t>0$$
$$u(x,0) = 2+x \qquad x>0$$
$$u(0,t) = 2-t \qquad t>0$$

At $t=0.5$, for $0\le x\le 1$, compare the numerical solution with the exact solution, $u = 2+x-t$.

A program listing is given in Fig. 10-7, and the results of a stable run and an unstable run are displayed in Fig. 10-8. The exact agreement in the stable run is explained by noting that the analytical solution is linear in x and t, and therefore the local truncation error is zero. Thus, the only errors in the calculation are rounding errors. The unstable run illustrates the growth of these errors even in the absence of any truncation errors.

```
      PROGRAM LLAXW
C     TITLE:  DEMO PROGRAM FOR LAX-WENDROFF METHOD
C             FOR LINEAR EQUATION UX + UT = 0
C     INPUT:  H, X-GRID SPACING
C             K, TIME STEP (K/H < 1 FOR STABILITY)
C             TMAX, MAXIMUM COMPUTATION TIME
C             F(X), INITIAL CONDITION ON U, U(X,0) = F(X)
C             P(T), BOUNDARY CONDITION, U(0,T) = P(T)
C             E(X,T), EXACT SOLUTION
C     OUTPUT: NUMERICAL AND EXACT SOLUTION AT T=TMAX
      COMMON U(0:500),V(0:500)
      REAL K
      F(X) = 2 + X
      P(T) = 2 - T
      E(X,T) = 2 + X - T
      PRINT*,'ENTER TMAX,X-GRID SPACING AND TIME STEP'
      READ*,TMAX,H,K
      T = 0
      S = K/H
```

Fig. 10-7 *(Program continues on next page)*

```
C     DEFINE SUFFICIENTLY LARGE NUMERICAL INITIAL INTERVAL
      NMAX = 1/H + TMAX/K + 1
C     SET INITIAL CONDITION
      DO 10 I = 0,NMAX
         X = I*H
         V(I) = F(X)
10    CONTINUE
C     ADVANCE SOLUTION TO TIME T+K AND SET BOUNDARY VALUES
15    DO 30 I = 1,NMAX-1
         U(I)=V(I)-.5*S*(V(I+1)-V(I-1))+.5*S*S*(V(I-1)-2*V(I)+V(I+1))
30    CONTINUE
      T = T + K
      NMAX = NMAX -1
      U(0) = P(T)
C     WRITE U OVER V TO PREPARE FOR ANOTHER TIME STEP
      DO 40 I = 0,NMAX
         V(I) = U(I)
40    CONTINUE
C     IF T IS LESS THAN TMAX, TAKE ANOTHER TIME STEP
      IF(ABS(TMAX-T).GT.K/2) GOTO 15
C     IF T EQUALS TMAX, PRINT RESULT
      WRITE(6,100)
      WRITE(6,110) H,K,S
      WRITE(6,120) TMAX
      ISTEP = .1/H
      IMAX = 1/H
      DO 50 I = 0,IMAX,ISTEP
         X = I*H
         EXACT = 2 + X - T
         WRITE(6,130) X,U(I),EXACT
50    CONTINUE
100   FORMAT(///,T9,'RESULTS FROM PROGRAM LLAXW',/)
110   FORMAT('H =',F5.2,T15,'K = ',F5.2,T30,'S =',F5.2,/)
120   FORMAT('T =',F5.2,T18,'NUMERICAL',T35,'EXACT',/)
130   FORMAT( 'X = ',F4.1,T13,F13.6,T30,F13.6)
      END
```

Fig. 10-7 (*Continued*)

```
H = 0.10        K =  0.10        S = 1.00

T = 0.50            NUMERICAL          EXACT

X =  0.             1.500000         1.500000
X =  0.1            1.600000         1.600000
X =  0.2            1.700000         1.700000
X =  0.3            1.800000         1.800000
X =  0.4            1.900000         1.900000
X =  0.5            2.000000         2.000000
X =  0.6            2.100001         2.100000
X =  0.7            2.200001         2.200000
X =  0.8            2.300001         2.300000
X =  0.9            2.400001         2.400000
X =  1.0            2.500001         2.500000

H = 0.02        K =  0.04        S = 2.00

T = 0.50            NUMERICAL          EXACT

X =  0.             1.480000         1.480000
X =  0.1         -228.278488         1.580000
X =  0.2         1824.329346         1.680000
X =  0.3         -314.192566         1.780000
X =  0.4         1201.677368         1.880000
X =  0.5        -1047.932617         1.980000
X =  0.6          449.913300         2.080000
X =  0.7        -1787.070679         2.180000
X =  0.8           70.098160         2.280000
X =  0.9         1202.175171         2.380000
X =  1.0        -1033.922974         2.480000
```

Fig. 10-8

10.8 Apply Wendroff's implicit method (*10.17*) to the initial–boundary value problem of Problem 10.7 with $s = 2.00$, the case in which the Lax–Wendroff method proved to be unstable.

See Figs. 10-9 and 10-10. As expected, this stable method produces the exact solution. Notice that even though the difference method is implicit, the calculations in the program proceed from left to right in x, in an explicit manner.

```
      PROGRAM WENDI
C     TITLE:  DEMO PROGRAM FOR WENDROFF'S IMPLICIT
C             METHOD FOR EQUATION UX + UT = 0
C     INPUT:  N, NUMBER OF X-SUBINTERVALS
C             K, TIME STEP (K/H < 1 FOR STABILITY)
C             TMAX, MAXIMUM COMPUTATION TIME
C             (X1,X2), X-INTERVAL
C             F(X), INITIAL CONDITION ON U, U(X,0) = F(X)
C             P(T), BOUNDARY CONDITION, U(0,T) = P(T)
C             E(X,T), EXACT SOLUTION
C     OUTPUT: NUMERICAL AND EXACT SOLUTION AT T=TMAX
      COMMON U(0:500),V(0:500)
      REAL K
      DATA T,X1,X2/0,0,1/
      F(X) = 2 + X
      P(T) = 2 - T
      E(X,T) = 2 + X - T
C     SET TMAX AND X AND T STEP SIZES
      PRINT*,'ENTER TMAX,NUMBER OF X-SUBINTERVALS AND TIME STEP'
      READ*,TMAX,N,K
      H = (X2-X1)/N
      S = K/H
C     SET INITIAL CONDITION AND BOUNDARY CONDITION
      DO 10 I = 0,N
         X = X1 + I*H
         V(I) = F(X)
10    CONTINUE
      U(0) = P(K)
C     ADVANCE SOLUTION TO TIME T+K AND SET BOUNDARY VALUE
15    T = T + K
      U(0) = P(T)
      DO 30 I = 1,N
         U(I)=V(I-1)+(1-S)*(V(I) - U(I-1))/(1+S)
30    CONTINUE
C     WRITE U OVER V TO PREPARE FOR ANOTHER TIME STEP
      DO 40 I = 0,N
         V(I) = U(I)
40    CONTINUE
C     IF T IS LESS THAN TMAX, TAKE ANOTHER TIME STEP
      IF(ABS(TMAX-T).GT.K/2) GOTO 15
C     IF T EQUALS TMAX, PRINT RESULT
      WRITE(6,100)
      WRITE(6,110) N,K,S
      WRITE(6,120) TMAX
      ISTEP = .1/H
      IMAX = 1/H
      DO 50 I = 0,IMAX,ISTEP
         X = X1 + I*H
         EXACT = 2 + X - T
         WRITE(6,130) X,U(I),EXACT
50    CONTINUE
100   FORMAT(///,T9,'RESULTS FROM PROGRAM WENDI',/)
110   FORMAT('N =',I4,T15,'K = ',F5.2,T30,'S =',F5.2,/)
120   FORMAT('T =',F5.2,T18,'NUMERICAL',T35,'EXACT',/)
130   FORMAT( 'X = ',F4.1,T13,F13.6,T30,F13.6)
      END
```

Fig. 10-9

```
N = 100        K =  0.02        S = 2.00

T = 0.50        NUMERICAL        EXACT

X =  0.         1.500000         1.500000
X =  0.1        1.600000         1.600000
X =  0.2        1.700000         1.700000
X =  0.3        1.800000         1.800000
X =  0.4        1.900000         1.900000
X =  0.5        2.000000         2.000000
X =  0.6        2.100000         2.100000
X =  0.7        2.200000         2.200000
X =  0.8        2.300000         2.300000
X =  0.9        2.400000         2.400000
X =  1.0        2.500000         2.500000
```

Fig. 10-10

```
      PROGRAM CLAXW
C     TITLE:  DEMO PROGRAM FOR LAX-WENDORFF METHOD
C             FOR CONSERVATION EQUATION (U*U/2)X+UT=0
C     INPUT:  H, X-GRID SPACING
C             K, TIME STEP
C             TMAX, MAXIMUM COMPUTATION TIME
C             F(X), INITIAL CONDITION ON U, U(X,0) = F(X)
C             P(T), BOUNDARY CONDITION, U(0,T) = P(T)
C             E(X,T), EXACT SOLUTION
C     OUTPUT: NUMERICAL AND EXACT SOLUTION AT T=TMAX
      COMMON U(0:500),V(0:500)
      REAL K
      F(X) = X
      P(T) = 0
      E(X,T) = X/(1+T)
C     SET TMAX AND X AND T STEP SIZES
      PRINT*,'ENTER TMAX,X-GRID SPACING AND TIME STEP'
      READ*,TMAX,H,K
      T = 0
      S = K/H
C     DEFINE SUFFICIENTLY LARGE NUMERICAL INITIAL INTERVAL
      NMAX = 1/H + TMAX/K + 1
C     SET INITIAL CONDITION
      DO 10 I = 0,NMAX
         X = I*H
         V(I) = X
10    CONTINUE
C     ADVANCE SOLUTION TO TIME T+K AND SET BOUNDARY VALUES
15    DO 30 I = 1,NMAX-1
         U(I)  =V(I) - .5*S*(V(I+1)**2-V(I-1)**2)/2
         U(I) = U(I) + S*S*(V(I+1)+V(I))*(V(I+1)**2-V(I)**2)/8
         U(I) = U(I) - S*S*(V(I)+V(I-1))*(V(I)**2-V(I-1)**2)/8
30    CONTINUE
      T = T + K
      NMAX = NMAX -1
      U(0) = 0
C     WRITE U OVER V TO PREPARE FOR ANOTHER TIME STEP
      DO 40 I = 0,NMAX
         V(I) = U(I)
40    CONTINUE
C     IF T IS LESS THAN TMAX, TAKE ANOTHER TIME STEP
      IF(ABS(TMAX-T).GT.K/2) GOTO 15
C     IF T EQUALS TMAX, PRINT RESULT
      WRITE(6,100)
      WRITE(6,110) H,K,S
      WRITE(6,120) TMAX
      ISTEP = .1/H
      IMAX = 1/H
      DO 50 I = 0,IMAX,ISTEP
         X = I*H
         EXACT = X/(1 + T)
         WRITE(6,130) X,U(I),EXACT
50    CONTINUE
100   FORMAT(///,T9,'RESULTS FROM PROGRAM CLAXW',/)
110   FORMAT('H =',F5.2,T15,'K = ',F5.2,T30,'S =',F5.2,/)
120   FORMAT('T =',F5.2,T18,'NUMERICAL',T35,'EXACT',/)
130   FORMAT( 'X = ',F4.1,T13,F13.6,T30,F13.6)
      END
```

Fig. 10-11

10.9 Use the Lax–Wendroff method (*10.19*) to approximate the solution of

$$\begin{aligned} (u^2/2)_x + u_t &= 0 \qquad x>0,\ t>0 \\ u(x,0) &= x \qquad x>0 \\ u(0,t) &= 0 \qquad t>0 \end{aligned}$$

At $t=1$, for $0 \le x \le 1$, compare the numerical solution with the exact solution, $u = x/(1+t)$.

See Figs. 10-11 and 10-12. A "sufficiently large numerical initial interval" is one that includes the numerical domain of dependence of the interval on which it is desired to approximate the solution.

H = 0.10 K = 0.10 S = 1.00

T = 1.00	NUMERICAL	EXACT
X = 0.	0.	0.
X = 0.1	0.050146	0.050000
X = 0.2	0.100292	0.100000
X = 0.3	0.150438	0.150000
X = 0.4	0.200585	0.200000
X = 0.5	0.250731	0.250000
X = 0.6	0.300877	0.300000
X = 0.7	0.351023	0.350000
X = 0.8	0.4C1169	0.400000
X = 0.9	0.451315	0.450000
X = 1.0	0.501461	0.500000

Fig. 10-12

10.10 Show that a necessary condition for the linear Lax–Wendroff method (*10.25*) to be stable is that $|s\lambda| \le 1$ for each of the eigenvalues λ of the matrix $\mathbf{A}$.

Making a von Neumann analysis (see Problem 9.5), we substitute

$$\mathbf{E}_{nj} = e^{in\beta}[\xi_1^j, \xi_2^j, \dots, \xi_M^j]^T$$

in (*10.25*), obtaining $[\xi_1^{j+1}, \dots, \xi_M^{j+1}]^T = \mathbf{G}[\xi_1^j, \dots, \xi_M^j]^T$, where

$$\mathbf{G} \equiv \mathbf{I} - (is \sin\beta)\mathbf{A} - \left(2s^2 \sin^2\frac{\beta}{2}\right)\mathbf{A}^2 \tag{1}$$

For stability, all eigenvalues μ of the *amplification matrix* $\mathbf{G}$ must satisfy $|\mu| \le 1$. But the eigenvalues of the matrix polynomial (*1*) are the values of the polynomial at the eigenvalues λ of $\mathbf{A}$:

$$\mu = 1 - (is\sin\beta)\lambda - \left(2s^2\sin^2\frac{\beta}{2}\right)\lambda^2 \tag{2}$$

From (*2*), since λ is real,

$$\begin{aligned} |\mu|^2 &= \left[1 - \left(2s^2\sin^2\frac{\beta}{2}\right)\lambda^2\right]^2 + [(s\sin\beta)\lambda]^2 \\ &= 1 - \left(4\sin^4\frac{\beta}{2}\right)\rho(1-\rho) \qquad (\rho \equiv s^2\lambda^2) \end{aligned} \tag{3}$$

It is clear from (*3*) that $|\mu|^2 \le 1$ for all β only if $0 \le \rho \le 1$; i.e., only if $|s\lambda| \le 1$. This condition is (it can be shown) also sufficient for stability.

10.11 Use the linear Lax–Wendroff method (*10.25*) to approximate the solution to the initial value problem

$$\begin{bmatrix} 4 & -6 \\ 1 & -3 \end{bmatrix}\begin{bmatrix} u \\ v \end{bmatrix}_x + \begin{bmatrix} u \\ v \end{bmatrix}_t = \begin{bmatrix} 0 \\ 0 \end{bmatrix}$$

$$u(x,0) = \sin x \qquad v(x,0) = \cos x$$

At $t = 0.5$, for $0 \le x \le 1$, compare the numerical solution with the exact solution,

```
      PROGRAM SLAXW
C     TITLE:  DEMO PROGRAM FOR LAX-WENDORFF METHOD
C             FOR SYSTEM OF TWO EQUATIONS, AUX + UT = 0
C     INPUT:  H, X-GRID SPACING
C             K, TIME STEP
C             TMAX, MAXIMUM COMPUTATION TIME
C             FU(X) & FV(X),INITIAL CONDITION ON U & V
C             EU(X,T) & EV(X,T), EXACT U & V SOLUTIONS
C             A & B, COEFFICIENT MATRIX AND ITS SQUARE
C     OUTPUT: NUMERICAL AND EXACT SOLUTION AT T=TMAX
      COMMON U(0:500),V(0:500),UN(0:500),VN(0:500),A(2,2),B(2,2)
      REAL K
      FU(X) = SIN(X)
      FV(X) = COS(X)
      EU(X,T) = (6*SIN(X-3*T)-6*COS(X-3*T))/5
     1         + (6*COS(X+2*T)-SIN(X+2*T))/5
      EV(X,T) = (SIN(X-3*T)-COS(X-3*T))/5
     1         + (6*COS(X+2*T)-SIN(X+2*T))/5
C     SET COEFFICIENT MATRIX AND ITS SQUARE
      DATA A(1,1),A(1,2),A(2,1),A(2,2)/4,-6,1,-3/
      DATA B(1,1),B(1,2),B(2,1),B(2,2)/10,-6,1,3/
C     SET TMAX AND X AND T STEP SIZES
      PRINT*,'ENTER TMAX,X-GRID SPACING AND TIME STEP'
      READ*,TMAX,H,K
      T = 0
      S = K/H
C     DEFINE SUFFICIENTLY LARGE NUMERICAL INITIAL INTERVAL
      NLOW = TMAX/K +1
      NHIGH = NLOW + 1/H
      NMAX = NHIGH + NLOW
C     SET INITIAL CONDITION
      DO 10 I = 0,NMAX
         X = (-NLOW + I)*H
         U(I) = FU(X)
         V(I) = FV(X)
10    CONTINUE
C     ADVANCE SOLUTION TO TIME T+K
      ILOW = 1
15    DO 30 I = ILOW,NMAX-1
         UN(I)=U(I)-.5*S*A(1,1)*(U(I+1)-U(I-1))
     1             -  .5*S*A(1,2)*(V(I+1)-V(I-1))
     2             +   S*S*B(1,1)*(U(I-1)-2*U(I)+U(I+1))/2
     3             +   S*S*B(1,2)*(V(I-1)-2*V(I)+V(I+1))/2
         VN(I)=V(I)-.5*S*A(2,1)*(U(I+1)-U(I-1))
     1             -  .5*S*A(2,2)*(V(I+1)-V(I-1))
     2             +   S*S*B(2,1)*(U(I-1)-2*U(I)+U(I+1))/2
     3             +   S*S*B(2,2)*(V(I-1)-2*V(I)+V(I+1))/2
30    CONTINUE
      T = T + K
      ILOW = ILOW + 1
      NMAX = NMAX - 1
C     WRITE U OVER UN AND VN OVER V TO PREPARE FOR NEXT TIME STEP
      DO 40 I = ILOW,NMAX
         U(I) = UN(I)
         V(I) = VN(I)
40    CONTINUE
C     IF T IS LESS THAN TMAX, TAKE ANOTHER TIME STEP
      IF(ABS(TMAX-T).GT.K/2) GOTO 15
C     IF T EQUALS TMAX, PRINT RESULT
      WRITE(6,100)
      WRITE(6,110) H,K,S
      WRITE(6,120) TMAX
      ISTEP = .1/H
      IMAX = 1/H + ILOW
      DO 50 I = ILOW,IMAX,ISTEP
         X = (I - ILOW)*H
         EXACTU = EU(X,T)
         EXACTV = EV(X,T)
         WRITE(6,130) X,U(I),EXACTU
         WRITE(6,140) V(I),EXACTV
50    CONTINUE
100   FORMAT(///,T9,'RESULTS FROM PROGRAM SLAXW',/)
110   FORMAT('H =',F5.2,T15,'K = ',F7.4,T30,'S =',F5.2,/)
120   FORMAT('T =',F5.2,T18,'NUMERICAL',T35,'EXACT',/)
130   FORMAT( 'X = ',F4.1,T13,'U=',F10.6,T30,'U=',F10.6)
140   FORMAT( T13,'V=',F10.6,T30,'V=',F10.6,/)
      END
```

Fig. 10-13

$$u(x, t) = [6\sin(x-3t) - 6\cos(x-3t) - \sin(x+2t) - 6\cos(x+2t)]/5$$
$$v(x, t) = [\sin(x-3t) - \cos(x-3t) - \sin(x+2t) + 6\cos(x+2t)]/5$$

Since the eigenvalues of

$$\mathbf{A} = \begin{bmatrix} 4 & -6 \\ 1 & -3 \end{bmatrix}$$

are $\lambda_1 = -2$, $\lambda_2 = 3$, it follows from Problem 10.10 that the stability condition in this case is $3k/h \le 1$. In the program of Fig. 10-13, note that to obtain a numerical solution on the line $\mathscr{L}$: $t = t_{max}$, $0 \le x \le 1$, the initial interval in the finite-difference calculation must be large enough to include the numerical domain of dependence of $\mathscr{L}$. Comparison of the numerical and exact solutions is made in Fig. 10-14.

```
H = 0.05          K =  0.0125      S = 0.25

T = 0.50             NUMERICAL           EXACT

X =  0.         U= -0.801758        U= -0.801810
                V=  0.266720        V=  0.266422

X =  0.1        U= -1.020323        U= -1.020427
                V=  0.135308        V=  0.134991

X =  0.2        U= -1.228694        U= -1.228847
                V=  0.002545        V=  0.002211

X =  0.3        U= -1.424788        U= -1.424989
                V= -0.130245        V= -0.130592

X =  0.4        U= -1.606646        U= -1.606893
                V= -0.261732        V= -0.262090

X =  0.5        U= -1.772451        U= -1.772742
                V= -0.390605        V= -0.390969

X =  0.6        U= -1.920546        U= -1.920878
                V= -0.515574        V= -0.515941

X =  0.7        U= -2.049451        U= -2.049821
                V= -0.635392        V= -0.635758

X =  0.8        U= -2.157879        U= -2.158283
                V= -0.748862        V= -0.749224

X =  0.9        U= -2.244747        U= -2.245180
                V= -0.854849        V= -0.855203

X =  1.0        U= -2.309186        U= -2.309645
                V= -0.952295        V= -0.952637
```

Fig. 10-14

10.12 Use the numerical method of characteristics to estimate, at $y = 0.01, 0.1, 0.2, 0.3$, the solution to

$$uu_x + u_y = -2u^3$$
$$u(x, 0) = x$$

on the characteristic through $Q \equiv (1, 0)$. Then compare the numerical results with the exact solution.

In the notation of Section 10.4, we have $a = u$, $b = 1$, $c = -2u^3$, $x(Q) = 1$, $y(Q) = 0$. The system (*10.27*)–(*10.28*) becomes

$$[x(P) - x(Q)] - u[y(P) - y(Q)] = 0$$
$$[u(P) - u(Q)] + 2u^3[y(P) - y(Q)] = 0$$

which is to be solved for $x(P)$ and $u(P)$ corresponding to the four given values of $y(P)$.

From the initial condition $u(x, 0) = x$, we have $u(Q) = 1$; so, an initial estimate $x(P_0)$, $u(P_0)$, can be determined by solving

$$[x(P_0) - 1] - u(Q)y(P) = 0$$
$$[u(P_0) - 1] + 2u(Q)^3 y(P) = 0$$

to obtain $x(P_0) = 1 + y(P)$, $u(P_0) = 1 - 2y(P)$. Now successive approximations can be defined by solving

$$[x(P_j) - 1] - \frac{u(Q) + u(P_{j-1})}{2} y(P) = 0$$

$$[u(P_j) - 1] + \frac{2u(Q)^3 + 2u(P_{j-1})^3}{2} y(P) = 0$$

for $j = 1, 2, \ldots$, until two approximations agree to within a set tolerance. For a tolerance of 10^{-8}, fewer than 20 iterations will be required.

By the method of Problem 5.5, the exact solution along the characteristic $s = 1$ is found to be

$$u(r, 1) = (1 + 4r)^{-1/2} \qquad x(r, 1) = \frac{1}{2}[(1 + 4r)^{1/2} + 1] \qquad y(r, 1) = r$$

The numerical and exact results for x and u at $y = r = 0.01, 0.1, 0.2, 0.3$, are compared in Table 10-2.

Table 10-2

	Numerical		Exact	
$y(P)$	$x(P)$	$u(P)$	$x(P)$	$u(P)$
0.01	1.00990286	0.980571601	1.00990195	0.98058069
0.1	1.09203010	0.840602064	1.09160798	0.84515425
0.2	1.17240756	0.724075551	1.17082039	0.74535599
0.3	1.24394498	0.626299874	1.24161985	0.67419986

10.13 By Problem 5.12, the open-channel flow equations can be expressed in characteristic $\alpha\beta$-coordinates as

$$2c_\alpha + v_\alpha = S(x, v)t_\alpha \tag{1}$$
$$2c_\beta - v_\beta = -S(x, v)t_\beta \tag{2}$$
$$x_\alpha = \lambda_1(v, c)t_\alpha \tag{3}$$
$$x_\beta = \lambda_2(v, c)t_\beta \tag{4}$$

where $c \equiv \sqrt{gu}$, $\lambda_1(v, c) \equiv v + c$, $\lambda_2(v, c) \equiv v - c$, and $S(x, v) \equiv g[S_0(x) - S_f(v^2)]$. Assuming that $c(x, 0)$ and $v(x, 0)$ are prescribed, obtain a numerical solution of (*1*)–(*4*) by *Hartree's method*, which uses a rectangular grid, $(x_n, t_j) = (nh, jk)$.

It is sufficient to show how the solution is advanced from level j to level $j + 1$ (Fig. 10-15). Assume c and v are known at the grid points on level j and let P have coordinates (x_n, t_{j+1}). The α- and β-characteristics through P intersect the line $t = t_j$ at Q and R, respectively. For error control, $k = t_{j+1} - t_j$ is chosen small enough to locate $x(Q)$ and $x(R)$ between x_{n-1} and x_{n+1}.

If (*1*)–(*4*) is discretized using averages of the coefficients at t_j and t_{j+1}, there results

$$2[c(P) - c(Q)] + [v(P) - v(Q)] = \frac{S(P) + S(Q)}{2} k \tag{5}$$

$$2[c(P) - c(R)] - [v(P) - v(R)] = -\frac{S(P) + S(R)}{2} k \tag{6}$$

$$x(P) - x(Q) = \frac{\lambda_1(P) + \lambda_1(Q)}{2} k \tag{7}$$

$$x(P) - x(R) = \frac{\lambda_2(P) + \lambda_2(R)}{2} k \tag{8}$$

where $S(P) \equiv S(x(P), v(P))$, etc. (5)–(8) constitutes four nonlinear equations in the four unknowns $x(Q)$, $x(R)$, $c(P)$, $v(P)$; the quantities $c(Q)$, $v(Q)$, $c(R)$, $v(R)$ are evaluated by interpolation between the grid points on $t = t_j$. The system may be solved by an iterative procedure similar to the one outlined in Problem 10.5.

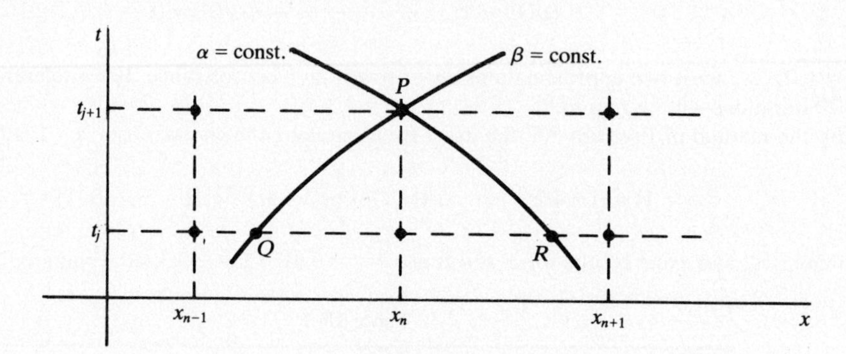

Fig. 10-15

Supplementary Problems

10.14 Demonstrate the von Neumann stability of (10.17).

10.15 (a) On a grid $(x_n, y_m, t_j) = (nh, mh, jk)$, derive an explicit difference equation for the wave equation

$$u_{tt} - c^2(u_{xx} + u_{yy}) = 0$$

(b) Use the von Neumann method to derive a stability condition.

10.16 Consider the initial value problem

$$\begin{aligned} u_{tt} - 4u_{xx} &= 0 & -\infty < x < \infty,\ t > 0 \\ u(x, 0) &= \cos x & -\infty < x < \infty \\ u_t(x, 0) &= 0 & -\infty < x < \infty \end{aligned}$$

(a) Find the D'Alembert solution and evaluate it at $x = 0$, $t = 0.04$. (b) With $h = 0.1$, $k = 0.02$, use (10.2) and the starting formula (2) of Problem 10.1 to calculate $U_{02} \approx u(0, 0.04)$. (c) Repeat (b) for the (cruder) starting formula

$$U_{n1} = U_{n0} = \cos nh$$

10.17 (a) Construct a centered-difference approximation to the damped wave equation

$$u_{tt} = c^2 u_{xx} - 2bu_t \qquad (b > 0)$$

(b) Make a von Neumann stability analysis of your method. [*Hint*: The analysis is similar to Problem 9.20.] (c) Show that in the limiting case $c^2 s^2 = 1$, the method becomes the DuFort–Frankel method (9.19). [*Hint*: In Problem 9.10, set $h^2/k^2 \equiv c^2$ and $a^2 \equiv c^2/2b$.]

10.18 (*a*) Solve analytically the mixed boundary value problem

$$\begin{aligned} x^2 u_{xx} - y^2 u_{yy} &= 1 \qquad x>1,\ y>1 \\ u(x,1) &= \log x \qquad x>1 \\ u_y(x,1) &= 2 \qquad x>1 \\ u(1,y) &= 2\log y \qquad y>1 \end{aligned}$$

[*Hint*: Assume $u(x, y) = X(x) + Y(y)$.] (*b*) Determine the characteristics of the PDE.

10.19 (*a*) Write out the characteristic equations (*10.7*)–(*10.11*) for Problem 10.18. (*b*) Integrate the first two characteristic equations to obtain the first characteristic grid point, P, between $Q = (2, 1)$ and $R = (2.1, 1)$. (*c*) Difference the remaining characteristic equations and estimate $u_x(P)$, $u_y(P)$, and $u(P)$; compare these values with those furnished by the analytical solution.

10.20 Show that to apply the method of characteristics to (*10.4*) in the region Ω: $x>0, y>0$, it is necessary that u and u_y be given on $y = 0$ (y is the timelike variable). Moreover, show that (i) if $\lambda_+>0$ and $\lambda_-<0$ in Ω, then u or u_x or a linear combination of u and u_x must be specified on $x = 0$; (ii) if $\lambda_+>0$ and $\lambda_->0$, then u and u_x must be specified on $x = 0$; (iii) if $\lambda_+<0$ and $\lambda_-<0$, then neither u nor u_x can be specified on $x = 0$ independently of the values of $u(x, 0)$ and $u_y(x, 0)$.

10.21 (*a*) Verify that $u = xy$ solves

$$\begin{aligned} u_{xx} - u^2 u_{yy} &= 0 \qquad x>0,\ y>2 \\ u(x,2) &= 2x \qquad x>0 \\ u_y(x,2) &= x \qquad x>0 \end{aligned}$$

(*b*) Determine the characteristics of the PDE and the location of the first characteristic grid point, P, between $Q = (1, 2)$ and $R = (2, 2)$.

10.22 Use the numerical method of characteristics to obtain the *initial* approximation to the solution of Problem 10.21 at grid point P. Compare the numerical and the exact results.

10.23 (*a*) With $h = 0.5$ and $k = 0.2$, apply (*10.13*) to approximate the solution to $u_t - 2u_x = u$, $u(x, 0) = \cos x$, at $(x, t) = (1, k)$. Compare the numerical solution with the exact solution, $u = e^t \cos(x + 2t)$. (*b*) Repeat with $h = 0.1$ and $k = 0.04$.

10.24 (*a*) With $h = 0.5$ and $k = 0.2$, apply (*10.14*) to approximate the solution to $u_t + 2u_x = 1$, $u(x, 0) = \sin x$, at $(x, t) = (1, k)$. Compare the numerical solution with the exact solution, $u = t + \sin(x - 2t)$. (*b*) Repeat with $h = 0.1$ and $k = 0.04$.

10.25 With $h = 0.1$ and $k = 0.1$, apply Wendroff's implicit method (*10.17*) to approximate the solution to the initial–boundary value problem

$$\begin{aligned} u_t + 2u_x &= 1 \qquad x>0,\ t>0 \\ u(x,0) &= \sin x \qquad x>0 \\ u(0,t) &= t - \sin 2t \qquad t>0 \end{aligned}$$

at $(x, t) = (0.1, 0.1)$. Compare the numerical solution with the exact solution, $u = t + \sin(x - 2t)$.

10.26 With $h = 0.1$ and $k = 0.2$, apply Wendroff's implicit method (*10.17*) to approximate the solution to

$$\begin{aligned} u_t - u_x &= x - t \qquad x<1,\ t>0 \\ u(x,0) &= 0 \qquad x<1 \\ u(1,t) &= t \qquad t>0 \end{aligned}$$

at $(x, t) = (0.9, 0.2)$. Compare the numerical solution with the exact solution, $u = xt$.

10.27 Use the change of variable $v = e^{-t}u$ to transform $u_t - 2u_x = u$, $u(x, 0) = \cos x$ to $v_t - 2v_x = 0$, $v(x, 0) = \cos x$. With $h = 0.5$ and $k = 0.2$, apply the Lax–Wendroff method (*10.19*), where $F(v) = -2v$, to approximate v at $(x, t) = (0.5, 0.2)$. Recover u and compare the numerical result with the exact solution from Problem 10.23.

10.28 With $h = 0.1$ and $k = 0.2$, apply the Lax–Wendroff method (*10.19*) to approximate the solution to

$$\left(\frac{u^2}{2}\right)_x + u_t = 0, \quad u(x, 0) = x$$

at $(x, t) = (0.2, 0.2)$. Compare the numerical solution with the exact solution, $u = x(1 + t)^{-1}$.

10.29 Use the numerical method of characteristics to obtain an initial approximation to the solution of

$$\begin{aligned} yu_x + xu_y &= x^2 + y^2 \qquad -\infty < x < \infty, \ y > 0 \\ u(x, 0) &= 0 \qquad -\infty < x < \infty \end{aligned}$$

at $y = 0.3$ on the characteristic through $(1, 0)$. Show analytically that the equation of this characteristic is $y = (x^2 - 1)^{1/2}$ and that $u = xy$ is the exact solution to the problem. Compare numerical and exact solutions.

10.30 Use the numerical method of characteristics to estimate the solution of

$$uu_x - yu_y = x, \quad u(x, 1) = x - 2$$

at $y = 1.5$ on the characteristic through $(2, 0)$. Compare the numerical results with the exact solution, $u = x - 2y$.

Chapter 11

Difference Methods for Elliptic Equations

11.1 LINEAR ALGEBRAIC EQUATIONS

In a *linear* elliptic boundary value problem, if all derivatives are replaced by their corresponding difference quotients, a system of *linear* algebraic equations results.

EXAMPLE 11.1 On the square Ω: $0<x<\ell$, $0<y<\ell$, consider the Dirichlet problem for Poisson's equation,

$$u_{xx}+u_{yy}=f(x, y) \quad \text{in } \Omega \tag{11.1}$$

$$u=g(x, y) \quad \text{on } S \tag{11.2}$$

Choosing a mesh spacing $h=\ell/4$, define on Ω the grid points

$$(x_m, y_n)=(mh, nh) \qquad (m, n=0, 1, 2, 3, 4)$$

Using central differences to approximate u_{xx} and u_{yy}, one obtains the difference equation

$$\frac{U_{m+1,n}-2U_{mn}+U_{m-1,n}}{h^2}+\frac{U_{m,n+1}-2U_{nm}+U_{m,n-1}}{h^2}=f_{mn} \tag{11.3}$$

where $f_{mn}\equiv f(x_m, y_n)$, and the boundary values are $U_{mn}=g_{mn}$ for m or n equal to 0 or 4.

A system like (*11.3*) is more clearly displayed as a system of linear equations if the grid points are labeled by a single index. Thus, if the interior nodes are ordered left-to-right and bottom-to-top, as in Fig. 11-1, then we can write $U_{11}\equiv U_1$, $U_{21}\equiv U_2, \ldots$, and can reindex f similarly. With this indexing, multiplication of (*11.3*) by $-h^2$ produces the following nine linear equations in the nine unknowns U_j:

$$\begin{bmatrix} 4 & -1 & 0 & -1 & 0 & 0 & 0 & 0 & 0 \\ -1 & 4 & -1 & 0 & -1 & 0 & 0 & 0 & 0 \\ 0 & -1 & 4 & 0 & 0 & -1 & 0 & 0 & 0 \\ -1 & 0 & 0 & 4 & -1 & 0 & -1 & 0 & 0 \\ 0 & -1 & 0 & -1 & 4 & -1 & 0 & -1 & 0 \\ 0 & 0 & -1 & 0 & -1 & 4 & 0 & 0 & -1 \\ 0 & 0 & 0 & -1 & 0 & 0 & 4 & -1 & 0 \\ 0 & 0 & 0 & 0 & -1 & 0 & -1 & 4 & -1 \\ 0 & 0 & 0 & 0 & 0 & -1 & 0 & -1 & 4 \end{bmatrix} \begin{bmatrix} U_1 \\ U_2 \\ U_3 \\ U_4 \\ U_5 \\ U_6 \\ U_7 \\ U_8 \\ U_9 \end{bmatrix} = \begin{bmatrix} B_1 \\ B_2 \\ B_3 \\ B_4 \\ B_5 \\ B_6 \\ B_7 \\ B_8 \\ B_9 \end{bmatrix}$$

where, in the case $g=0$, $B_j\equiv -h^2 f_j$.

By the scheme of Example 11.1 the difference equations for the general, two-dimensional, linear, elliptic boundary value problem (see Problem 11.7) can be put in the form

$$\mathbf{AU}=\mathbf{B} \tag{11.4}$$

The following remarks are pertinent to the general problem.

(1) The dimension of the vectors **U** and **B** is equal to the number of nodes at which the solution is to be approximated.

(2) The vector **B** is determined by the boundary conditions and the u-independent terms in the PDE.

(3) The matrix **A** is square and contains at most five nonzero entries per row. With Ω fixed, the order of **A**, given in (1) above, is a certain decreasing function of the mesh spacing h. Thus,

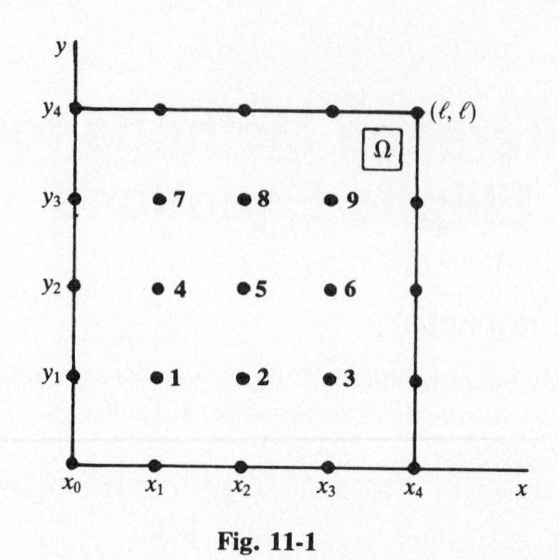

Fig. 11-1

even if the entries of **A** do not involve h, the eigenvalues of **A**—and those of various iteration matrices to be derived from **A**—will be functions of h.

(4) Provided the boundary value problem has a unique solution and h is sufficiently small, **A** is nonsingular, so that the system *(11.4)* has exactly one solution.

11.2 DIRECT SOLUTION OF LINEAR EQUATIONS

A method for solving *(11.4)* is called a *direct method* if it produces the exact solution to *(11.4)* (up to rounding errors) by a finite number of algebraic operations. *Gaussian elimination* is an example of a direct method. Direct methods are generally restricted to problems such that *(11.4)* can be accommodated in the central memory of the available computer.

If *(11.4)* is to be solved for a given nonsingular matrix **A** and several vectors **B**, the ***LU*-*decomposition method*** is more economical than Gaussian elimination. This method is based on a factorization of **A** of the form **A** = **LU**, where **L** is a lower-triangular matrix and **U** is an upper-triangular matrix (see Problem 11.8).

The matrix **A** is usually *sparse* (most entries are zero), *banded* (some set of contiguous diagonals contains all the nonzero entries), symmetric, and/or block tridiagonal; efficient direct methods exploit any such special properties.

11.3 ITERATIVE SOLUTION OF LINEAR EQUATIONS

Iterative methods (or *indirect methods*) generate a sequence of approximations to the solution of a system of algebraic equations. In contrast to direct methods, they do not produce the exact solution to the system in a finite number of algebraic operations. Iterative methods generally require less computer storage and are easier to program than direct methods.

In most linear algebra and numerical analysis literature, iterative methods are stated for a system of equations in single-index form; e.g., *(11.4)*. In computational applications we shall find it easier to use multiple indices to identify the unknowns; single indexing will be employed only in discussions of the convergence of iterative methods.

The iterative methods, which below are stated for *(11.1)*–*(11.2)*, extend to the general, linear, elliptic boundary value problem.

Jacobi Point Iteration

$$U_{mn}^{k+1} = (U_{m-1,n}^{k} + U_{m,n-1}^{k} + U_{m+1,n}^{k} + U_{m,n+1}^{k} + F_{mn})/4 \tag{11.5}$$

Gauss–Seidel Point Iteration

$$U_{mn}^{k+1} = (U_{m-1,n}^{k+1} + U_{m,n-1}^{k+1} + U_{m+1,n}^{k} + U_{m,n+1}^{k} + F_{mn})/4 \tag{11.6}$$

Successive Overrelaxation (SOR) Point Iteration

$$\begin{aligned} \bar{U}_{mn}^{k+1} &= (U_{m-1,n}^{k+1} + U_{m,n-1}^{k+1} + U_{m+1,n}^{k} + U_{m,n+1}^{k} + F_{mn})/4 \\ U_{mn}^{k+1} &= \omega \bar{U}_{mn}^{k+1} + (1-\omega)U_{mn}^{k} \qquad (0 < \omega < 2) \end{aligned} \tag{11.7}$$

Some properties of these methods, in the present application, follow.

(1) The boundary condition (*11.2*) is accounted for by setting $U_{mn}^k = g_{mn}$ at all boundary nodes, for $k = 0, 1, 2, \ldots$.

(2) In all three methods the initial estimate, U_{mn}^0, can be chosen arbitrarily and, as $k \to \infty$, the U_{mn}^k will converge to the solution of the difference equations.

(3) F_{mn} is determined by the right-hand side of the PDE (*11.1*): $F_{mn} = -h^2 f_{mn}$.

(4) If (*11.1*) is Laplace's equation ($f = 0$), Jacobi's method consists in successively replacing the U-value at a node by the average of the U-values at the four neighboring nodes.

(5) Jacobi's method is independent of the order in which the nodes are scanned.

(6) The Gauss–Seidel and SOR methods are stated here for a scanning of nodes in the numerical order of Fig. 11-1. If the nodes are scanned in a different order, (*11.6*) and (*11.7*) must be modified.

(7) The Gauss–Seidel method differs from the Jacobi method only in that new information about U is used as soon as it becomes available.

(8) The SOR method takes note of the direction in which the Gauss–Seidel iterates are proceeding and (for $\omega > 1$) extrapolates in that direction in an effort to accelerate convergence.

(9) In (*11.7*), ω is called the *relaxation parameter*; the method is characterized as *underrelaxation* or *overrelaxation* according as $0 < \omega < 1$ or $1 < \omega < 2$. With $\omega = 1$, the SOR method reduces to the Gauss–Seidel method.

The methods (*11.5*), (*11.6*), (*11.7*) are *point iterative* methods because they update U one grid point at a time. Improved convergence rates can be obtained by using *block iterative* methods which update U at several grid points simultaneously. This improved convergence is gained at the expense of having to solve a system of linear equations each time a block of nodes is updated. In most block iterative methods the calculations are arranged so that the linear system is tridiagonal and therefore easy to solve. For instance, choosing as the block the horizontal line of nodes $n =$ const., we obtain the following *row iterative* counterparts of (*11.5*), (*11.6*), (*11.7*):

Jacobi Row Iteration

$$U_{mn}^{k+1} = (U_{m-1,n}^{k+1} + U_{m,n-1}^{k} + U_{m+1,n}^{k+1} + U_{m,n+1}^{k} + F_{mn})/4 \tag{11.8}$$

Gauss–Seidel Row Iteration

$$U_{mn}^{k+1} = (U_{m-1,n}^{k+1} + U_{m,n-1}^{k+1} + U_{m+1,n}^{k+1} + U_{m,n+1}^{k} + F_{mn})/4 \tag{11.9}$$

SOR Row Iteration (or ***LSOR Iteration***)

$$\begin{aligned} \bar{U}_{mn}^{k+1} &= (\bar{U}_{m-1,n}^{k+1} + U_{m,n-1}^{k+1} + \bar{U}_{m+1,n}^{k+1} + U_{m,n+1}^{k} + F_{mn})/4 \\ U_{mn}^{k+1} &= \omega \bar{U}_{mn}^{k+1} + (1-\omega)U_{mn}^{k} \qquad (0 < \omega < 2) \end{aligned} \tag{11.10}$$

Some properties of (*11.8*), (*11.9*), (*11.10*) are listed below.

(1) In Jacobi's method (*11.8*) the rows can be updated in any order.

(2) In the Gauss–Seidel and LSOR methods (*11.9*) and (*11.10*) the rows must be updated in the order $n = 1, 2, \ldots$, or else the iteration formulas must be modified.

(3) Equations (*11.8*) and (*11.9*) give rise to tridiagonal systems in the row of unknowns $U_{0n}^{k+1}, U_{1n}^{k+1}, \ldots, U_{Mn}^{k+1}$; (*11.10*) does the same for $\bar{U}_{0n}^{k+1}, \bar{U}_{1n}^{k+1}, \ldots, \bar{U}_{Mn}^{k+1}$.

(4) Column iteration methods similar to (*11.8*), (*11.9*), (*11.10*) are also available (see Problem 11.24). By alternating row and column iterations a variety of ADI methods can be devised (cf. Section 9.4).

11.4 CONVERGENCE OF POINT ITERATIVE METHODS

To investigate the convergence of (*11.5*), (*11.6*), (*11.7*), suppose the difference equation (*11.3*) written in singly indexed form (*11.4*). Next, write the coefficient matrix $\mathbf{A}$ as $\mathbf{A} = -\mathbf{L} + \mathbf{D} - \mathbf{U}$, where $\mathbf{L}$, $\mathbf{D}$, and $\mathbf{U}$ are, respectively, strictly lower triangular (zeros on the main diagonal), diagonal, and strictly upper triangular matrices. Assume that $\det(\mathbf{D}) \neq 0$.

EXAMPLE 11.2 For the matrix $\mathbf{A}$ of Example 11.1, $\mathbf{D}$ is the 9×9 diagonal matrix with 4s along the main diagonal; $\mathbf{L}$ is the 9×9 matrix with 1s along the third subdiagonal, the pattern two 1s, 0, two 1s, 0, ... along the first subdiagonal, and zeros elsewhere; and $\mathbf{U} = \mathbf{L}^T$.

Methods (*11.5*), (*11.6*), (*11.7*) can be expressed in matrix form as follows:

Jacobi Point Iteration

$$\mathbf{U}^{k+1} = \mathbf{T}_J\mathbf{U}^k + \mathbf{C}_J \qquad (\mathbf{T}_J \equiv \mathbf{D}^{-1}(\mathbf{L}+\mathbf{U}), \quad \mathbf{C}_J \equiv \mathbf{D}^{-1}\mathbf{B}) \tag{11.11}$$

Gauss–Seidel Point Iteration

$$\mathbf{U}^{k+1} = \mathbf{T}_G\mathbf{U}^k + \mathbf{C}_G \qquad (\mathbf{T}_G \equiv (\mathbf{D}-\mathbf{L})^{-1}\mathbf{U}, \quad \mathbf{C}_G \equiv (\mathbf{D}-\mathbf{L})^{-1}\mathbf{B}) \tag{11.12}$$

SOR Point Iteration

$$\mathbf{U}^{k+1} = \mathbf{T}_\omega\mathbf{U}^k + \mathbf{C}_\omega \qquad (\mathbf{T}_\omega \equiv (\mathbf{D}-\omega\mathbf{L})^{-1}[(1-\omega)\mathbf{D}+\mathbf{U}], \quad \mathbf{C}_\omega \equiv (\mathbf{D}-\omega\mathbf{L})^{-1}\mathbf{B}) \tag{11.13}$$

Theorem 11.1: In (*11.11*), (*11.12*), or (*11.13*), if $\{\mathbf{U}^k\}$ converges to $\mathbf{U}^*$, then $\mathbf{A}\mathbf{U}^* = \mathbf{B}$.

As in Section 9.3, let the spectral radius of a square matrix $\mathbf{T}$ be denoted $\rho(\mathbf{T})$.

Theorem 11.2: The sequence $\{\mathbf{U}^k\}$ defined by $\mathbf{U}^{k+1} = \mathbf{T}\mathbf{U}^k + \mathbf{C}$, with $\mathbf{U}^0$ arbitrary, converges to a unique vector, $\mathbf{U}^*$, independent of $\mathbf{U}^0$, if and only if $\rho(\mathbf{T}) < 1$.

Theorem 11.3 (*Stein–Rosenberg*): If, for the matrix $\mathbf{A}$ of (*11.4*), $a_{ij} \leq 0$ for $i \neq j$ and $a_{ii} > 0$, then exactly one of the following statements holds:

(1) $0 < \rho(\mathbf{T}_G) < \rho(\mathbf{T}_J) < 1$ (3) $\rho(\mathbf{T}_J) = \rho(\mathbf{T}_G) = 0$
(2) $1 < \rho(\mathbf{T}_J) < \rho(\mathbf{T}_G)$ (4) $\rho(\mathbf{T}_J) = \rho(\mathbf{T}_G) = 1$

Theorem 11.4: For the system $\mathbf{A}\mathbf{U} = \mathbf{B}$ if (i) $a_{ij} \leq 0$ for $i \neq j$ and $a_{ii} > 0$,

$$\text{(ii)} \quad a_{ii} \geq \sum_{\substack{j \\ j \neq i}} |a_{ij}| \quad \text{with strict inequality for some } i$$

and (iii) a change in any component of $\mathbf{B}$ affects every component of $\mathbf{U}$, then both the Jacobi and the Gauss–Seidel point iterative methods converge (conclusion (1) or (3) of Theorem 11.3).

Theorem 11.5: For the system $\mathbf{A}\mathbf{U} = \mathbf{B}$ with $a_{ii} \neq 0$, a necessary condition for the convergence of the SOR point iterative method is $0 < \omega < 2$. If $\rho(\mathbf{T}_J) < 1$, the condition is also sufficient.

11.5 CONVERGENCE RATES

Let $\mathbf{U}^*$ represent the exact solution to $\mathbf{AU}=\mathbf{B}$ and let $\mathbf{U}^k$ represent an approximation to $\mathbf{U}^*$ obtained by k applications of an iterative method

$$\mathbf{U}^k = \mathbf{TU}^{k-1} + \mathbf{C} \tag{11.14}$$

The *residual vector*, $\mathbf{R}^k \equiv \mathbf{B} - \mathbf{AU}^k$, is a measure of the amount by which $\mathbf{U}^k$ fails to satisfy the system $\mathbf{AU}=\mathbf{B}$, if $\mathbf{U}^k = \mathbf{U}^*$, then $\mathbf{R}^k = 0$. The *maximum residual* after k applications of (*11.14*) is the maximum of the magnitudes of the components of $\mathbf{R}^k$.

The *convergence rate* of (*11.14*) is defined to be $-\log_{10} \rho(\mathbf{T})$, where $\rho(\mathbf{T})$ is the spectral radius of the iteration matrix. For large k the reciprocal of the convergence rate is roughly the number of further iterations of (*11.14*) required to reduce the maximum residual by a factor of ten. For a square mesh, the *asymptotic convergence rate* is the dominant term in the convergence rate as the mesh spacing approaches zero (cf. remark (3) of Section 11.1).

To compare the convergence rates of the three point-iterative methods, some way of relating $\rho(\mathbf{T}_J)$, $\rho(\mathbf{T}_G)$, and $\rho(\mathbf{T}_\omega)$ is needed. This relationship is given in Theorem 11.6, which involves two new notions.

Definition: Matrix $\mathbf{A}$ is *2-cyclic* if there exists a permutation matrix $\mathbf{P}$ such that

$$\mathbf{PAP}^T = \begin{bmatrix} \mathbf{D}_1 & \mathbf{F} \\ \mathbf{G} & \mathbf{D}_2 \end{bmatrix}$$

where $\mathbf{D}_1$ and $\mathbf{D}_2$ are square diagonal matrices.

Definition: The 2-cyclic matrix $\mathbf{A} = -\mathbf{L} + \mathbf{D} - \mathbf{U}$ is *consistently ordered* if $\det(-\beta\mathbf{L} + \alpha\mathbf{D} - \beta^{-1}\mathbf{U})$ is independent of the scalar β.

Theorem 11.6: If $\mathbf{A}$ is 2-cyclic and consistently ordered, then the eigenvalues μ of $\mathbf{T}_J$ and the eigenvalues λ of $\mathbf{T}_\omega$ satisfy

$$(\lambda + \omega - 1)^2 = \lambda\omega^2\mu^2 \qquad (\omega \neq 0, \lambda \neq 0) \tag{11.15}$$

Since $\mathbf{T}_G = \mathbf{T}_\omega$ when $\omega = 1$, (*11.15*) relates the eigenvalues of $\mathbf{T}_J$ to the eigenvalues of both $\mathbf{T}_G$ and $\mathbf{T}_\omega$.

Table 11-1

Method	Convergence Rate	Asymptotic Convergence Rate
Point Jacobi	$-\log(\cos h)$	$h^2/2$
Point Gauss–Seidel	$-\log(\cos^2 h)$	h^2
Optimal Point SOR	$-\log\dfrac{1-\sin h}{1+\sin h}$	$2h$
Row Jacobi	$-\log\dfrac{\cos h}{2-\cos h}$	h^2
Row Gauss–Seidel	$-\log\left(\dfrac{\cos h}{2-\cos h}\right)^2$	$2h^2$
Optimal Row SOR	$-\log\left[\dfrac{1-\sqrt{2}\sin(h/2)}{1+\sqrt{2}\sin(h/2)}\right]^2$	$2\sqrt{2}\,h$

The relationship (*11.15*) can be used to find the value of the relaxation parameter, $\bar{\omega}$, which maximizes the convergence rate (minimizes $\rho(\mathbf{T}_\omega)$) for the SOR method.

Theorem 11.7: If **A** is 2-cyclic and consistently ordered,

$$\bar{\omega} = \frac{2}{1+\sqrt{1-\rho(\mathbf{T}_J)^2}} \tag{11.16}$$

When the relaxation parameter is given by (*11.16*) the SOR method is called *optimal SOR*. Table 11-1 displays the convergence rates for the iterative methods of Section 11.3 on a square of side π. For an arbitrary side ℓ, replace h in the table by $\pi h/\ell$.

Solved Problems

11.1 Determine the truncation error associated with using centered differences to approximate the Laplacian operator, $u_{xx}+u_{yy}$, on a rectangular grid, $(x_m, y_n) = (mh, nk)$.

By (*9.3*),

$$\text{T.E.} = -\frac{h^2}{12}u_{xxxx}(\bar{x}, y_n) - \frac{k^2}{12}u_{yyyy}(x_m, \bar{y}) = O(h^2+k^2)$$

provided u_{xxxx} and u_{yyyy} are bounded.

If a solution to a boundary value problem for Poisson's equation has identically zero fourth derivatives, e.g., $u = xy$, then the exact solution to the difference equation gives the exact solution to the boundary value problem. Such solutions are valuable when comparing different numerical methods.

11.2 Formulate difference equations with truncation error $O(h^2)$, together with discrete boundary conditions, for the Neumann problem

$$u_{xx}+u_{yy} = f(x, y) \qquad \text{in } \Omega \tag{1}$$

$$\frac{\partial u}{\partial n} = g(x, y) \qquad \text{on } S \tag{2}$$

where Ω is the rectangle $0<x<a$, $0<y<b$. Choose grid points (mh, nh) such that $Mh = a$, $Nh = b$.

By (*11.3*) and Problem 11.1, (*1*) can be approximated with truncation error $O(h^2)$ by

$$-U_{m-1,n} - U_{m,n-1} + 4U_{mn} - U_{m+1,n} - U_{m,n+1} = -h^2 f_{mn} \tag{3}$$

in which $m = 0, 1, \ldots, M$ and $n = 0, 1, \ldots, N$. Note the tacit assumption that f is also defined on S.

To approximate $\partial u/\partial n$ by a centered difference requires the introduction of ghost points (open dots in Fig. 11-2). At those grid points on S that are not corner points, the boundary conditions are:

$$U_{M+1,n} - U_{M-1,n} = 2h\, g_{Mn} \qquad (n = 1, 2, \ldots, N-1) \tag{4}$$

$$U_{m,N+1} - U_{m,N-1} = 2h\, g_{mN} \qquad (m = 1, 2, \ldots, M-1) \tag{5}$$

$$U_{-1,n} - U_{1n} = 2h\, g_{0n} \qquad (n = 1, 2, \ldots, N-1) \tag{6}$$

$$U_{m,-1} - U_{m1} = 2h\, g_{m0} \qquad (m = 1, 2, \ldots, M-1) \tag{7}$$

At a corner grid point, where **n** is undefined, let us take as the "normal derivative" the average of the two derivatives along the two outer normals to the sides meeting at the corner. This leads to the final four boundary conditions

$$\begin{aligned} U_{-1,0}+U_{0,-1}&=U_{10}+U_{01}+4h\,g_{00}\\ U_{M,-1}+U_{M+1,0}&=U_{M1}+U_{M-1,0}+4h\,g_{M0}\\ U_{M+1,N}+U_{M,N+1}&=U_{M-1,N}+U_{M,N-1}+4h\,g_{MN}\\ U_{0,N+1}+U_{-1,N}&=U_{0,N-1}+U_{1,N}+4h\,g_{0N} \end{aligned} \tag{8}$$

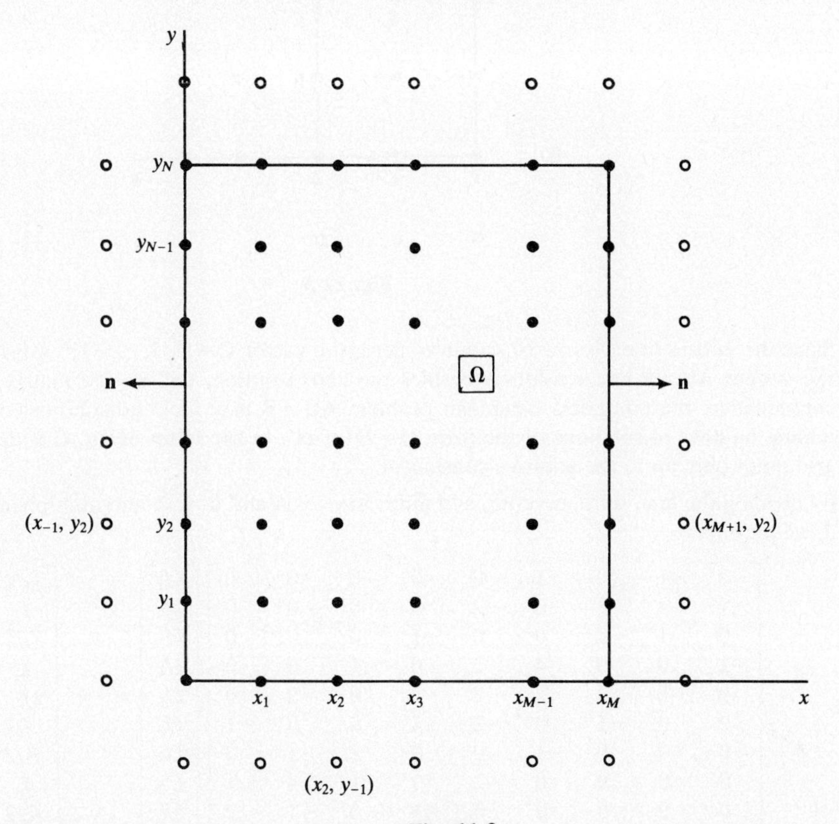

Fig. 11-2

11.3 In Problem 11.2, let $M=N=2$ and let $g\equiv 0$. (*a*) Write out in matrix form, $\mathbf{AU}=\mathbf{B}$, the difference system (*3*)–(*8*). (*b*) Show that $\mathbf{A}$ is singular. (*c*) By elementary row operations (which do not alter the system) obtain a representation $\mathbf{A'U}=\mathbf{B'}$ with $\mathbf{A'}$ symmetric. (*d*) Show that $\mathbf{A'U}=\mathbf{B'}$ can be solved (nonuniquely) only if f satisfies a consistency condition similar to

$$\int_\Omega f\,d\Omega=\int_S g\,dS\ (=0)$$

(*a*) For the single-indexing indicated in Fig. 11-3. we obtain, since $g\equiv 0$, the representation

$$\begin{bmatrix} 4 & -2 & 0 & -2 & 0 & 0 & 0 & 0 & 0\\ -1 & 4 & -1 & 0 & -2 & 0 & 0 & 0 & 0\\ 0 & -2 & 4 & 0 & 0 & -2 & 0 & 0 & 0\\ -1 & 0 & 0 & 4 & -2 & 0 & -1 & 0 & 0\\ 0 & -1 & 0 & -1 & 4 & -1 & 0 & -1 & 0\\ 0 & 0 & -1 & 0 & -2 & 4 & 0 & 0 & -1\\ 0 & 0 & 0 & -2 & 0 & 0 & 4 & -2 & 0\\ 0 & 0 & 0 & 0 & -2 & 0 & -1 & 4 & -1\\ 0 & 0 & 0 & 0 & 0 & -2 & 0 & -2 & 4 \end{bmatrix} \begin{bmatrix} U_1\\ U_2\\ U_3\\ U_4\\ U_5\\ U_6\\ U_7\\ U_8\\ U_9 \end{bmatrix} = -h^2 \begin{bmatrix} f_1\\ f_2\\ f_3\\ f_4\\ f_5\\ f_6\\ f_7\\ f_8\\ f_9 \end{bmatrix}$$

or $\mathbf{AU}=\mathbf{B}$.

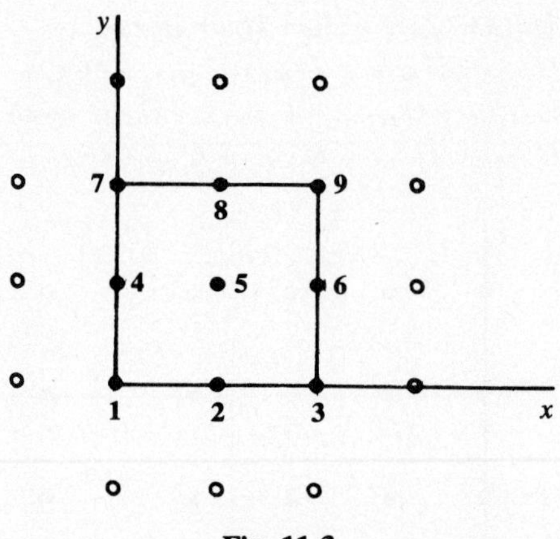

Fig. 11-3

(b) Since the entries in each row of $\mathbf{A}$ sum to zero, the vector $\mathbf{C} \equiv [1, 1, \ldots, 1]^T$ satisfies $\mathbf{AC} = \mathbf{0}$. Thus, the system $\mathbf{AU} = \mathbf{0}$ has a solution besides the zero solution, and so the matrix $\mathbf{A}$ is singular. In consequence, the numerical Neumann problem $\mathbf{AU} = \mathbf{B}$ may have no solution, or it may have an infinite number of solutions of the form $\mathbf{U} = \mathbf{U}^* + \alpha\mathbf{C}$. In the latter event, $\mathbf{U}$ is determined at each grid point only up to the additive constant α.

(c) By dividing the first, third, seventh, and ninth rows of $\mathbf{A}$ and $\mathbf{B}$ by 2, and multiplying the fifth row by 2, we produce

$$\begin{bmatrix} 2 & -1 & 0 & -1 & 0 & 0 & 0 & 0 & 0 \\ -1 & 4 & -1 & 0 & -2 & 0 & 0 & 0 & 0 \\ 0 & -1 & 2 & 0 & 0 & -1 & 0 & 0 & 0 \\ -1 & 0 & 0 & 4 & -2 & 0 & -1 & 0 & 0 \\ 0 & -2 & 0 & -2 & 8 & -2 & 0 & -2 & 0 \\ 0 & 0 & -1 & 0 & -2 & 4 & 0 & 0 & -1 \\ 0 & 0 & 0 & -1 & 0 & 0 & 2 & -1 & 0 \\ 0 & 0 & 0 & 0 & -2 & 0 & -1 & 4 & -1 \\ 0 & 0 & 0 & 0 & 0 & -1 & 0 & -1 & 2 \end{bmatrix} \begin{bmatrix} U_1 \\ U_2 \\ U_3 \\ U_4 \\ U_5 \\ U_6 \\ U_7 \\ U_8 \\ U_9 \end{bmatrix} = -h^2 \begin{bmatrix} f_1/2 \\ f_2 \\ f_3/2 \\ f_4 \\ 2f_5 \\ f_6 \\ f_7/2 \\ f_8 \\ f_9/2 \end{bmatrix}$$

or $\mathbf{A'U} = \mathbf{B'}$.

(d) Because $\mathbf{A'} = \mathbf{A'}^T$ and $\mathbf{A'C} = \mathbf{0}$ (since $\mathbf{AC} = \mathbf{0}$), we have, if a solution $\mathbf{U}$ exists,

$$\mathbf{B'}^T\mathbf{C} = (\mathbf{A'U})^T\mathbf{C} = \mathbf{U}^T(\mathbf{A'C}) = \mathbf{U}^T\mathbf{0} = 0$$

which means that the entries in the vector $\mathbf{B'}$ sum to zero. But this condition is equivalent to

$$\int_\Omega f(x, y)\, dx\, dy = 0$$

if the integral is evaluated by the trapezoidal rule using the nine grid points.

11.4 Show how to apply finite differences to

$$\begin{aligned} u_{xx} + u_{yy} &= f(x, y) && \text{in } \Omega \\ u &= g(x, y) && \text{on } S \end{aligned}$$

in the case that Ω has a curved boundary.

At any grid point in Ω whose four neighboring grid points are also in Ω the usual difference expressions for u_{xx} and u_{yy} can be used. Consider a grid point in Ω with at least one neighboring node not in Ω; e.g., $P = (x_m, y_n)$ in Fig. 11-4. The coordinates of the intermediary points q and r on S are respectively $(x_m + \alpha h, y_n)$ and $(x_m, y_n + \beta h)$, where $0 < \alpha, \beta < 1$. Since u is specified on S, $u(q)$ and $u(r)$ are known.

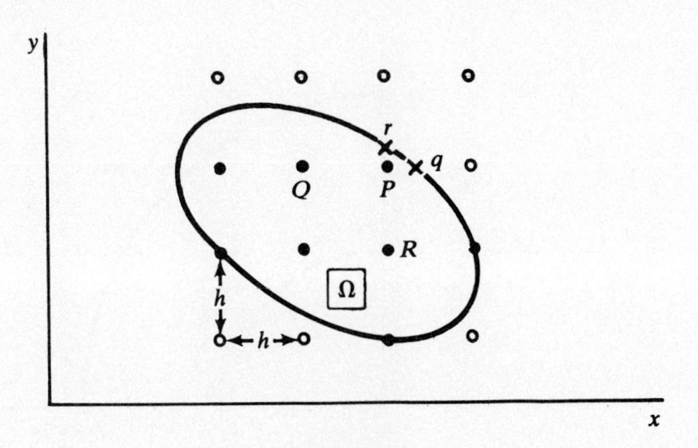

Fig. 11-4

By Taylor's theorem,

$$u(q) = u(P) + \alpha h u_x(P) + \frac{(\alpha h)^2}{2} u_{xx}(P) + O(h^3)$$

$$u(Q) = u(P) - h u_x(P) + \frac{h^2}{2} u_{xx}(P) + O(h^3)$$

Eliminating $u_x(P)$ from the above pair of equations, we have

$$u_{xx}(P) = \frac{\alpha u(Q) - (1+\alpha)u(P) + u(q)}{h^2\alpha(\alpha+1)/2} + O(h)$$

Similarly,

$$u_{yy}(P) = \frac{\beta u(R) - (1+\beta)u(P) + u(r)}{h^2\beta(\beta+1)/2} + O(h)$$

Thus, an $O(h)$ approximation to Poisson's equation at P is

$$\frac{U(Q)}{\alpha+1} + \frac{U(R)}{\beta+1} - \left(\frac{1}{\alpha} + \frac{1}{\beta}\right) U(P) + \frac{U(q)}{\alpha(\alpha+1)} + \frac{U(r)}{\beta(\beta+1)} = \frac{h^2}{2} f(P) \quad (1)$$

11.5 Making use of Problem 11.4, approximate the solution to

$$\begin{aligned} u_{xx} + u_{yy} &= 0 & x^2+y^2<1,\ y>0 \\ u(x, y) &= 100 & x^2+y^2=1,\ y>0 \\ u(x, y) &= 0 & y=0,\ -1<x<1 \end{aligned}$$

Choose a square grid with $h = 0.5$.

Symmetry about the y-axis allows us to reduce the number of unknowns in the difference system from three to two: we need only consider Laplace's equation on the quarter-disk, with boundary conditions as indicated in Fig. 11-5. From these boundary conditions,

$$U_{00} = 0 \qquad U_{10} = 0 \qquad U_{02} = 100 \qquad U(q) = 100 \qquad U(r) = 100 \qquad U_{11} - U_{-1,1} = 0$$

The only grid points at which u must be estimated are $P = (x_1, y_1)$ and $Q = (x_0, y_1)$.

The difference equation centered at Q is

$$U_{11} + U_{02} - 4U_{01} + U_{-1,1} + U_{00} = 0$$

which, by the boundary conditions, simplifies to

$$2U_{01} - U_{11} = 50 \quad (1)$$

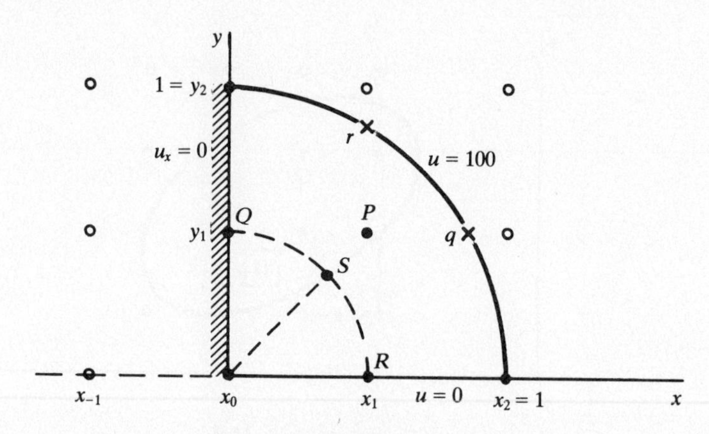

Fig. 11-5

The coordinates of q and r are $(\sqrt{3}h, h)$ and $(h, \sqrt{3}h)$; hence, in the notation of Problem 11.4, $\alpha = \beta = \sqrt{3} - 1$. Now, by (*1*) of Problem 11.4, the difference equation centered at P is

$$\frac{U_{01}}{\sqrt{3}} + \frac{U_{10}}{\sqrt{3}} - \frac{2U_{11}}{\sqrt{3}-1} + \frac{U(q)}{3-\sqrt{3}} + \frac{U(r)}{3-\sqrt{3}} = 0$$

which, in view of the boundary conditions, simplifies to

$$(1-\sqrt{3})U_{01} + 2\sqrt{3}\,U_{11} = 200 \tag{2}$$

Solving (*1*)–(*2*), we find

$$U_{01} \equiv U(Q) = \frac{100(2+\sqrt{3})}{1+3\sqrt{3}} \approx 60.2 \qquad U_{11} \equiv U(P) = \frac{50(7+\sqrt{3})}{1+3\sqrt{3}} \approx 70.5$$

11.6 (*a*) Show how to apply finite differences to Laplace's equation in polar coordinates,

$$\frac{\partial^2 u}{\partial r^2} + \frac{1}{r}\frac{\partial u}{\partial r} + \frac{1}{r^2}\frac{\partial^2 u}{\partial \theta^2} = 0$$

(*b*) Rework Problem 11.5 in polar coordinates, on a mesh with $\Delta r = 0.5$ and $\Delta\theta = \pi/4$.

(*a*) Define the grid $(r_m, \theta_n) = (m\,\Delta r, n\,\Delta\theta)$, where $m, n = 0, 1, 2, \ldots$, and let U_{mn} be an approximation to $u(r_m, \theta_n)$. Using centered differences to approximate each derivative in Laplace's equation, we obtain, after grouping like terms,

$$\left(1-\frac{1}{2m}\right)U_{m-1,n} + \frac{1}{(m\,\Delta\theta)^2}U_{m,n-1} - 2\left[1+\frac{1}{(m\,\Delta\theta)^2}\right]U_{mn} + \left(1+\frac{1}{2m}\right)U_{m+1,n} + \frac{1}{(m\,\Delta\theta)^2}U_{m,n+1} = 0 \tag{1}$$

(*b*) On the polar grid, the sole unknowns are $U(S) \equiv U_{11}$ and $U(Q) \equiv U_{12}$. (See Fig. 11-5; the symmetry condition along the vertical axis is now $u_\theta = 0$, which yields the numerical condition $U_{13} = U_{11}$.) Application of (*1*) of part (*a*), centered at S and Q, and substitution of boundary values, gives the two equations

$$2[1+(\Delta\theta)^2]U_{11} - U_{12} = 150(\Delta\theta)^2$$
$$-U_{11} + [1+(\Delta\theta)^2]U_{12} = 75(\Delta\theta)^2$$

These yield ($\Delta\theta = \pi/4$):

$$U_{11} \equiv U(S) \approx 46.3 \qquad U_{12} \equiv U(Q) \approx 57.3$$

Now, the boundary value problem under consideration can readily be solved analytically (by conformal mapping or by letting $a \to 0$ in Problem 7.39). The exact solution yields

$$u(Q) = \frac{400}{\pi} \arctan \frac{1}{2} \approx 59.0$$

which shows that the coarse meshes used above and in Problem 11.5 have produced quite accurate results.

11.7 (*a*) Formulate difference equations for

$$au_{xx} + bu_{yy} + cu_x + du_y + eu = f(x, y) \qquad \text{in } \Omega \tag{1}$$

$$u = g(x, y) \qquad \text{on } S \tag{2}$$

where Ω is the rectangle $0 < x < Mh$, $0 < y < Nh$. The coefficients are allowed to depend on x and y, provided $a, b > 0$ (elliptic PDE), and $e \leq 0$, in Ω. (*b*) Show that if the mesh spacing h is chosen sufficiently small (while M and N are made correspondingly large), the system of difference equations has a unique solution.

(*a*) With $(x_m, y_n) = (mh, nh)$,

$$(au_{xx})_{mn} = a_{mn} \frac{u_{m-1,n} - 2u_{mn} + u_{m+1,n}}{h^2} + O(h^2) \qquad (cu_x)_{mn} = c_{mn} \frac{u_{m+1,n} - u_{m-1,n}}{2h} + O(h^2)$$

$$(bu_{yy})_{mn} = b_{mn} \frac{u_{m,n-1} - 2u_{mn} + u_{m,n+1}}{h^2} + O(h^2) \qquad (du_y)_{mn} = d_{mn} \frac{d_{m,n+1} - u_{m,n-1}}{2h} + O(h^2)$$

These approximations, substituted in (*1*), yield the difference equation

$$\alpha_0 U_{mn} - \alpha_1 U_{m-1,n} - \alpha_2 U_{m,n-1} - \alpha_3 U_{m+1,n} - \alpha_4 U_{m,n+1} = -h^2 f_{mn} \tag{3}$$

where $\alpha_0 \equiv (2a + 2b - h^2 e)_{mn}$,

$$\alpha_1 \equiv \left(a - \frac{ch}{2}\right)_{mn} \qquad \alpha_3 \equiv \left(a + \frac{ch}{2}\right)_{mn}$$

$$\alpha_2 \equiv \left(b - \frac{dh}{2}\right)_{mn} \qquad \alpha_4 \equiv \left(b + \frac{dh}{2}\right)_{mn}$$

Equation (*3*) is required to hold for $m = 1, 2, \ldots, M-1$; $n = 1, 2, \ldots, N-1$. The boundary values for U_{mn} are obtained from (*2*).

(*b*) Since $e \leq 0$, we have, for all m and n,

$$\alpha_0 \geq \alpha_1 + \alpha_2 + \alpha_3 + \alpha_4 \tag{4}$$

Also, since $a > 0$ and $b > 0$, it follows that for h sufficiently small,

$$\alpha_i > 0 \qquad (i = 0, \ldots, 4) \tag{5}$$

Now, system (*3*) has a unique solution if and only if its homogeneous version, obtained by taking $f \equiv 0$ and $g \equiv 0$ in (*1*)–(*2*), has only the zero solution. If, contrary to what we wish to prove, the homogeneous (*3*) has a nonzero solution, we may suppose that the largest component, $U_{\mu\nu}$, of this solution is positive. Then, from (*3*) with $f_{\mu\nu} = 0$,

$$\alpha_0 U_{\mu\nu} = \alpha_1 U_{\mu-1,\nu} + \alpha_2 U_{\mu,\nu-1} + \alpha_3 U_{\mu+1,\nu} + \alpha_4 U_{\mu,\nu+1}$$

which, together with (*4*) and $U_{\mu\nu} > 0$, implies

$$\alpha_1(U_{\mu\nu} - U_{\mu-1,\nu}) + \alpha_2(U_{\mu\nu} - U_{\mu,\nu-1}) + \alpha_3(U_{\mu\nu} - U_{\mu+1,\nu}) + \alpha_4(U_{\mu\nu} - U_{\mu,\nu+1}) \leq 0 \tag{6}$$

In view of (*5*), (*6*) can hold only if $U = U_{\mu\nu} = \max$ at all four neighbors of (x_μ, y_ν). Repetition of this argument leads to a boundary node at which $U = U_{\mu\nu} > 0$, which contradicts the assumption $g \equiv 0$. The proof is now complete.

11.8 (*a*) Show how to express an invertible matrix **A** of order N as the product of a lower-triangular matrix **L** and an upper-triangular matrix **U**. (*b*) Carry out the factorization of the matrix **A** that would result if h were taken to be $\ell/3$ in Example 11.1.

(*a*) The matrix factorization **LU** = **A** may be written componentwise as

$$\sum_{k=1}^{N} l_{ik}u_{kj} = a_{ij} \qquad (i, j = 1, 2, \ldots, N) \tag{1}$$

For **L** and **U** to be respectively lower- and upper-triangular, $l_{ij} = 0$ for $j > i$ and $u_{ij} = 0$ for $j < i$. Setting $i = j = 1$ in (*1*), we find $l_{11}u_{11} = a_{11}$; thus the diagonal elements of **L** and **U** are not uniquely determined. We shall choose all $l_{ii} = 1$. From (*1*) it then follows that the rows of **U** and the columns of **L** can be found by applying the pair of formulas

$$u_{ij} = a_{ij} - \sum_{k=1}^{i-1} l_{ik}u_{kj} \qquad (j = i, i+1, \ldots, N) \tag{2}$$

$$l_{ij} = \left(a_{ij} - \sum_{k=1}^{j-1} l_{ik}u_{kj}\right) \Big/ u_{jj} \qquad (i = j+1, j+2, \ldots, N) \tag{3}$$

in the order $i = 1, j = 1, i = 2, j = 2, \ldots, j = N-1, i = N$. The sums in (*2*) and (*3*) are understood to be zero whenever the upper limit of summation is less than one. Because det (**A**) $= \Pi u_{jj} \neq 0$, the right side of (*3*) is always well-defined.

(*b*) The choice $h = \ell/3$ produces the 4×4 matrix

$$\mathbf{A} = \begin{bmatrix} 4 & -1 & -1 & 0 \\ -1 & 4 & 0 & -1 \\ -1 & 0 & 4 & -1 \\ 0 & -1 & -1 & 4 \end{bmatrix}$$

From (*2*), with $i = 1$: $u_{11} = 4$, $u_{12} = -1$, $u_{13} = -1$, $u_{14} = 0$.

From (*3*), with $j = 1$: $l_{21} = -1/4$, $l_{31} = -1/4$, $l_{41} = 0$.

From (*2*), with $i = 2$: $u_{22} = 15/4$, $u_{23} = -1/4$, $u_{24} = -1$.

From (*3*), with $j = 2$: $l_{32} = -1/15$, $l_{42} = -4/15$.

From (*2*), with $i = 3$: $u_{33} = 56/15$, $u_{34} = -16/15$.

From (*3*), with $j = 3$: $l_{43} = -2/7$.

From (*2*), with $i = 4$: $u_{44} = 24/7$.

Now we have obtained the desired factorization:

$$\begin{bmatrix} 1 & 0 & 0 & 0 \\ -\frac{1}{4} & 1 & 0 & 0 \\ -\frac{1}{4} & -\frac{1}{15} & 1 & 0 \\ 0 & -\frac{4}{15} & -\frac{2}{7} & 1 \end{bmatrix} \begin{bmatrix} 4 & -1 & -1 & 0 \\ 0 & \frac{15}{4} & -\frac{1}{4} & -1 \\ 0 & 0 & \frac{56}{15} & -\frac{16}{15} \\ 0 & 0 & 0 & \frac{24}{7} \end{bmatrix} = \begin{bmatrix} 4 & -1 & -1 & 0 \\ -1 & 4 & 0 & 0 \\ -1 & 0 & 4 & -1 \\ 0 & -1 & -1 & 4 \end{bmatrix}$$

11.9 Once a system **AV** = **B** has been expressed in the form **LUV** = **B**, show how it can be solved for **V** by a forward substitution followed by a backward substitution.

In the system **LUV** = **B**, let **W** = **UV**. Then **AV** = **B** can be written **LW** = **B**, or

$$\sum_{j=1}^{N} l_{ij}W_j = B_i \qquad (i = 1, 2, \ldots, N)$$

Since $l_{ij} = 0$ for $i < j$, we can easily solve for **W** as follows:

$$
\begin{aligned}
W_1 &= B_1/l_{11} \\
W_2 &= (B_2 - l_{21}W_1)/l_{22} \\
W_3 &= (B_3 - l_{31}W_1 - l_{32}W_2)/l_{33} \\
&\cdots\cdots\cdots\cdots\cdots \\
W_i &= \left(B_i - \sum_{j=1}^{i-1} l_{ij}W_j\right)\Big/ l_{ii}
\end{aligned}
$$

Now that $\mathbf{W}$ is known, the system $\mathbf{UV} = \mathbf{W}$, or

$$\sum_{j=1}^{N} u_{ij}V_j = W_i \qquad (i = 1, 2, \ldots, N)$$

with $u_{ij} = 0$ for $i > j$, can be solved for $\mathbf{V}$ as follows:

$$
\begin{aligned}
V_N &= W_N/u_{NN} \\
V_{N-1} &= (W_{N-1} - u_{N-1,N}W_N)/u_{N-1,N-1} \\
V_{N-2} &= (W_{N-2} - u_{N-2,N}W_N - u_{N-2,N-1}W_{N-1})/u_{N-1,N-1} \\
&\cdots\cdots\cdots\cdots\cdots \\
V_{N-i} &= \left(W_{N-i} - \sum_{j=1}^{i} u_{N-i,N-1-j}W_{N-1-j}\right)\Big/ u_{N-i,N-i}
\end{aligned}
$$

For the solution of a single linear system, the work required in an **LU**-decomposition is the same as that required in straightforward Gaussian elimination. One attractive feature of the **LU**-decomposition approach is that once **L** and **U** have been found for a given **A**, it is possible to solve $\mathbf{AV} = \mathbf{B}$ for any and all right-hand sides **B** just by using the forward-backward substitution method outlined above.

11.10 Write the matrix **A** of Example 11.1 in block tridiagonal form and obtain a block **LU**-decomposition of **A**.

With

$$\mathbf{H} \equiv \begin{bmatrix} 4 & -1 & 0 \\ -1 & 4 & -1 \\ 0 & -1 & 4 \end{bmatrix} \qquad \mathbf{I} \equiv \begin{bmatrix} 1 & 0 & 0 \\ 0 & 1 & 0 \\ 0 & 0 & 1 \end{bmatrix} \qquad \mathbf{0} = \begin{bmatrix} 0 & 0 & 0 \\ 0 & 0 & 0 \\ 0 & 0 & 0 \end{bmatrix}$$

we have

$$\mathbf{A} = \begin{bmatrix} \mathbf{H} & -\mathbf{I} & \mathbf{0} \\ -\mathbf{I} & \mathbf{H} & -\mathbf{I} \\ \mathbf{0} & -\mathbf{I} & \mathbf{H} \end{bmatrix}$$

By multiplying out the left side of the desired decomposition,

$$\begin{bmatrix} \mathbf{I} & \mathbf{0} & \mathbf{0} \\ \mathbf{L}_{21} & \mathbf{I} & \mathbf{0} \\ \mathbf{L}_{31} & \mathbf{L}_{32} & \mathbf{I} \end{bmatrix} \begin{bmatrix} \mathbf{U}_{11} & \mathbf{U}_{12} & \mathbf{U}_{13} \\ \mathbf{0} & \mathbf{U}_{22} & \mathbf{U}_{23} \\ \mathbf{0} & \mathbf{0} & \mathbf{U}_{33} \end{bmatrix} = \begin{bmatrix} \mathbf{H} & -\mathbf{I} & \mathbf{0} \\ -\mathbf{I} & \mathbf{H} & -\mathbf{I} \\ \mathbf{0} & -\mathbf{I} & \mathbf{H} \end{bmatrix}$$

we obtain, in succession, $\mathbf{U}_{11} = \mathbf{H}$, $\mathbf{U}_{12} = -\mathbf{I}$, $\mathbf{U}_{13} = \mathbf{0}$, $\mathbf{L}_{21} = -\mathbf{H}^{-1}$, $\mathbf{L}_{31} = \mathbf{0}$, $\mathbf{U}_{22} = \mathbf{H} - \mathbf{H}^{-1}$, $\mathbf{U}_{23} = -\mathbf{I}$, $\mathbf{L}_{32} = -(\mathbf{H} - \mathbf{H}^{-1})^{-1}$, $\mathbf{U}_{33} = \mathbf{H} + (\mathbf{H} - \mathbf{H}^{-1})^{-1}$.

11.11 Determine the eigenvalues and the corresponding eigenfunctions of the $N \times N$ tridiagonal matrix

$$\begin{bmatrix} b & c & & & & 0 \\ a & b & c & & & \\ & a & b & c & & \\ & & \cdot & \cdot & \cdot & \\ & & & a & b & c \\ 0 & & & & a & b \end{bmatrix}$$

If we set $U_0 = U_{N+1} = 0$, then the eigenvalue problem $\mathbf{AU} = \lambda\mathbf{U}$ can be expressed as

$$aU_{n-1} + (b-\lambda)U_n + cU_{n+1} = 0 \qquad (n = 1, 2, \ldots, N) \tag{1}$$

If we look for a solution of (*1*) with U_n proportional to r^n, then (*1*) implies that r must satisfy the quadratic equation

$$cr^2 + (b-\lambda)r + a = 0 \tag{2}$$

With r_1 and r_2 the solutions to (*2*), set $U_n = \alpha r_1^n + \beta r_2^n$, where α and β are constants to be determined. The end conditions $U_0 = U_{N+1} = 0$ require

$$\alpha + \beta = 0 \qquad \text{and} \qquad \alpha r_1^{N+1} + \beta r_2^{N+1} = 0 \tag{3}$$

from which $(r_1/r_2)^{N+1} = 1$, or

$$\frac{r_1}{r_2} = e^{i2k\pi/(N+1)} \qquad (k = 1, 2, \ldots, N) \tag{4}$$

where we have disallowed $r_1 = r_2$. The product of the roots of (*2*) is

$$r_1 r_2 = \frac{a}{c} \qquad \left(\frac{a}{c} \neq 0 \text{ assumed}\right) \tag{5}$$

Together, (*4*) and (*5*) yield

$$r_1 = \sqrt{a/c}\, e^{ik\pi/(N+1)} \qquad r_2 = \sqrt{a/c}\, e^{-ik\pi/(N+1)} \tag{6}$$

which, substituted in the expression for the sum of the roots of (*2*),

$$r_1 + r_2 = -\frac{b-\lambda}{c}$$

determine the eigenvalues λ_j as

$$\lambda_j = b + 2c\sqrt{\frac{a}{c}}\cos\frac{j\pi}{N+1} \qquad (j = 1, 2, \ldots, N)$$

Note that if a and c are of like sign (the usual case), the λ_j are all real.

Using (*6*) and (*3*) to determine α and β, we find that the nth component of an eigenvector $\mathbf{U}^k$ corresponding to λ_k is given by

$$U_n^k = \left(\sqrt{\frac{a}{c}}\right)^n \sin\frac{nk\pi}{N+1}$$

11.12 Let $\mathbf{H}$ be an $M \times M$ matrix with M distinct eigenvalues, $\lambda_1, \lambda_2, \ldots, \lambda_M$, and consider the eigenvalue problem $\mathbf{AV} = \gamma\mathbf{V}$, where $\mathbf{A}$ is the $N \times N$ block tridiagonal matrix

$$\begin{bmatrix} \mathbf{H} & -\mathbf{I} & & & \mathbf{0} \\ -\mathbf{I} & \mathbf{H} & -\mathbf{I} & & \\ & -\mathbf{I} & \mathbf{H} & -\mathbf{I} & \\ & \cdot & \cdot & \cdot & \\ & & -\mathbf{I} & \mathbf{H} & -\mathbf{I} \\ \mathbf{0} & & & -\mathbf{I} & \mathbf{H} \end{bmatrix}$$

Here, $\mathbf{0}$ and $\mathbf{I}$ are the $M \times M$ null and identity matrices. Calculate the eigenvalues of $\mathbf{A}$.

Let $\mathbf{U}^k$ be an eigenvector of $\mathbf{H}$ corresponding to the eigenvalue λ_k. In the eigenvalue problem $\mathbf{AV} = \gamma\mathbf{V}$ we will look for an eigenvector $\mathbf{V}$ in the form

$$\mathbf{V} = [\alpha_1(\mathbf{U}^k)^T, \alpha_2(\mathbf{U}^k)^T, \ldots, \alpha_N(\mathbf{U}^k)^T]^T$$

for scalars α_i not all zero. From $\mathbf{AV} = \gamma\mathbf{V}$,

$$\begin{aligned}
\alpha_1 \mathbf{H}\mathbf{U}^k - \alpha_2 \mathbf{I}\mathbf{U}^k \quad &= \alpha_1 \mathbf{U}^k \\
-\alpha_1 \mathbf{I}\mathbf{U}^k + \alpha_2 \mathbf{H}\mathbf{U}^k - \alpha_3 \mathbf{I}\mathbf{U}^k \quad &= \alpha_2 \mathbf{U}^k \\
\cdots\cdots\cdots\cdots\cdots\cdots\cdots\cdots\cdots \\
\alpha_{N-1}\mathbf{I}\mathbf{U}^k + \alpha_N \mathbf{H}\mathbf{U}^k &= \alpha_N \mathbf{U}^k
\end{aligned}$$

which, after using the condition $\mathbf{H}\mathbf{U}^k = \lambda_k \mathbf{U}^k$ ($\mathbf{U}^k \neq \mathbf{0}$), leads to

$$\begin{bmatrix} \lambda_k & -1 & & & & 0 \\ -1 & \lambda_k & -1 & & & \\ & -1 & \lambda_k & -1 & & \\ & & \cdot & \cdot & \cdot & \\ & & & -1 & \lambda_k & -1 \\ 0 & & & & -1 & \lambda_k \end{bmatrix} \begin{bmatrix} \alpha_1 \\ \alpha_2 \\ \alpha_3 \\ \cdots \\ \alpha_{N-1} \\ \alpha_N \end{bmatrix} = \begin{bmatrix} \alpha_1 \\ \alpha_2 \\ \alpha_3 \\ \cdots \\ \alpha_{N-1} \\ \alpha_N \end{bmatrix} \tag{1}$$

By Problem 11.1, the eigenvalues γ_{jk} of (*1*) are given by

$$\gamma_{jk} = \lambda_k - 2\cos\frac{j\pi}{N+1} \tag{2}$$

As k ranges from 1 to M and j ranges from 1 to N, (*2*) yields the MN eigenvalues of $\mathbf{A}$.

11.13 Find the eigenvalues of the matrix $\mathbf{A}$ of Example 11.1.

Problem 11.10 gives

$$\mathbf{A} = \begin{bmatrix} \mathbf{H} & -\mathbf{I} & \mathbf{0} \\ -\mathbf{I} & \mathbf{H} & -\mathbf{I} \\ \mathbf{0} & -\mathbf{I} & \mathbf{H} \end{bmatrix} \qquad \text{where} \qquad \mathbf{H} = \begin{bmatrix} 4 & -1 & 0 \\ -1 & 4 & -1 \\ 0 & -1 & 4 \end{bmatrix}$$

By Problem 11.12, the eigenvalues of $\mathbf{A}$ are given by

$$\gamma_{jk} = \lambda_k - 2\cos\frac{j\pi}{4} \qquad (j = 1, 2, 3)$$

where λ_k, the kth eigenvalue of $\mathbf{H}$, is given by Problem 11.11 as

$$\lambda_k = 4 - 2\cos\frac{k\pi}{4} \qquad (k = 1, 2, 3)$$

Thus, the eigenvalues of $\mathbf{A}$ are given by

$$\gamma_{jk} = 4 - 2\left(\cos\frac{j\pi}{4} + \cos\frac{k\pi}{4}\right) \qquad (j, k = 1, 2, 3)$$

11.14 Calculate the spectral radius, $\rho(\mathbf{T}_J)$, of the Jacobi iteration matrix corresponding to the matrix $\mathbf{A}$ of Example 11.1.

With $\mathbf{A} = -\mathbf{L} + \mathbf{D} - \mathbf{U}$ and $\mathbf{T}_J = \mathbf{D}^{-1}(\mathbf{L} + \mathbf{U})$,

$$\mathbf{AV} = \gamma\mathbf{V} \Rightarrow (-\mathbf{L} + \mathbf{D} - \mathbf{U})\mathbf{V} = \gamma\mathbf{V} \Rightarrow \mathbf{DV} = (\mathbf{L} + \mathbf{U})\mathbf{V} + \gamma\mathbf{V} \Rightarrow \mathbf{T}_J\mathbf{V} = (\mathbf{I} - \gamma\mathbf{D}^{-1})\mathbf{V}$$

Since $d_{ii} = 4$, $\mathbf{D}^{-1} = (1/4)\mathbf{I}$ and the last equation becomes

$$\mathbf{T}_J\mathbf{V} = \left(1 - \frac{\gamma}{4}\right)\mathbf{V} \tag{1}$$

Now, the eigenvalues of $\mathbf{A}$ were found in Problem 11.13 to be

$$\gamma_{jk} = 4 - 2\left(\cos\frac{j\pi}{4} + \cos\frac{k\pi}{4}\right) \qquad (j, k = 1, 2, 3)$$

which, together with (*1*), implies that the eigenvalues of $\mathbf{T}_J$ are given by

$$\mu_{jk} = \frac{1}{2}\left(\cos\frac{j\pi}{4} + \cos\frac{k\pi}{4}\right) \qquad (j, k = 1, 2, 3) \tag{2}$$

The largest in magnitude of these eigenvalues is μ_{11}; hence, $\rho(\mathbf{T}_J) = \sqrt{2}/2$.

11.15 For the system of Example 11.1, write the Jacobi row iteration method, (*11.8*), in matrix form and determine the spectral radius of the resulting iteration matrix, $\mathbf{T}$.

By Problem 11.10, the system can be represented as

$$\begin{bmatrix} \mathbf{H} & -\mathbf{I} & \mathbf{0} \\ -\mathbf{I} & \mathbf{H} & -\mathbf{I} \\ \mathbf{0} & -\mathbf{I} & \mathbf{H} \end{bmatrix} \begin{bmatrix} \mathbf{v}_1 \\ \mathbf{v}_4 \\ \mathbf{v}_7 \end{bmatrix} = \begin{bmatrix} \mathbf{c}_1 \\ \mathbf{c}_4 \\ \mathbf{c}_7 \end{bmatrix} \tag{1}$$

where $\mathbf{v}_i = [U_i, U_{i+1}, U_{i+2}]^T$ and $\mathbf{c}_i = [B_i, B_{i+1}, B_{i+2}]^T$ for $i = 1, 4, 7$. Let

$$L \equiv \begin{bmatrix} \mathbf{0} & \mathbf{0} & \mathbf{0} \\ \mathbf{I} & \mathbf{0} & \mathbf{0} \\ \mathbf{0} & \mathbf{I} & \mathbf{0} \end{bmatrix} \qquad D \equiv \begin{bmatrix} \mathbf{H} & \mathbf{0} & \mathbf{0} \\ \mathbf{0} & \mathbf{H} & \mathbf{0} \\ \mathbf{0} & \mathbf{0} & \mathbf{H} \end{bmatrix} \qquad U \equiv \begin{bmatrix} \mathbf{0} & \mathbf{I} & \mathbf{0} \\ \mathbf{0} & \mathbf{0} & \mathbf{I} \\ \mathbf{0} & \mathbf{0} & \mathbf{0} \end{bmatrix}$$

where D is invertible (because $\mathbf{H}$ is invertible). Then (*1*) may be written

$$(-L + D - U)\mathbf{V} = \mathbf{c} \qquad \text{or} \qquad \mathbf{V} = D^{-1}(L + U)\mathbf{V} + D^{-1}\mathbf{c} \tag{2}$$

where $\mathbf{V} = [\mathbf{v}_1^T, \mathbf{v}_4^T, \mathbf{v}_7^T]^T = \mathbf{U}$ and $\mathbf{c} = [\mathbf{c}_1^T, \mathbf{c}_4^T, \mathbf{c}_7^T]^T = \mathbf{B}$.

The second equation (*2*) is equivalent to the fixed-point iteration

$$\mathbf{V}^{k+1} = D^{-1}(L + U)\mathbf{V}^k + D^{-1}\mathbf{c} \tag{3}$$

and (*3*) is identical to (*11.8*). In fact, since

$$D^{-1}(L + U) = \begin{bmatrix} \mathbf{0} & \mathbf{H}^{-1} & \mathbf{0} \\ \mathbf{H}^{-1} & \mathbf{0} & \mathbf{H}^{-1} \\ \mathbf{0} & \mathbf{H}^{-1} & \mathbf{0} \end{bmatrix} \equiv \mathbf{T} \tag{4}$$

(*3*) yields

$$\begin{aligned} \mathbf{v}_1^{k+1} &= \mathbf{H}^{-1}\mathbf{v}_4^k + \cdots \\ \mathbf{v}_4^{k+1} &= \mathbf{H}^{-1}(\mathbf{v}_1^k + \mathbf{v}_7^k) + \cdots \\ \mathbf{v}_7^{k+1} &= \mathbf{H}^{-1}\mathbf{v}_4^k + \cdots \end{aligned}$$

which is just the "solved form" of (*11.8*) in single-index notation.

To determine the eigenvalues μ of $\mathbf{T}$, let $\mathbf{w}$ be an eigenvector of $\mathbf{H}$ corresponding to the eigenvalue λ_k and look for eigenvectors of $\mathbf{T}$ of the form $\mathbf{V} = [\alpha_1\mathbf{w}^T, \alpha_4\mathbf{w}^T, \alpha_7\mathbf{w}^T]^T$. $\mathbf{TV} = \mu\mathbf{V}$ implies $(L + U)\mathbf{V} = \mu D\mathbf{V}$, or

$$\begin{aligned} \alpha_4\mathbf{w} \quad &= \alpha_1\mu\mathbf{H}\mathbf{w} = \mu\lambda_k\alpha_1\mathbf{w} \\ \alpha_1\mathbf{w} + \alpha_7\mathbf{w} &= \alpha_4\mu\mathbf{H}\mathbf{w} = \mu\lambda_k\alpha_4\mathbf{w} \\ \alpha_4\mathbf{w} \quad &= \alpha_7\mu\mathbf{H}\mathbf{w} = \mu\lambda_k\alpha_7\mathbf{w} \end{aligned}$$

Since $\mathbf{w} \neq \mathbf{0}$, we now have

$$\begin{bmatrix} 0 & 1 & 0 \\ 1 & 0 & 1 \\ 0 & 1 & 0 \end{bmatrix} \begin{bmatrix} \alpha_1 \\ \alpha_4 \\ \alpha_7 \end{bmatrix} = \mu\lambda_k \begin{bmatrix} \alpha_1 \\ \alpha_4 \\ \alpha_7 \end{bmatrix}$$

By Problem 11.11, the eigenvalues of the above tridiagonal matrix are

$$\mu\lambda_k = 2\cos\frac{j\pi}{4} \qquad (j = 1, 2, 3)$$

and the eigenvalues of $\mathbf{H}$ are

$$\lambda_k = 4 - 2\cos\frac{k\pi}{4} \qquad (k = 1, 2, 3)$$

Thus, the eigenvalues of the matrix **T** are

$$\mu_{jk} = \frac{\cos(j\pi/4)}{2 - \cos(k\pi/4)} \qquad (j, k = 1, 2, 3) \tag{5}$$

From (*5*),

$$\rho(\mathbf{T}) = \mu_{11} = \frac{\sqrt{2}}{4 - \sqrt{2}}$$

Note that this is smaller than $\sqrt{2}/2$, the spectral radius of the point Jacobi iteration matrix for the same problem, as found in Problem 11.14. Equation (*11.15*) can be established for block tridiagonal matrices of the form (*1*). It follows that the spectral radius of the row Gauss–Seidel method for this problem is $2/(4-\sqrt{2})^2$.

11.16 Write a computer program which uses the SOR method (*11.13*) to approximate the solution to the boundary value problem

$$\begin{aligned} u_{xx} + u_{yy} &= 0 \qquad \text{in } \Omega\text{: } 0 < x, y < 1 \\ u(x, y) &= e^{2\pi x} \sin 2\pi y \qquad \text{on } S \end{aligned}$$

Choosing a mesh spacing $h = 0.1$, run the program at $\omega = \bar{\omega}$ (optimal SOR) and at $\omega = 1$ (Gauss–Seidel method).

Figure 11-6 lists a program. From Table 11-1,

$$\rho(\mathbf{T}_J) = \cos\frac{\pi h}{\ell} = \cos\frac{\pi}{10}$$

for this problem. It follows from (*11.16*) that $\bar{\omega} = 1.528$ is the optimal relaxation parameter for SOR. The comparisons with the exact solution ($u = e^{2\pi x}\sin 2\pi y$ in Ω) given in Table 11-2 were obtained by iterating until the maximum residual was less than 0.005. Note that the Gauss–Seidel method required 67 iterations, while SOR required only 24. For small choices of mesh spacing h, the superior convergence

Table 11-2

	Gauss–Seidel (K = 67)	Optimal SOR (K = 24)	Exact
M, N = 1, 1	1.249327	1.246912	1.101777
M, N = 2, 1	2.386185	2.381623	2.065233
M, N = 2, 2	3.862067	3.852746	3.341618
M, N = 3, 1	4.431387	4.424993	3.871189
M, N = 3, 2	7.171626	7.159750	6.263716
M, N = 3, 3	7.175988	7.159992	6.263716
M, N = 4, 1	8.165332	8.158091	7.256373
M, N = 4, 2	13.213451	13.200061	11.741060
M, N = 4, 3	13.218328	13.200035	11.741060
M, N = 4, 4	8.179705	8.158471	7.256376
M, N = 5, 1	15.013905	15.006215	13.601753
M, N = 5, 2	24.294676	24.280386	22.008101
M, N = 5, 3	24.299553	24.280504	22.008101
M, N = 5, 4	15.028279	15.006590	13.601758
M, N = 5, 5	0.023098	0.001065	0.000003

```
      PROGRAM SOR
C     TITLE:  DEMO PROGRAM FOR GAUSS-SEIDEL OR
C             SOR METHOD FOR POISSON'S EQUATION
C             ON A RECTANGLE WITH A SQUARE GRID
C     INPUT:  MMAX, NUMBER OF X-SUBINTERVALS
C             NMAX, NUMBER OF Y-SUBINTERVALS
C             OMEGA, RELAXATION PARAMETER
C             KMAX, MAXIMUM NUMBER OF ITERATIONS
C             TOL, CONVERGENCE CRITERION FOR RESIDUALS
C             (X1,X2), X-INTERVAL
C             (Y1,Y2), Y-INTERVAL
C             G1(X), LOWER BOUNDARY CONDITION
C             G2(Y), RIGHT BOUNDARY CONDITION
C             G3(X), UPPER BOUNDARY CONDITION
C             G4(Y), LEFT BOUNDARY CONDITION
C             F(X,Y), RIGHT SIDE OF POISSON'S EQ.
C             E(X,Y), EXACT SOLUTION
C     OUTPUT: NUMERICAL AND EXACT SOLUTION
      COMMON U(0:51,0:51),V(0:51,0:51)
      DATA X1,X2,Y1,Y2/0,1,0,1/
      PI = 4*ATAN(1.)
      G1(X) = 0
      G2(Y) = EXP(2*PI)*SIN(2*PI*Y)
      G3(X) = 0
      G4(Y) = SIN(2*PI*Y)
      F(X,Y) = 0
      E(X,Y) = EXP(2*PI*X)*SIN(2*PI*Y)
      PRINT*,'ENTER GRID SPACING, H, AND RELAXATION PARAMETER'
      READ*,H,OMEGA
      PRINT*,'ENTER MAXIMUM ITERATION NUMBER, RESIDUAL TOLERANCE'
      READ*,KMAX,TOL
      MMAX =(X2-X1)/H
      NMAX =(Y2-Y1)/H
C     SET BOUNDARY VALUES AND INITIAL ESTIMATE TO SOLUTION
      DO 10 M = 1,MMAX-1
      DO 10 N = 1,NMAX-1
         X = X1 + M*H
         Y = Y1 + N*H
         U(M,N) = 0
         U(M,0) = G1(X)
         U(MMAX,N) = G2(Y)
         U(M,NMAX) = G3(X)
         U(0,N) = G4(Y)
10    CONTINUE
      DO 15 K = 1,KMAX
C     CALCULATE K-TH ITERATE
      DO 20 M = 1,MMAX-1
      DO 20 N = 1,NMAX-1
         X = X1 + M*H
         Y = Y1 + N*H
         UOLD = U(M,N)
         U(M,N) = (U(M+1,N)+U(M,N+1)+U(M-1,N)+U(M,N-1))/4
     1            - H*H*F(X,Y)/4
         U(M,N) = OMEGA*U(M,N) + (1-OMEGA)*UOLD
20    CONTINUE
C     CALCULATE THE MAXIMUM RESIDUAL
      RMAX = 0
      DO 30 M = 1,MMAX-1
      DO 30 N = 1,NMAX-1
         X = X1 + M*H
         Y = Y1 + N*H
         RES=-H*H*F(X,Y)+U(M+1,N)+U(M,N+1)+U(M-1,N)+U(M,N-1)-4*U(M,N)
         IF(ABS(RES).GT.RMAX) RMAX = ABS(RES)
30    CONTINUE
C     IF RMAX SUFFICIENTLY SMALL PRINT ANSWER
      IF(RMAX.LT.TOL) GOTO 40
C     IF RMAX EXCEEDS TOLERANCE AND K < KMAX PERFORM ANOTHER ITERATION
15    CONTINUE
      PRINT*,'CONVERGENCE CRITEREON WAS NOT MET'
40    CONTINUE
      WRITE(6,100)
      WRITE(6,110) H,KMAX,TOL,OMEGA
      WRITE(6,120) K,RMAX
      DO 50 M = 1,MMAX/2
      DO 50 N = 1,M
         X = X1 + M*H
         Y = Y1 + N*H
         WRITE(6,130) M,N,U(M,N),E(X,Y)
50    CONTINUE
100   FORMAT(///,T12,'RESULTS FROM PROGRAM SOR',/)
110   FORMAT('H='F5.2,'  KMAX=',I4,'  TOL=',E8.2,'  OMEGA=',F6.3,/)
120   FORMAT('K=',I4,'  RMAX=',E8.2,T25,'NUMERICAL',T42,'EXACT',/)
130   FORMAT( 'M,N =',I1,',',I1,T20,F13.6,T37,F13.6)
      END
```

Fig. 11-6

rate of the SOR method as against the Gauss–Seidel method becomes even more striking. It is seen that the numerical results are not in very good agreement with the analytical solution. To improve the numerical solution one might (i) drive the maximum residual below a smaller tolerance, or (ii) use a finer mesh. An analysis of the truncation error (see Problem 11.1) shows it to be the villain here: to improve the accuracy, a finer mesh is needed.

Supplementary Problems

11.17 Using a rectangular grid $(x_m, y_n) = (mh, nk)$, write a difference equation for the quasilinear PDE

$$(au_x)_x + (bu_y)_y + cu = f$$

where $a(x, y, u)$ and $b(x, y, u)$ are positive functions.

11.18 Let a flow field be given in Ω by $\mathbf{q} = -(au_x, bu_y)$, where $a(x, y, u) > 0$, $b(x, y, u) > 0$. Give a numerical method for finding u, if the net flux across the boundary of any subregion of Ω is zero.

11.19 Let Ω denote the square $0 < x < 1, 0 < y < 1$ and consider the boundary value problem

$$(au_x)_x + (bu_y)_y = 0 \qquad \text{in } \Omega$$
$$u = xy \qquad \text{on S}$$

Introduce a square grid, $(x_m, y_n) = (mh, nh)$, with $h = 0.25$, and center on each grid point a region

$$R_{mn}: \ x_m - \frac{h}{2} < x < x_m + \frac{h}{2}, \ y_n - \frac{h}{2} < y < y_n + \frac{h}{2}$$

Suppose that $a = b = 1$ except in R_{22}, where $a = b = 0$. (*a*) Using the result of Problem 11.18 and harmonic means for the coefficients—e.g.,

$$a_{m-1/2,n} = \frac{2a_{m-1,n}a_{mn}}{a_{m-1,n} + a_{mn}}$$

—write a difference system for U_{mn} ($m, n = 1, 2, 3$; U_{22} excepted). (*b*) Write out the Gauss–Seidel iteration equations (*11.6*) for the system of (*a*), assuming the nodes are scanned bottom-to-top, left-to-right. (*c*) Use the iteration equations to estimate the U_{mn}, and compare the values with those of the solution, $u = xy$, of the boundary value problem in which $a = b = 1$ throughout Ω.

11.20 Consider the boundary value problem defined in Fig. 11-7 (see page 186). With $h = 1/3$ and $(x_m, y_n) = (mh, nh)$, write the difference equations centered at (*a*) (x_2, y_2), (*b*) (x_3, y_5), (*c*) (x_5, y_1). (*d*) Obtain the remaining 15 difference equations and solve the system.

11.21 Consider the system of linear equations $\mathbf{AU} = \mathbf{B}$, where

$$\mathbf{A} = [a_{ij}] \qquad \mathbf{U} = (U_1, U_2, \ldots, U_n)^T \qquad \mathbf{B} = (B_1, B_2, \ldots, B_n)^T$$

If $a_{ii} \neq 0$, show that (*a*) Jacobi's point iterative method can be expressed by

$$U_i^{k+1} = \left(B_i - \sum_{j=1}^{i-1} a_{ij}U_j^k - \sum_{j=i+1}^{n} a_{ij}U_j^k\right) a_{ii}^{-1} \tag{1}$$

(*b*) the Gauss–Seidel point iterative method is given by

$$U_i^{k+1} = \left(B_i - \sum_{j=1}^{i-1} a_{ij}U_j^{k+1} - \sum_{j=i+1}^{n} a_{ij}U_j^k\right) a_{ii}^{-1} \tag{2}$$

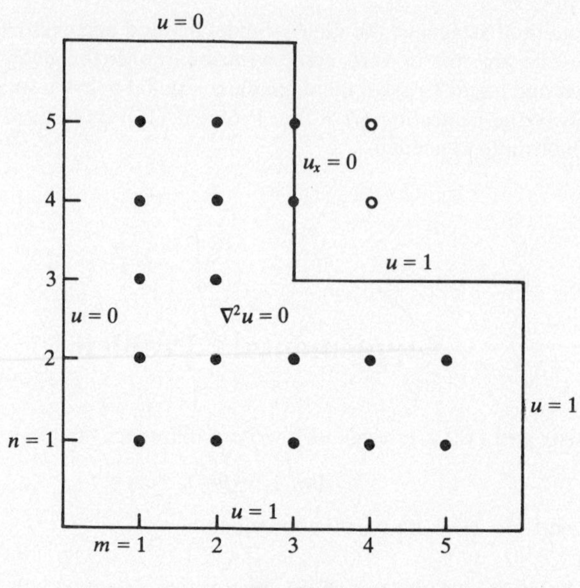

Fig. 11-7

11.22 Assume that (2) of Problem 11.21 has been used to obtain the approximate solution

$$(U_1^{k+1}, U_2^{k+1}, \ldots, U_{i-1}^{k+1}, U_i^k, U_{i+1}^k, \ldots, U_n^k)^T$$

to $\mathbf{AU} = \mathbf{B}$. The residual vector, $\mathbf{R}_i^{k+1}$, associated with this approximation depends on both the iteration counter, k, and the index, i, of the first "unimproved" component. Rewrite (2) in terms of R_{ii}^{k+1}, the ith component of $\mathbf{R}_i^{k+1}$.

11.23 Find an expression for the SOR point iterative method like that found for the Gauss–Seidel in Problem 11.22.

11.24 Formulate the column iteration counterparts of (11.8) and (11.9).

11.25 Determine the **LU**-factorization of

$$\mathbf{A} = \begin{bmatrix} 1 & 0 & 1 \\ 2 & 3 & 4 \\ 1 & 9 & 9 \end{bmatrix}$$

with $l_{ii} = 1$.

11.26 Given the elliptic boundary value problem

$$u_{xx} + u_{yy} + cu = f(x, y) \quad \text{in } \Omega\text{: } 0 < x, y < 3h \qquad (1)$$

$$u = g(x, y) \quad \text{on } S \qquad (2)$$

where c is a constant distinct from 4, (a) write a difference equation for (1) on the grid $(x_m, y_n) = (mh, nh)$; (b) with $\mathbf{U} \equiv (U_{11}, U_{21}, U_{12}, U_{22})^T$, determine the matrix $\mathbf{T}_J$ for the Jacobi point iteration method; (c) find the eigenvalues of $\mathbf{T}_J$ [*Hint*: solve the characteristic equation]; (d) determine the range of c-values for which $\rho(\mathbf{T}_J) < 1$.

11.27 Consider the *parabolic* problem

$$u_t = u_{xx} + u_{yy} \qquad x, y \text{ in } \Omega, t > 0 \tag{1}$$
$$u(x, y, 0) = f(x, y) \qquad x, y \text{ in } \Omega \tag{2}$$
$$u(x, y, t) = g(x, y, t) \qquad x, y \text{ on } S, t > 0 \tag{3}$$

Introduce a grid $(x_m, y_n, t_j) = (mh, nh, jk)$ and define

$$w(x, y) \equiv u(x, y, t_{j+1}) \qquad v(x, y) \equiv u(x, y, t_j)$$

Verify that the result of approximating (*1*) at time t_{j+1} using centered space differences and a backward time difference is the *elliptic* system (cf. Problem 11.26(*a*))

$$(e-4)W_{mn} + W_{m,n+1} + W_{m-1,n} + W_{m,n-1} + W_{m+1,n} = eV_{mn}$$

with $e \equiv -h^2/k$. Conclude that Jacobi's method or another technique of this chapter can be used to advance the solution of the parabolic problem from one time level to the next.

Chapter 12

Variational Formulation of Boundary Value Problems

12.1 INTRODUCTION

In certain cases, the solution of a boundary value problem for a PDE is also a solution of an associated calculus of variations problem. A typical problem in the calculus of variations is to find, for functions u belonging to a prescribed set $\mathscr{A}$, the extreme values of the integral expression

$$J[u(\mathbf{x})] = \int_\Omega F(\mathbf{x}, u(\mathbf{x}), \nabla u(\mathbf{x}))\, d\Omega$$

where F denotes a given function. Hence we shall begin by describing some of the structure of the domain $\mathscr{A}$ of J.

12.2 THE FUNCTION SPACE $L^2(\Omega)$

In Chapter 6 we considered the space $L^2(a, b)$ of functions $f(x)$ that are defined and square integrable on (a, b) in $\mathbf{R}^1$. More generally, let Ω denote a bounded region in $\mathbf{R}^n$ and consider the set $L^2(\Omega)$ of all real-valued functions $u(\mathbf{x})$ defined on Ω which satisfy

$$\int_\Omega u(\mathbf{x})^2\, d\Omega < \infty$$

Like $L^2(a, b)$, $L^2(\Omega)$ is a vector space over the real numbers, and the expected definition

$$\langle u, v\rangle \equiv \int_\Omega u(\mathbf{x})\, v(\mathbf{x})\, d\Omega$$

makes it an inner product space.

A subset of $L^2(\Omega)$ is said to be a *subspace* of $L^2(\Omega)$ if the subset is closed under the operation of forming linear combinations.

EXAMPLE 12.1 (*a*) For k a nonnegative integer, the subset $C^k(\bar{\Omega})$ of all u in $L^2(\Omega)$ which, together with all derivatives of order k or less, are continuous on $\bar{\Omega}$ is a subspace of $L^2(\Omega)$. (*b*) Let $u_1, \ldots, u_N$ denote N elements of $L^2(\Omega)$. The subset $\mathscr{M}$ of all linear combinations of $u_1, \ldots, u_N$ is a subspace of $L^2(\Omega)$—the subspace *spanned by* the u_i. (*c*) For m a positive integer, the subset $H^m(\Omega)$ of all u in $L^2(\Omega)$ whose derivatives of order m or less are also in $L^2(\Omega)$ is a subspace of $L^2(\Omega)$.

A subspace $\mathscr{M}$ of $L^2(\Omega)$ is *dense in* $L^2(\Omega)$ if for any $\epsilon > 0$ and any f in $L^2(\Omega)$ there exists a v in $\mathscr{M}$ such that

$$\|v - f\|^2 \equiv \int_\Omega (v - f)^2\, d\Omega < \epsilon$$

i.e., if any f in $L^2(\Omega)$ can be approximated with arbitrary precision in the least-squares sense by a function from $\mathscr{M}$.

Theorem 12.1: For each positive integer m, the following subspaces are dense in $L^2(\Omega)$: $C^m(\bar{\Omega})$, $H^m(\Omega)$, and the set of all u in $C^m(\bar{\Omega})$ such that $u = 0$ on S, the boundary of Ω.

Theorem 12.2: If $\mathscr{M}$ is a dense subspace in $L^2(\Omega)$ and if an element u of $L^2(\Omega)$ satisfies $\langle u, v\rangle = 0$ for all v in $\mathscr{M}$, then $u = 0$.

12.3 THE CALCULUS OF VARIATIONS

A real-valued function J whose domain of definition is a subset of $L^2(\Omega)$ will be called a *functional.* The fundamental problem of the calculus of variations is, then, to find the extreme values of a given functional J over a specified domain $\mathscr{A}$ in $L^2(\Omega)$.

EXAMPLE 12.2 (*a*) For Ω a bounded set in $\mathbf{R}^2$ having smooth boundary S and for given functions f in $C(\Omega)$ and g in $C(S)$, let $\mathscr{A} \equiv \{u(x, y) \text{ in } H^1(\Omega) : u = g \text{ on } S\}$. Minimize

$$J[u] = \int_\Omega (u_x^2 + u_y^2 - 2fu)\, dx\, dy$$

over $\mathscr{A}$. (*b*) For Ω a bounded set in $\mathbf{R}^n$ with smooth boundary S and for given functions $a_{ij}(\mathbf{x})$ and $f(\mathbf{x})$ in $C(\Omega)$, let $A \equiv \{u \text{ in } H^1(\Omega) : u = 0 \text{ on } S\}$. Minimize

$$J[u] = \int_\Omega \left(\sum_{i,j=1}^{n} a_{ij} \frac{\partial u}{\partial x_i} \frac{\partial u}{\partial x_j} - 2fu \right) d\Omega$$

Such problems can be treated in much the same way as extreme-value problems in elementary calculus, if the notion of the derivative is suitably generalized.

Variation of a Functional

First, we associate with the domain $\mathscr{A}$ of the functional J, a set $\mathscr{M}$ of *comparison functions* such that, for any u in $\mathscr{A}$ and any v in $\mathscr{M}$, $u + \epsilon v$ belongs to $\mathscr{A}$ for every real number ϵ. $\mathscr{M}$ is necessarily a subspace of $L^2(\Omega)$.

EXAMPLE 12.3 (*a*) For $\mathscr{A}$ as in Example 12.2(*a*), we may take $\mathscr{M} \equiv \{v \text{ in } H^1(\Omega) : v = 0 \text{ on } S\}$. (*b*) For $\mathscr{A}$ as in Example 12.2(*b*), we may take $\mathscr{M}$ to be identical to $\mathscr{A}$. In general, whenever $\mathscr{A}$ is a *subspace* of $L^2(\Omega)$ and not just a subset as in the first example, we may take $\mathscr{M} \equiv \mathscr{A}$ (for u, v in $\mathscr{A}$, the linear combination $u + \epsilon v$ is also in $\mathscr{A}$).

For J a functional on domain $\mathscr{A}$, to which corresponds a set of comparison functions $\mathscr{M}$, let u belong to $\mathscr{A}$ and v belong to $\mathscr{M}$. Then

$$\phi(\epsilon) \equiv J[u + \epsilon v] \tag{12.1}$$

is a real-valued function of the real variable ϵ.

Definition: If the limit

$$\phi'(0) = \lim_{\epsilon \to 0} \frac{J[u + \epsilon v] - J[u]}{\epsilon} \tag{12.2}$$

exists for every v in $\mathscr{M}$ (the value of the limit will generally depend on the "direction" v), we write

$$\phi'(0) = \delta J[u; v]$$

and call this the *variation of J at u (in the direction v).*

Theorem 12.3: Let J denote a functional on domain $\mathscr{A}$ with associated set of comparison functions $\mathscr{M}$, and suppose that u_0 in $\mathscr{A}$ is a local extreme point for J. If J has a variation at u_0, it must vanish; i.e.,

$$\delta J[u_0; v] = 0 \qquad \text{for all } v \text{ in } \mathscr{M}$$

Gradient of a Functional

Theorem 12.3 shows that the variation of a functional J is analogous to the directional derivative of a function on $\mathbf{R}^n$, which is obtained by computing the scalar product of the gradient of the function with a unit vector in the given direction. Under certain conditions, we can define what is meant by "the gradient of a functional."

Consider functional J with domain $\mathscr{A}$ contained in $L^2(\Omega)$. Suppose that the comparison functions comprise a *dense* subspace $\mathscr{M}$ of $L^2(\Omega)$. Finally, let $\mathscr{D}$ consist of all u in $\mathscr{A}$ such that J has a variation at u and, moreover, such that there exists a G in $L^2(\Omega)$ having the property

$$\delta J[u; v] = \langle G, v\rangle \qquad \text{for all } v \text{ in } \mathscr{M}$$

If $\mathscr{D}$ is nonempty, we call the function G the *gradient of J at u*, and write $G = \nabla J[u]$; the subset $\mathscr{D}$ of $\mathscr{A}$ is called the *domain of the gradient.*

If u_0 in $\mathscr{D}$ furnishes a local extremum for J, Theorem 12.3 and the definition of the gradient of J imply that $\langle \nabla J[u_0], v\rangle = 0$ for all v in $\mathscr{M}$. But $\mathscr{M}$ is dense in $L^2(\Omega)$, and so, by Theorem 12.2, $\nabla J[u_0] = 0$. We can, in fact, prove

Theorem 12.4: If the subspace $\mathscr{M}$ of comparison functions is dense in $L^2(\Omega)$ and if u_0 in $\mathscr{A}$ is a local extreme point for J, then u_0 necessarily belongs to $\mathscr{D}$ and

$$\nabla J[u_0] = 0$$

(the *Euler equation* for J).

The point of Theorem 12.4 is that to solve a PDE which is the Euler equation of a suitably constructed functional J, subject to whatever boundary conditions are part of the definition of the domain $\mathscr{D}$, it is sufficient to minimize J over the class $\mathscr{A}$ of admissible functions. As the smoothness conditions incorporated into the definition of $\mathscr{D}$ appear to be weaker than what is required for a classical solution, the solution of the variational problem is considered to be a *generalized solution* of the boundary value problem. In Section 12.5 we shall develop an even broader notion of the solution of a boundary value problem.

12.4 VARIATIONAL PRINCIPLES FOR EIGENVALUES AND EIGENFUNCTIONS

Consider the Sturm–Liouville problem (*6.13*), with $C_2 = C_4 = 0$. Theorem 6.4 describes the eigenvalues and eigenfunctions of this problem. Define

$$\mathscr{A}_0 \equiv \{\phi \text{ in } H^1(a, b) : \phi(a) = \phi(b) = 0\} \tag{12.3}$$

and define a functional J on $\mathscr{A}_0$ by

$$J[\phi] \equiv \frac{\displaystyle\int_a^b [p(x)\phi'(x)^2 + q(x)\phi(x)^2]\,dx}{\displaystyle\int_a^b r(x)\phi(x)^2\,dx} \tag{12.4}$$

It is easy to show that $\lambda_1 \le J[\phi]$ for all ϕ in $\mathscr{A}_0$; indeed,

$$\lambda_1 = \min_{\phi \in \mathscr{A}_0} J[\phi] \tag{12.5}$$

In addition, for $k = 1, 2, \ldots,$ define

$$\mathscr{A}_k \equiv \left\{\phi \text{ in } \mathscr{A}_0 : \int_a^b r(x)\phi(x)\,u_j(x)\,dx = 0 \quad (j = 1, 2, \ldots, k)\right\} \tag{12.6}$$

where $u_j(x)$ denotes an eigenfunction of (*6.13*) corresponding to the eigenvalue λ_j. Then

$$\lambda_{k+1} = \min_{\phi \in \mathscr{A}_k} J[\phi] \qquad (k = 1, 2, \ldots) \tag{12.7}$$

The functional (*12.4*) is called the *Rayleigh quotient* of the Sturm–Liouville problem (*6.13*), whose eigenvalues are characterized by the *minimum principles* (*12.5*) and (*12.7*).

More generally, consider the elliptic boundary value problem

$$-\nabla\cdot(p(\mathbf{x})\nabla u(\mathbf{x}))+q(\mathbf{x})u(\mathbf{x})=\lambda r(\mathbf{x})u(\mathbf{x}) \qquad \mathbf{x} \text{ in } \Omega \tag{12.8}$$

$$u(\mathbf{x})=0 \qquad \mathbf{x} \text{ on } S \tag{12.9}$$

where $p(\mathbf{x})>0$, $q(\mathbf{x})$, and $r(\mathbf{x})>0$ are all in $C^1(\bar{\Omega})$. As in the case of the one-dimensional Sturm–Liouville problem, all eigenvalues of (*12.8*)–(*12.9*) are real and can be arranged in a countably infinite, increasing sequence. Moreover, weighted eigenfunctions $r^{1/2}u$ belonging to distinct eigenvalues are orthogonal; i.e.,

$$\langle r^{1/2}u_m, r^{1/2}u_n\rangle \equiv \int_\Omega r(\mathbf{x})u_m(\mathbf{x})u_n(\mathbf{x})\,d\Omega=0 \qquad (m\neq n)$$

The smallest eigenvalue, λ_1, satisfies

$$\lambda_1=\min_{\phi\in\mathcal{A}_0} J[\phi] \tag{12.10}$$

where

$$\mathcal{A}_0=\{\phi \text{ in } H^1(\Omega) : \phi=0 \text{ on } S\} \tag{12.11}$$

and where the Rayleigh quotient is given by

$$J[\phi]=\frac{\displaystyle\int_\Omega [p(\mathbf{x})\nabla\phi(\mathbf{x})\cdot\nabla\phi(\mathbf{x})+q(\mathbf{x})\phi(\mathbf{x})^2]\,d\Omega}{\displaystyle\int_\Omega r(\mathbf{x})\phi(\mathbf{x})^2\,d\Omega} \tag{12.12}$$

Moreover, for $k=1,2,\ldots,$

$$\lambda_{k+1}=\min_{\phi\in\mathcal{A}_k} J[\phi] \tag{12.13}$$

where

$$\mathcal{A}_k=\{\phi \text{ in } \mathcal{A}_0 : \langle r^{1/2}\phi, r^{1/2}u_j\rangle=0 \quad (1\le j\le k)\} \tag{12.14}$$

and u_j denotes an eigenfunction belonging to the eigenvalue λ_j. It is readily shown (cf. Problem 12.7(*b*)) that the minima in (*12.10*) and (*12.13*) are assumed at the eigenfunctions; i.e., $\lambda_k = J[u_k]$ $(k=1,2,\ldots)$.

Minimum principles analogous to (*12.10*) and (*12.13*) hold when the differential operator in (*12.8*) is replaced by a general linear, elliptic, differential operator.

12.5 WEAK SOLUTIONS OF BOUNDARY VALUE PROBLEMS

Let Ω denote a bounded region in $\mathbf{R}^2$ with smooth boundary S consisting of complementary arcs S_1 and S_2. Let p, q, f, g_1, g_2 denote given functions which are defined and smooth on Ω and/or on S. Finally, define a linear partial differential operator by

$$Lu(x,y)=-\nabla^2u+pu_x+qu_y \tag{12.15}$$

and consider the mixed boundary value problem

$$Lu=f \qquad \text{in } \Omega \tag{12.16}$$

$$u=g_1 \qquad \text{on } S_1 \tag{12.17}$$

$$\frac{\partial u}{\partial n}=g_2 \qquad \text{on } S_2 \tag{12.18}$$

For arbitrary u and v in $C^2(\Omega)$, Green's first identity, (*1.7*), gives

$$\langle Lu, v\rangle = \int_\Omega \nabla u \cdot \nabla v \, d\Omega + \int_\Omega (pu_x + qu_y)v \, d\Omega - \int_S v \frac{\partial u}{\partial n} dS \tag{12.19}$$

Now define

$$\begin{aligned} \mathscr{A} &\equiv \{u \text{ in } H^1(\Omega) : u = g_1 \text{ on } S_1\} \\ \mathscr{M} &\equiv \{v \text{ in } H^1(\Omega) : v = 0 \text{ on } S_1\} \end{aligned} \tag{12.20}$$

and note that for u in $\mathscr{A}$ and v in $\mathscr{M}$,

$$\int_S v \frac{\partial u}{\partial n} dS = \int_{S_2} v \frac{\partial u}{\partial n} dS$$

Moreover, if u in $\mathscr{A}$ satisfies (*12.18*),

$$\int_{S_2} v \frac{\partial u}{\partial n} dS = \int_{S_2} vg_2 \, dS \qquad \text{for all } v \text{ in } \mathscr{M}$$

Thus, if u is a classical solution of the boundary value problem (*12.16*)–(*12.18*), u must satisfy

$$K[u, v] = F[v] \qquad \text{for all } v \text{ in } \mathscr{M} \tag{12.21}$$

where, for u in $\mathscr{A}$ and v in $\mathscr{M}$,

$$K[u, v] \equiv \int_\Omega \nabla u \cdot \nabla v \, d\Omega + \int_\Omega (pu_x + qu_y)v \, d\Omega \tag{12.22}$$

$$F[v] \equiv \int_\Omega fv^2 \, d\Omega + \int_{S_2} g_2 v \, dS \tag{12.23}$$

Definition: A *weak solution* of (*12.16*)–(*12.18*) is any function $u(x, y)$ that belongs to $\mathscr{A}$ and satisfies (*12.21*).

A notion of weak solution has already been encountered in Problem 5.15. Evidently, every classical solution of the boundary value problem is a weak solution. However, a weak solution u need not be a classical solution: it need only be sufficiently regular to allow definition of $K[u, v]$ for all v in $\mathscr{M}$.

In the special (self-adjoint) case that p and q in (*12.15*) vanish on Ω [i.e., when (*12.16*) is Poisson's equation], then

$$\begin{aligned} K[u, v] &= K[v, u] \qquad \text{for all } u, v \text{ in } \mathscr{M} \\ K[u, u] &\geq 0 \qquad \text{for all } u \text{ in } \mathscr{M} \end{aligned} \tag{12.24}$$

Whenever (*12.24*) holds, it follows that for u in $\mathscr{A}$, v in $\mathscr{M}$,

$$2\{K[u, v] - F[v]\} = \delta J[u; v] \tag{12.25}$$

where

$$J[u] \equiv K[u, u] - 2F[u] \tag{12.26}$$

If u_0 is a weak solution of the boundary value problem (for Poisson's equation), it follows from (*12.25*) that u_0 is a stationary point for J. In fact, it can be shown that u_0 minimizes J over $\mathscr{A}$, which leads to

Theorem 12.5: Let $K[u, v]$ satisfy (*12.24*) and let $J[u]$ be given by (*12.26*). Then u_0 minimizes J over $\mathscr{A}$ if and only if $K[u_0, v] = F[v]$ for all v in $\mathscr{M}$.

According to Theorem 12.5, the weak formulation of a boundary value problem is, provided (*12.24*) holds, the same thing as the variational formulation guaranteed by Theorem 12.4. However, when (*12.24*) does not hold, only the weak formulation is possible. Thus, the notion of a weak solution to a boundary value problem is more comprehensive than that of a variational solution, which in turn is more comprehensive than the notion of a classical solution.

Solved Problems

12.1 Let Ω denote a bounded region in $\mathbf{R}^2$ with smooth boundary S. For $\mathscr{A} = \mathscr{M} = H^1(\Omega)$, let the following functionals be defined on $\mathscr{A}$:

$$J_1[u] = \int_\Omega (u_x^2 + u_y^2 - 2fu)\,dx\,dy$$

$$J_2[u] = \int_\Omega (u_x^2 + u_y^2 - 2fu)\,dx\,dy + \int_S pu^2\,dS$$

$$J_3[u] = \int_\Omega (u_x^2 + u_y^2 - 2fu)\,dx\,dy + \int_S (pu^2 - 2gu)\,dS$$

Here p and g are in $C(\bar{\Omega})$ and f is in $L^2(\Omega)$. Compute $\delta J[u; v]$ for each functional.

It suffices to calculate $\delta J_3[u; v]$; setting $p = g = 0$ or $g = 0$ will yield the other two variations. For u, v in $\mathscr{A}$ and arbitrary real ϵ, we have

$$\begin{aligned}\phi_3(\epsilon) &\equiv J_3[u + \epsilon v]\\ &= J_3[u] + 2\epsilon\int_\Omega (u_xv_x + u_yv_y - fv)\,dx\,dy + 2\epsilon\int_S (puv - gv)\,dS\\ &\quad + \epsilon^2\int_\Omega (v_x^2 + v_y^2)\,dx\,dy + \epsilon^2\int_S pv^2\,dS\end{aligned}$$

and so

$$\delta J_3[u; v] \equiv \phi_3'(0) = 2\int_\Omega (u_xv_x + u_yv_y - fv)\,dx\,dy + 2\int_S (pu - g)v\,dS \tag{1}$$

12.2 Let Ω denote a bounded region in $\mathbf{R}^n$ with smooth boundary S. On $\mathscr{A} = \mathscr{M} = H^1(\Omega)$ define the functional

$$J_4[u] = \int_\Omega \left[\sum_{i,j=1}^{n} a_{ij}(\mathbf{x})\frac{\partial u}{\partial x_i}\frac{\partial u}{\partial x_j} + c(\mathbf{x})u^2 - 2f(\mathbf{x})u\right] d\Omega$$

where the functions $a_{ij} = a_{ji}$ (see Section 2.1), c, and f are all in $C[\Omega]$. Find $\delta J_4[u; v]$.

For u, v in $\mathscr{A}$ and arbitrary real ϵ,

$$\begin{aligned}\phi_4(\epsilon) &\equiv J_4(u + \epsilon v)\\ &= J_4(u) + \epsilon\int_\Omega \left[\sum a_{ij}\left(\frac{\partial u}{\partial x_i}\frac{\partial v}{\partial x_j} + \frac{\partial u}{\partial x_j}\frac{\partial v}{\partial x_i}\right) + 2cuv - 2fv\right] d\Omega + \epsilon^2\int_\Omega \left[\sum a_{ij}\frac{\partial v}{\partial x_i}\frac{\partial v}{\partial x_j} + cv^2\right] d\Omega\end{aligned}$$

Hence, from the symmetry of the a_{ij},

$$\delta J_4[u; v] \equiv \phi_4'(0) = 2\int_\Omega \left[\sum a_{ij}\frac{\partial u}{\partial x_i}\frac{\partial v}{\partial x_j} + (cu - f)v\right] d\Omega$$

12.3 Let Ω denote a bounded region in $\mathbf{R}^2$ with smooth boundary S. With

$$\mathscr{A} = \{u \text{ in } H^1(\Omega) : u = g \text{ on } S\} \qquad \mathscr{M} = \{v \text{ in } H^1(\Omega) : v = 0 \text{ on } S\}$$

and f, g in $C[\bar{\Omega}]$, consider the functional

$$J_1[u] = \int_\Omega (u_x^2 + u_y^2 - 2fu)\,dx\,dy \qquad (u \text{ in } \mathscr{A})$$

Find $\mathscr{D}_1$ and $\nabla J_1[u]$. Note that $\mathscr{M}$ is dense in $L^2(\Omega)$.

Problem 12.1 gives (for any $\mathscr{A}$ and $\mathscr{M}$)

$$\delta J_1[u; v] = 2\int_\Omega (u_x v_x + u_y v_y - fv)\,dx\,dy \tag{1}$$

Green's first identity, (*1.7*), implies

$$\int_\Omega (u_x v_x + u_y v_y)\,dx\,dy = \int_S v\frac{\partial u}{\partial n}\,dS - \int_\Omega v\nabla^2 u\,dx\,dy \tag{2}$$

Since $v = 0$ on S for v in $\mathscr{M}$, (*2*) used in (*1*) leads to

$$\delta J_1[u; v] = -2\int_\Omega (\nabla^2 u + f)v\,dx\,dy = \langle G, v\rangle$$

where
$$G(x, y) \equiv -2[\nabla^2 u(x, y) + f(x, y)] \qquad (x, y) \text{ in } \Omega$$

If $\nabla^2 u$ is in $L^2(\Omega)$, then G belongs to $L^2(\Omega)$ and we conclude that

$$\nabla J_1[u] = G = -2(\nabla^2 u + f)$$

for u belonging to $\mathscr{D}_1 = \{u \text{ in } H^2(\Omega) : u = g \text{ on } S\}$, where $H^2(\Omega)$ denotes the class of functions for which $\nabla^2 u$ belongs to $L^2(\Omega)$. (A more precise characterization of the domain $\mathscr{D}$ of the gradient of a functional can be given in the context of functional analysis.)

12.4 For each functional of Problem 12.1, find $\mathscr{D}$ and $\nabla J[u]$.

Again it suffices to treat J_3 and then to specialize the results to the other two functionals. Applying (*2*) of Problem 12.3 to (*1*) of Problem 12.1, we obtain

$$\delta J_3[u; v] = 2\int_S \left(\frac{\partial u}{\partial n} + pu - g\right) v\,dS - 2\int_\Omega (\nabla^2 u + f)v\,dx\,dy \tag{1}$$

Thus, if u in $\mathscr{A}$ satisfies the additional conditions

$$\text{(i)}\quad u \text{ belongs to } H^2(\Omega) \qquad\qquad \text{(ii)}\quad \frac{\partial u}{\partial n} + pu = g \quad \text{on } S \tag{2}$$

(*1*) shows that $\nabla J_3[u] = -2(\nabla^2 u + f)$ on the domain $\mathscr{D}_3$ defined by (*2*). (∇J_3 is in $L^2(\Omega)$ because of (*2*)(i).)

The functional J_3 is seen to be the energy integral for Poisson's equation with an inhomogeneous mixed boundary condition:

$$-\nabla^2 u = f \qquad \text{in } \Omega$$

$$\frac{\partial u}{\partial n} + pu = g \qquad \text{on } S$$

Setting $p = g = 0$, we see that J_1, the energy integral for Poisson's equation with a homogeneous Neumann condition, has gradient

$$\nabla J_1[u] = -2(\nabla^2 u + f) \qquad \text{for } u \text{ in } \mathscr{D}_1 = \left\{u \text{ in } H^2(\Omega) : \frac{\partial u}{\partial n} = 0 \text{ on } S\right\}$$

In Problem 12.3, a different domain $\mathscr{A}$ for J_1 produced a different domain $\mathscr{D}_1$ for ∇J_1.

12.5 Let Ω denote a bounded region in $\mathbf{R}^2$ with smooth boundary S. Let S_1 denote a connected subset of S and let S_2 denote the complement of S_1 in S. Let p, f, g_1, g_2 be functions in $C(\bar{\Omega})$. If

$$\mathscr{A} = \{u \text{ in } H^1(\Omega) : u = g_1 \text{ on } S_1\} \qquad \mathscr{M} = \{v \text{ in } H^1(\Omega) : v = 0 \text{ on } S_1\} \tag{0}$$

find $\mathscr{D}$ and ∇J for the functionals

$$J_1[u] = \int_{\Omega} (u_x^2 + u_y^2 - 2fu)\, dx\, dy$$

$$J_2[u] = \int_{\Omega} (u_x^2 + u_y^2 - 2fu)\, dx\, dy + \int_{S_2} pu^2\, dS$$

$$J_3[u] = \int_{\Omega} (u_x^2 + u_y^2 - 2fu)\, dx\, dy + \int_{S_2} (pu^2 - 2g_2 u)\, dS$$

This problem illustrates further the relation between $\mathscr{A}$ and $\mathscr{D}$. Note that the admissible class $\mathscr{A}$ of *(0)* is "between" the classes $\mathscr{A}$ of Problems 12.3 and 12.4, in the sense that

$$\{u \text{ in } H^1(\Omega) : u = g \text{ on } S\} \subset \{u \text{ in } H^1(\Omega) : u = g \text{ on } S_1 \subset S\} \subset \{u \text{ in } H^1(\Omega)\}$$

As usual, we may restrict attention initially to J_3. Since $v = 0$ on S_1, we have, analogous to *(1)* of Problem 12.4,

$$\delta J_3[u; v] = 2\int_{S_2} \left(\frac{\partial u}{\partial n} + pu - g_2\right) v\, dS - 2\int_{\Omega} (\nabla^2 u + f) v\, dx\, dy \tag{1}$$

We conclude that if u in $\mathscr{A}$ satisfies the additional conditions

$$\text{(i)}\quad u \text{ belongs to } H^2(\Omega) \qquad \text{(ii)}\quad \frac{\partial u}{\partial n} + pu = g_2 \quad \text{on } S_2 \tag{2}$$

then $\nabla J_3[u] = -2(\nabla^2 u + f)$ on the domain

$$\mathscr{D}_3 = \left\{u \text{ in } H^2(\Omega) : u = g_1 \text{ on } S_1, \frac{\partial u}{\partial n} + pu = g_2 \text{ on } S_2\right\}$$

Now setting $g_2 = 0$ and $g_2 = p = 0$, respectively, we obtain:

$$\nabla J_2[u] = -2(\nabla^2 u + f) \qquad \text{on } \mathscr{D}_2 = \left\{u \text{ in } H^2(\Omega) : u = g_1 \text{ on } S_1, \frac{\partial u}{\partial n} + pu = 0 \text{ on } S_2\right\}$$

$$\nabla J_1[u] = -2(\nabla^2 u + f) \qquad \text{on } \mathscr{D}_1 = \left\{u \text{ in } H^2(\Omega) : u = g_1 \text{ on } S_1, \frac{\partial u}{\partial n} = 0 \text{ on } S_2\right\}$$

In the case of each functional, the definition *(0)* of $\mathscr{A}$ specifies how u is to behave over the part S_1 of the boundary. Then the definition of $\mathscr{D}$ imposes a condition over the remainder of the boundary; this condition involves the normal derivative of u. Boundary conditions incorporated in the definition of $\mathscr{A}$ are called *stable* boundary conditions; those included in the definition of $\mathscr{D}$ are called *natural* boundary conditions.

12.6 Let Ω denote a bounded region in $\mathbf{R}^2$ with smooth boundary S. Let f, g, and p be functions in $C(\bar{\Omega})$. Consider Poisson's equation

$$-\nabla^2 u(x, y) = f(x, y) \qquad \text{in } \Omega \tag{1}$$

and the Dirichlet, Neumann, and mixed boundary conditions

$$u = g \qquad \text{on } S \tag{2}$$

$$\frac{\partial u}{\partial n} = g \qquad \text{on } S \tag{3}$$

$$\frac{\partial u}{\partial n} + pu = g \qquad \text{on } S \tag{4}$$

Give variational formulations of the problems *(1)*–*(2)*, *(1)*–*(3)*, and *(1)*–*(4)*.

All three boundary value problems are covered by the functional $J_3[u]$ of Problem 12.5, with $\mathscr{A}$ and $\mathscr{M}$ as given in *(0)* of that problem.

(i) In Problem 12.5, take $S_1 = S$, $g_1 = g$. Then, by Theorem 12.4, if u_0 minimizes

$$\int_\Omega (u_x^2 + u_y^2 - 2fu)\,dx\,dy \qquad \text{over } \{u \text{ in } H^1(\Omega) : u = g \text{ on } S\}$$

u_0 solves (1)–(2).

(ii) In Problem 12.5, take $S_2 = S$, $p = 0$, $g_2 = g$. Then, by Theorem 12.4, if u_0 minimizes

$$\int_\Omega (u_x^2 + u_y^2 - 2fu)\,dx\,dy - 2\int_S gu\,dS \qquad \text{over } H^1(\Omega)$$

u_0 solves (1)–(3).

(iii) In Problem 12.5, take $S_2 = S$, $g_2 = g$. Then, by Theorem 12.4, if u_0 minimizes

$$\int_\Omega (u_x^2 + u_y^2 - 2fu)\,dx\,dy + \int_S (pu^2 - 2gu)\,dS \qquad \text{over } H^1(\Omega)$$

u_0 solves (1)–(4).

It is seen from (i) that the Dirichlet condition (2) is a stable boundary condition for the corresponding variational problem, whereas the Neumann condition (3) and the mixed condition (4) figure as natural boundary conditions.

12.7 Consider the boundary value problem

$$-\nabla^2 u + qu = \lambda ru \qquad \text{in } \Omega \tag{1}$$

$$u = 0 \qquad \text{on } S \tag{2}$$

with $q \geq 0$ and $r > 0$ in Ω. Let $\lambda_1 < \lambda_2 \leq \cdots$ denote the eigenvalues of (1)–(2) arranged in increasing order, and let u_n denote the eigenfunction corresponding to λ_n. With $J[\phi]$ as given by (12.12), with $p \equiv 1$, and $\mathscr{A}_0$ as given by (12.11), prove: (a) if ϕ_* minimizes J over $\mathscr{A}_0$, then ϕ_* satisfies (1)–(2) with $\lambda = J[\phi_*]$; (b) for $n = 1, 2, \ldots, \lambda_n = J[u_n]$; (c) $\lambda_1 = \min_{\mathscr{A}_0} J[\phi]$.

(a) On $\mathscr{A}_0 \equiv \mathscr{M} \equiv \{\phi \text{ in } H^1(\Omega) : \phi = 0 \text{ on } S\}$ define

$$N[\phi] \equiv \int_\Omega (\nabla\phi \cdot \nabla\phi + q\phi^2)\,d\Omega \qquad D[\phi] \equiv \int_\Omega r\phi^2\,d\Omega \tag{3}$$

Then, for $J[\phi] = N[\phi]/D[\phi]$,

$$\delta J[\phi; v] = \frac{\delta N[\phi; v]D[\phi] - N[\phi]\delta D[\phi; v]}{D[\phi]^2} \tag{4}$$

If ϕ_* in $\mathscr{A}_0$ minimizes J, then $\delta J[\phi_*; v] = 0$, or

$$\delta N[\phi_*; v] - J[\phi_*]\delta D[\phi_*; v] = 0 \qquad \text{for all } v \text{ in } \mathscr{M} \tag{5}$$

But, since

$$\delta N[\phi; v] = 2\int_\Omega (\nabla\phi \cdot \nabla v + q\phi v)\,d\Omega \qquad \delta D[\phi; v] = 2\int_\Omega r\phi v\,d\Omega$$

(5) implies

$$\int_\Omega (\nabla\phi_* \cdot \nabla v + q\phi_* v - \lambda_* r\phi_* v)\,d\Omega = 0 \qquad \text{for all } v \text{ in } \mathscr{M} \tag{6}$$

where $\lambda_* \equiv J[\phi_*]$; or, by (1.7) and the boundary condition on v,

$$\langle -\nabla^2\phi_* + q\phi_* - \lambda_* r\phi_*, v\rangle = 0 \qquad \text{for all } v \text{ in } \mathscr{M} \tag{7}$$

Theorem 12.4 guarantees that the minimizing element ϕ_* is such as to render

$$\nabla J[\phi_*] = -\nabla^2\phi_* + q\phi_* - \lambda_* r\phi_*$$

an element of $L^2(\Omega)$; thus ϕ_* must belong to the subspace

$$\mathcal{D} \equiv \{\phi \text{ in } H^2(\Omega) : \phi = 0 \quad \text{on } S\}$$

of $\mathcal{A}_0$. Indeed, $\nabla J[\phi_*]$ must be the *zero element* of $L^2(\Omega)$, so that ϕ_* in $\mathcal{D}$ solves (*1*)–(*2*) with $\lambda = \lambda_*$.

(*b*) If u_n satisfies (*1*)–(*2*) with $\lambda = \lambda_n$, then multiplying (*1*) by u_n and integrating over Ω leads to

$$\int_\Omega (-\nabla^2 u_n + qu_n - \lambda_n r u_n)u_n \, d\Omega = 0$$

But, by Green's first identity and the boundary condition,

$$\int_\Omega (-\nabla^2 u_n)u_n \, d\Omega = \int_\Omega \nabla u_n \cdot \nabla u_n \, d\Omega$$

Hence,

$$\int_\Omega (\nabla u_n \cdot \nabla u_n + qu_n^2 - \lambda_n r u_n^2) \, d\Omega = 0 \qquad \text{or} \qquad \lambda_n = J[u_n]$$

(*c*) Since ϕ_* minimizes $J[\phi]$ over $\mathcal{A}_0$ and since each u_n belongs to $\mathcal{A}_0$, we have, by (*b*),

$$\lambda_* \equiv J[\phi_*] \le J[u_n] = \lambda_n \qquad (n = 1, 2, \ldots)$$

But λ_* is itself an eigenvalue, and so $\lambda_* = \lambda_1$. We have just proved *Rayleigh's principle*: the smallest eigenvalue of the boundary value problem (*1*)–(*2*) is identical to the smallest value of the functional $J[\phi]$. (Cf. Problem 9.8.)

12.8 Consider the Sturm–Liouville problem (*12.8*)–(*12.9*) and let $\phi_1, \ldots, \phi_{N-1}$ denote any $N-1$ elements of $\mathcal{A}_0$, as defined in (*12.11*). Let $\mathcal{D}_{N-1}$ denote the following subspace of $\mathcal{A}_0$:

$$\mathcal{D}_{N-1} \equiv \{u \text{ in } \mathcal{A}_0 : \langle r^{1/2}\phi_j, r^{1/2}u\rangle = 0 \quad (j = 1, \ldots, N-1)\} \tag{1}$$

For the Rayleigh quotient J of (*12.12*), write

$$C[\phi_1, \ldots, \phi_{N-1}] \equiv \min_{u \in \mathcal{D}_{N-1}} J[u] \tag{2}$$

Prove that $C[\phi_1, \ldots, \phi_{N-1}] \le \lambda_N$, with equality if $\phi_1, \ldots, \phi_{N-1}$ coincide with eigenfunctions of the Sturm–Liouville problem, corresponding to the first $N-1$ eigenvalues. This result, the *Courant minimax principle*, implies that λ_N, the Nth eigenvalue of the Sturm–Liouville problem, can be characterized as the maximum over all subsets $\{\phi_1, \ldots, \phi_{N-1}\}$ of $\mathcal{A}_0$ of the minimum of $J[u]$ on $\mathcal{D}_{N-1}$.

To establish the desired inequality it will suffice to construct a function w in $\mathcal{D}_{N-1}$ such that $J[w] \le \lambda_N$. Write

$$w = \sum_{i=1}^{N} c_i u_i(\mathbf{x}) \tag{3}$$

where $u_1(\mathbf{x}), \ldots, u_N(\mathbf{x})$ denote the eigenfunctions associated with $\lambda_1, \ldots, \lambda_N$. w will belong to $\mathcal{D}_{N-1}$ provided

$$0 = \langle r^{1/2}\phi_j, r^{1/2}w\rangle = \sum_{i=1}^{N} c_i \langle r^{1/2}\phi_j, r^{1/2}u_i\rangle \qquad (j = 1, 2, \ldots, N-1) \tag{4}$$

As (*4*) constitutes a system of $N-1$ linear equations in the N unknowns $c_1, \ldots, c_N$, there will always exist a nontrivial solution $(\hat{c}_1, \ldots, \hat{c}_N)$. [$w = 0$ certainly belongs to $\mathcal{D}_{N-1}$, but $J[0]$ is undefined.] Now

$$J[w]=\frac{\int_{\Omega}\left[p(\mathbf{x})\sum_{i,j=1}^{N}\hat{c}_i\hat{c}_j\,\nabla u_i(\mathbf{x})\cdot\nabla u_j(\mathbf{x})+q(\mathbf{x})\left(\sum_{i=1}^{N}\hat{c}_iu_i(\mathbf{x})\right)^2\right]d\Omega}{\int_{\Omega}\left[r(\mathbf{x})\left(\sum_{i=1}^{N}\hat{c}_iu_i(\mathbf{x})\right)^2\right]d\Omega}=\frac{\sum_{i,j=1}^{N}\theta_{ij}\hat{c}_i\hat{c}_j}{\sum_{i=1}^{N}\beta_i\hat{c}_i^2}$$

where

$$\theta_{ij}\equiv\int_{\Omega}[p(\mathbf{x})\nabla u_i(\mathbf{x})\cdot\nabla u_j(\mathbf{x})+q(\mathbf{x})u_i(\mathbf{x})u_j(\mathbf{x})]\,d\Omega\qquad \beta_i\equiv\int_{\Omega}r(\mathbf{x})u_i(\mathbf{x})^2\,d\Omega$$

and where we have used the weighted orthogonality of the eigenfunctions in evaluating the denominator. From Green's first identity and the fact that $u_i(\mathbf{x})$ *is* an eigenfunction of (*12.8*)–(*12.9*), we can show that

$$\theta_{ij}=\lambda_i\int_{\Omega}r(\mathbf{x})u_i(\mathbf{x})u_j(\mathbf{x})\,d\Omega=\begin{cases}0 & j\neq i\\ \lambda_i\beta_i & j=i\end{cases}$$

Therefore

$$J[w]=\frac{\sum_{i=1}^{N}\lambda_i\beta_i\hat{c}_i^2}{\sum_{i=1}^{N}\beta_i\hat{c}_i^2}\equiv\sum_{i=1}^{N}\kappa_i\lambda_i \tag{5}$$

where $\kappa_i\geq 0$, $\kappa_1+\kappa_2+\cdots+\kappa_N=1$. Thus, as a convex combination of the λ_i,

$$\lambda_1\leq J[w]\leq\lambda_N$$

The proof is completed by noting that, if $\phi_j=u_j$ $(j=1,2,\ldots,N-1)$, u_N will belong to $\mathscr{D}_{N-1}=\mathscr{A}_{N-1}$. We then have, using (*12.13*),

$$C[u_1,\ldots,u_{N-1}]\equiv\min_{u\in\mathscr{A}_{N-1}}J[u]=\lambda_N$$

12.9 Let p^*, q^*, r^* denote an alternate set of coefficient functions for the Sturm–Liouville problem (*12.8*)–(*12.9*); these are supposed to obey the same continuity and positivity conditions as do p, q, r. Let the alternate and original problems have Rayleigh quotients $J^*[\phi]$ and $J[\phi]$, and eigenvalues $\{\lambda_n^*\}$ and $\{\lambda_n\}$. Show that if

$$J[\phi]\leq J^*[\phi]\qquad\text{for all }\phi\text{ in }\mathscr{A}_0 \tag{1}$$

then $\lambda_n\leq\lambda_n^*$ $(n=1,2,\ldots)$.

Inequality (*1*) implies that

$$C[\phi_1,\ldots,\phi_{N-1}]\leq C^*[\phi_1,\ldots,\phi_{N-1}]$$

If, then, $\psi_1,\ldots,\psi_{N-1}$ is a set of functions in $\mathscr{A}_0$ that maximizes $C[\phi_1,\ldots,\phi_{N-1}]$ and if $\psi_1^*,\ldots,\psi_{N-1}^*$ in $\mathscr{A}_0$ maximizes $C^*[\phi_1,\ldots,\phi_{N-1}]$, the Courant minimax principle implies

$$\lambda_N=C[\psi_1,\ldots,\psi_{N-1}]\leq C^*[\psi_1,\ldots,\psi_{N-1}]\leq C^*[\psi_1^*,\ldots,\psi_{N-1}^*]=\lambda_N^*$$

12.10 For the one-dimensional Sturm–Liouville problem of Section 12.4, write $p_m>0$, q_m, and $r_m>0$ for the minimum values of the coefficient functions on $[a,b]$, and p_M, $q_M>0$, and r_M for the maximum values. Establish the following bounds for the eigenvalues:

$$\frac{q_m}{r_M}+\left(\frac{n\pi}{b-a}\right)^2\frac{p_m}{r_M}\leq\lambda_n\leq\frac{q_M}{r_m}+\left(\frac{n\pi}{b-a}\right)^2\frac{p_M}{r_m}\qquad(n=1,2,\ldots) \tag{1}$$

The Rayleigh quotient (*12.4*) clearly satisfies, for all ϕ in $\mathscr{A}_0$,

$$J[\phi] \leq \frac{p_M \int_a^b \phi'(x)^2\,dx + q_M \int_a^b \phi(x)^2\,dx}{r_m \int_a^b \phi(x)^2\,dx} \equiv J^*[\phi]$$

Now, $J^*[\phi]$ is the Rayleigh quotient for the Sturm–Liouville problem

$$-p_M w''(x) + q_M w(x) = \lambda^* r_m w(x) \qquad a < x < b$$
$$w(a) = w(b) = 0$$

which has eigenvalues

$$\lambda_n^* = \frac{q_M}{r_m} + \left(\frac{n\pi}{b-a}\right)^2 \frac{p_M}{r_m} \qquad (n = 1, 2, \ldots)$$

Therefore, by Problem 12.9, $\lambda_n \leq \lambda_n^*$, which is the upper bound asserted in (*1*). The lower bound is established similarly, from the inequality

$$J[\phi] \geq \frac{p_m \int_a^b \phi'(x)^2\,dx + q_m \int_a^b \phi(x)^2\,dx}{r_M \int_a^b \phi(x)^2\,dx}$$

12.11 Let Ω denote a bounded region in $\mathbf{R}^2$ with smooth boundary S composed of complementary arcs S_1 and S_2. Let p, q, r, f, g_1, g_2, and h denote given functions defined and smooth on $\bar{\Omega}$; in addition, suppose $p > 0$ on Ω. Give the weak formulation of the mixed boundary value problem

$$-\nabla\cdot(p\nabla u) + qu_x + ru_y = f \qquad \text{on } \Omega \tag{1}$$

$$u = g_1 \qquad \text{on } S_1 \tag{2}$$

$$\frac{\partial u}{\partial n} + hu = g_2 \qquad \text{on } S_2 \tag{3}$$

(Unless q and r vanish on Ω, this problem does not admit a variational formulation.)

For arbitrary u, v in $C^2(\Omega)$ we can use (*1.7*) to show that

$$\langle -\nabla\cdot(p\nabla u), v\rangle = \int_\Omega (\nabla u \cdot \nabla v)p\,d\Omega - \int_S v\frac{\partial u}{\partial n}p\,dS$$

Then, if u satisfies (*1*),

$$\int_\Omega (\nabla u\cdot\nabla v)p\,d\Omega + \int_\Omega (qu_x + ru_y)v\,d\Omega - \int_S v\frac{\partial u}{\partial n}p\,dS - \int_\Omega fv\,d\Omega = 0 \tag{4}$$

for all v in $H^1(\Omega)$.

Define

$$\mathcal{A} \equiv \{u \text{ in } H^1(\Omega) : u = g_1 \text{ on } S_1\} \qquad \mathcal{M} \equiv \{v \text{ in } H^1(\Omega) : v = 0 \text{ on } S_1\}$$

If u in $\mathcal{A}$ satisfies (*1*)–(*3*), then, for every v in $\mathcal{M}$, (*4*) gives

$$K[u, v] = F[v] \tag{5}$$

where

$$K[u, v] \equiv \int_\Omega (\nabla u\cdot\nabla v)p\,d\Omega + \int_\Omega (qu_x + ru_y)v\,d\Omega + \int_{S_2} phuv\,dS$$

$$F[v] \equiv \int_\Omega fv\,d\Omega + \int_{S_2} pg_2 v\,dS \tag{6}$$

The weak formulation of (*1*)–(*3*) is, therefore, to find a u in $\mathcal{A}$ that satisfies (*5*) for every v in $\mathcal{M}$.

The weak formulations of PDE (*1*) plus a Dirichlet or a Neumann condition on S are obtained, respectively, by taking $S_1 = S$ and taking $S_2 = S$ and $h = 0$, in the above formulation.

Supplementary Problems

12.12 Let $\Omega \equiv \{(x, y) : x^2 + y^2 < 1\}$ and $\mathscr{A} = \{u \text{ in } H^1(\Omega) : u = x^2 \text{ on } x^2 + y^2 = 1\}$. For

$$J[u] = \int_\Omega (y^2 u_x^2 + x^2 u_y^2)\, d\Omega \qquad (u \text{ in } \mathscr{A})$$

define $\mathscr{M}$ and find $\delta J[u; v]$.

12.13 Find $\mathscr{D}$ and $\nabla J[u]$ in Problem 12.12.

12.14 Let Ω be as in Problem 12.12. If u in $\mathscr{A} = H^1(\Omega)$ minimizes the functional

$$J[u] = \int_\Omega (y^2 u_x^2 + x^2 u_y^2 - 2Fu)\, d\Omega$$

over $\mathscr{A}$, where F is in $L^2(\Omega)$, what boundary value problem does $u(x, y)$ solve?

12.15 Suppose that Ω is a bounded region in $\mathbf{R}^2$ with smooth boundary S. Let F belong to $L^2(\Omega)$ and let

$$\mathscr{A}_1 \equiv \left\{u \text{ in } H^2(\Omega) : u = \frac{\partial u}{\partial n} = 0 \text{ on } S\right\}$$

$$J[u] \equiv \int_\Omega [(\nabla^2 u)^2 - 2Fu]\, d\Omega \qquad \text{for } u \text{ in } \mathscr{A}_1$$

Define $\mathscr{M}$ and find $\delta J[u; v]$.

12.16 Find $\mathscr{D}$ and $\nabla J[u]$ in Problem 12.15.

12.17 Repeat Problem 12.16 if $\mathscr{A}_1$ in Problem 12.15 is replaced by $\mathscr{A}_2 = H^2(\Omega)$.

12.18 With Ω, $\mathscr{A}_1$, and $J[u]$ as in Problem 12.15, define on $\mathscr{A}_1$ the functional

$$\tilde{J}[u] = \int_\Omega [(\nabla^2 u)^2 - 2(u_{xx}u_{yy} - 2u_{xy}^2) - 2Fu]\, d\Omega$$

Verify that $\nabla \tilde{J}[u] = \nabla J[u]$, on the domain $\tilde{\mathscr{D}} = \mathscr{D}$.

12.19 Let Ω denote a bounded region in $\mathbf{R}^2$ with smooth boundary S composed of complementary arcs S_1, S_2, and S_3. Give a variational formulation of the following boundary value problem:

$$\begin{aligned} -\nabla^2 u &= F && \text{in } \Omega \\ u &= g_1 && \text{on } S_1 \\ \frac{\partial u}{\partial n} &= g_2 && \text{on } S_2 \\ \frac{\partial u}{\partial n} + pu &= g_3 && \text{on } S_3 \end{aligned}$$

12.20 Let $J[\phi]$ be given by (*12.4*) and let (*a*) $\mathscr{A}_0 = H^1(a, b)$, (*b*) $\mathscr{A}_0 = \{\phi \text{ in } H^1(a, b) : \phi(a) = 0\}$. Show that if u^* minimizes J over $\mathscr{A}_0$, then u^* is an eigenfunction of (*6.13*) with (*a*) $C_1 = C_3 = 0$, (*b*) $C_2 = C_3 = 0$.

12.21 Consider

$$J[\phi] = \frac{\displaystyle\int_\Omega (p\nabla\phi \cdot \nabla\phi + q\phi^2)\, d\Omega + \int_S \alpha\phi^2\, dS}{\displaystyle\int_\Omega r\phi^2\, d\Omega} \qquad \text{on } \mathscr{A}_0 = H^1(\Omega)$$

Here, functions p, q, and r obey the usual conditions; function α is in $C^1(\bar{\Omega})$ and is nonnegative on S. Prove: (*a*) If u_* minimizes J over $\mathscr{A}_0$, then u_* satisfies

$$-\nabla\cdot(p\nabla u)+qu=\lambda ru \qquad \text{in } \Omega \tag{1}$$

$$\frac{\partial u}{\partial n}+\alpha u=0 \qquad \text{on } S \tag{2}$$

with $\lambda=\lambda_1=J[u_*]$; i.e., $u_*=u_1$, an eigenfunction of (*1*)–(*2*) belonging to the smallest eigenvalue. (*b*) $\lambda_n=J[u_n]$ $(n=1,2,\ldots)$.

12.22 For the one-dimensional Sturm–Liouville problem of Section 12.4, infer from Problem 12.10 that (*a*) if $q_m<0$, at most finitely many eigenvalues are negative; (*b*) if $q_m\geq 0$, all eigenvalues are positive; (*c*) $\sum_{n=1}^{\infty}\lambda_n^{-1}$ converges.

12.23 Give weak formulations of the problems

(*a*)
$$-u_{xx}-u_{yy}+2u_x=1 \qquad \text{in } \Omega:\ x,y>0,\ x+y<2$$
$$u=0 \qquad \text{on } S$$

(*b*)
$$-u_{xx}-u_{yy}+2u_x=1 \qquad \text{in } \Omega:\ x,y>0,\ x+y<2$$
$$u_x(0,y)=2-y \qquad 0<y<2$$
$$u_y(x,0)=x(2-x) \qquad 0<x<2$$
$$u(x,2-x)=0 \qquad 0<x<2$$

Chapter 13

Variational Approximation Methods

This chapter presents some techniques for constructing approximate solutions to boundary value problems. These techniques are based on the ideas of Chapter 12 and differ markedly from the finite-difference methods of Chapters 9, 10, and 11.

13.1 THE RAYLEIGH–RITZ PROCEDURE

This approximation procedure is limited to boundary value problems admitting the variational formulation "Find u^* in $\mathcal{A}$ such that functional $J[u]$ is minimized over $\mathcal{A}$ by u^*." It was seen in Chapter 12 that such boundary value problems arise in connection with self-adjoint elliptic PDEs.

We suppose $\mathcal{A}$ to be some subset of $L^2(\Omega)$, where Ω denotes a bounded set in $\mathbf{R}^n$ with smooth boundary S consisting of complementary portions S_1 and S_2. Specifically, we take $\mathcal{A} = \{u \text{ in } H^1(\Omega) : u = g \text{ on } S_1\}$ for a given g in $C(\bar{\Omega})$; the associated subspace of comparison functions is taken as $\mathcal{M} = \{v \text{ in } H^1(\Omega) : v = 0 \text{ on } S_1\}$.

Let ϕ_0 denote an arbitrary function from $\mathcal{A}$ (e.g., $\phi_0 = g$) and let $\phi_1, \ldots, \phi_N$ denote N linearly independent functions in $\mathcal{M}$. Then,

$$u_N(\mathbf{x}) \equiv \phi_0(\mathbf{x}) + \sum_{j=1}^{N} c_j \phi_j(\mathbf{x}) \tag{13.1}$$

belongs to $\mathcal{A}$ for *all* choices of the constants $c_1, \ldots, c_N$; we denote by $\mathcal{A}_N$ the subset of $\mathcal{A}$ consisting of all such functions u_N. Let u_N^* denote the function in $\mathcal{A}_N$ that minimizes J over $\mathcal{A}_N$. It can be shown that u_N^* represents the best approximation, in the least-squares sense, from $\mathcal{A}_N$ to the exact solution u^*. This function u_N^* is called the *Rayleigh–Ritz approximation* to the solution of the boundary value problem.

The minimization of J over $\mathcal{A}_N$ is tantamount to the minimization over all $\mathbf{c}$ in $\mathbf{R}^N$ of the ordinary function

$$H(c_1, \ldots, c_N) \equiv J\left[\phi_0 + \sum_{j=1}^{N} c_j \phi_j\right]$$

The minimizing constants $c_1^*, c_2^*, \ldots, c_N^*$ must satisfy

$$\frac{\partial H}{\partial c_m}(c_1, \ldots, c_N) = 0 \qquad (m = 1, \ldots, N) \tag{13.2}$$

which is a system of N equations in N unknowns.

The Rayleigh–Ritz procedure may also be applied to eigenvalue problems of the sort treated in Section 12.4. In any eigenvalue problem the boundary conditions are all homogeneous. Hence, we take $\phi_0 \equiv 0$ in (*13.1*) and minimize the Rayleigh quotient J over the subspace $\mathcal{A}_N = \mathcal{M}_N$.

13.2 THE GALERKIN PROCEDURE

This approximation method is employed when the boundary value problem admits only a weak formulation; e.g., in the case of a linear elliptic PDE containing odd-order derivatives. In the event that conditions (*12.24*) hold in the weak formulation, so that a variational formulation also exists, it can be shown (see Problem 13.4(*b*)) that the Galerkin and Rayleigh–Ritz procedures coincide.

Consider, then, a boundary value problem with the weak formulation "Find u^* in $\mathcal{A}$ such that $K[u^*, v] = F[v]$ for all v in $\mathcal{M}$." Here we suppose that Ω, S, $\mathcal{A}$, and $\mathcal{M}$ are as described in Section

13.1. Let $\phi_1, \ldots, \phi_N$ denote N linearly independent *trial functions* in $\mathcal{M}$ and let ϕ_0 denote an arbitrary function in $\mathcal{A}$. As in the Rayleigh–Ritz procedure, we seek an approximation u_N^* to the weak solution u^* of the form (*13.1*). In addition, let $\psi_1, \ldots, \psi_N$ denote N linearly independent *weight functions* in $\mathcal{M}$; these may or may not be the same as the trial functions. The Galerkin approximate solution is required to satisfy

$$K[u_N^*, \psi_j] = F[\psi_j] \qquad (j = 1, \ldots, N) \tag{13.3}$$

This is a set of N equations in the N unknowns $c_1^*, \ldots, c_N^*$.

Solved Problems

13.1 Let Ω denote a bounded region in $\mathbf{R}^n$ having smooth boundary S consisting of complementary pieces S_1 and S_2. For the boundary value problem

$$-\nabla \cdot (p(\mathbf{x})\nabla u) + q(\mathbf{x})u = f(\mathbf{x}) \qquad \text{in } \Omega \tag{1}$$

$$u = g_1(\mathbf{x}) \qquad \text{on } S_1 \tag{2}$$

$$\frac{\partial u}{\partial n} = g_2(\mathbf{x}) \qquad \text{on } S_2 \tag{3}$$

where, as usual, $p(\mathbf{x}) > 0$ and $q(\mathbf{x}) \geq 0$ in Ω, explicitly describe the construction of the Rayleigh–Ritz approximate solution.

The methods of Chapter 12 lead to the following variational formulation of (*1*)–(*2*)–(*3*): Find u^* in $\mathcal{A}$ minimizing J over $\mathcal{A}$, where

$$\mathcal{A} \equiv \{u \text{ in } H^1(\Omega) : u = g_1 \text{ on } S_1\}$$

$$J[u] \equiv \int_\Omega (p\nabla u \cdot \nabla u + qu^2)\, d\Omega - 2F[u]$$

$$F[u] \equiv \int_\Omega fu\, d\Omega + \int_{S_2} g_2 u\, dS$$

$$\mathcal{M} \equiv \{v \text{ in } H^1(\Omega) : v = 0 \text{ on } S_1\}$$

Let $\phi_1, \ldots, \phi_N$ denote N linearly independent functions from $\mathcal{M}$. The Rayleigh–Ritz approximate solution is of the form

$$u_N = \sum_{j=0}^{N} c_j \phi_j \tag{4}$$

where ϕ_0 denotes an arbitrary function from $\mathcal{A}$ and $c_0 = 1$. We then have

$$H(c_1, \ldots, c_N) \equiv J[u_N]$$

$$= \int_\Omega \left[p \sum_{j,k=0}^{N} (\nabla\phi_j \cdot \nabla\phi_k) c_j c_k + q \sum_{j,k=0}^{N} \phi_j \phi_k c_j c_k \right] d\Omega - 2 \sum_{j=0}^{N} c_j F[\phi_j]$$

in which the linearity of $F[\;]$ has been recognized. For $m = 1, 2, \ldots, N$,

$$\frac{\partial H}{\partial c_m} = \int_\Omega \left[2p \sum_{k=0}^{N} (\nabla\phi_m \cdot \nabla\phi_k) c_k + 2q \sum_{k=0}^{N} \phi_m \phi_k c_k \right] d\Omega - 2F[\phi_m]$$

$$\equiv 2\left[\sum_{k=1}^{N} A_{mk} c_k - F_m \right]$$

where, for $1 \leq m, k \leq N$,

$$A_{mk} \equiv \int_\Omega (p\nabla\phi_m \cdot \nabla\phi_k + q\phi_m\phi_k)\,d\Omega \qquad F_m \equiv F[\phi_m] - \int_\Omega (p\nabla\phi_m \cdot \nabla\phi_0 + q\phi_m\phi_0)\,d\Omega$$

For u_N to minimize J over $\mathscr{A}_N$ it is necessary (and, H being a convex function, sufficient) that

$$\sum_{k=1}^{N} A_{mk}c_k = F_m \tag{5}$$

The solution $c_1^*, \ldots, c_N^*$ of (5) produces the Rayleigh–Ritz approximation u_N^* via (4).

13.2 Using the trial functions $\phi_1(x, y) = (6 - 2x - 3y)y$ and $\phi_2(x, y) = (6 - 2x - 3y)y^2$ (and $\phi_0 \equiv 0$), construct the Rayleigh–Ritz approximate solution to the boundary value problem indicated in Fig. 13-1.

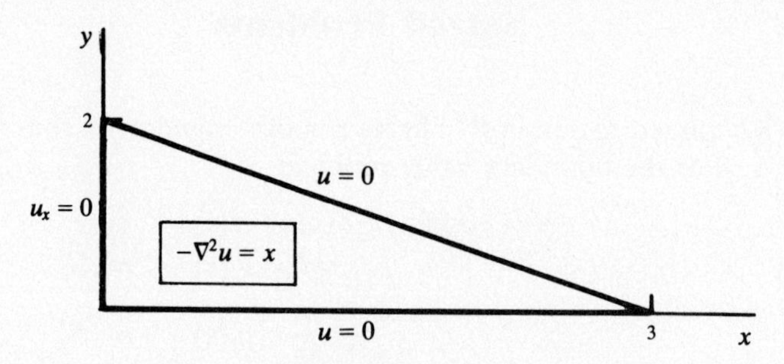

Fig. 13-1

According to Problem 13.1, we are seeking a solution of the form

$$u_2(x, y) = c_1\phi_1(x, y) + c_2\phi_2(x, y) \tag{1}$$

The constants c_1 and c_2 must satisfy

$$\begin{aligned} A_{11}c_1 + A_{12}c_2 &= F_1 \\ A_{21}c_1 + A_{22}c_2 &= F_2 \end{aligned} \tag{2}$$

where

$$A_{mk} = \int_\Omega \nabla\phi_m \cdot \nabla\phi_k\,dx\,dy \qquad F_m = \int_\Omega x\phi_m\,dx\,dy$$

Now,

$$A_{11} = \int_\Omega (36 - 24x - 72y + 4x^2 + 24xy + 40y^2)\,dx\,dy$$

$$A_{12} = \int_\Omega (72y - 48xy - 126y^2 + 8x^2y + 42xy^2 + 58y^3)\,dx\,dy = A_{21}$$

$$A_{22} = \int_\Omega (144y^2 - 96xy^2 - 216y^3 + 16x^2y^2 + 72xy^3 + 85y^4)\,dx\,dy$$

$$F_1 = \int_\Omega (6xy - 2x^2y - 3xy^2)\,dx\,dy \qquad F_2 = \int_\Omega (6xy^2 - 2x^2y^2 - 3xy^3)\,dx\,dy$$

For positive integers p and q,

$$\begin{aligned} \int_\Omega x^p y^q\,dx\,dy &= \int_0^3 \int_0^{2-(2x/3)} x^p y^q\,dx\,dy \\ &= \frac{3^{p+1}2^{q+1}}{q+1} \int_0^1 z^p(1-z)^{q+1}\,dz = \frac{3^{p+1}2^{q+1}}{q+1}\,\frac{p!(q+1)!}{(p+q+2)!} \end{aligned}$$

Thus, $A_{11}=26$, $A_{12}=A_{21}=163.2$, $A_{22}=-499.2$, $F_1=1.8$, $F_2=2.8$. Substituting these values in (2) and solving, we obtain

$$c_1^*=0.109 \qquad c_2^*=-0.001$$

13.3 Construct the Rayleigh–Ritz approximation to the solution of

$$-u''(x)+u(x)=1-x \qquad 0<x<1$$
$$u'(0)=u'(1)=0$$

using the trial functions (*a*) $\phi_1(x)=1$, $\phi_2(x)=x$, $\phi_3(x)=x^2$; (*b*) $\psi_1(x)=x^2(1-x)^2$, $\psi_2(x)=x^3(1-x)^2$, $\psi_3(x)=x^2(1-x)^3$. (*c*) Compare both approximate solutions with the exact solution,

$$u^*(x)=\frac{\cosh x-\cosh(1-x)}{\sinh 1}+1-x$$

(*a*) In this one-dimensional version of Problem 13.1, $\mathscr{A}=\mathscr{M}=H^1(0,1)$. Let

$$\hat{u}_3(x)=\sum_{i=1}^{3} c_i\phi_i(x)$$

Proceeding as in Problem 13.1, we find

$$\begin{bmatrix} 1 & \frac{1}{2} & \frac{1}{3} \\ \frac{1}{2} & \frac{4}{3} & \frac{5}{4} \\ \frac{1}{3} & \frac{5}{4} & \frac{23}{15} \end{bmatrix}\begin{bmatrix} c_1 \\ c_2 \\ c_3 \end{bmatrix}=\begin{bmatrix} \frac{1}{2} \\ \frac{1}{6} \\ \frac{1}{12} \end{bmatrix}$$

whence $\hat{c}_1=0.5384$, $\hat{c}_2=-0.0769$, $\hat{c}_3=-2\times10^{-8}$.

(*b*) Letting

$$\tilde{u}_3=\sum_{i=1}^{3} d_i\psi_i(x)$$

we find

$$\begin{bmatrix} 0.0206 & 0.0103 & 0.0098 \\ 0.0103 & 0.0068 & 0.0083 \\ 0.0098 & 0.0083 & 0.0115 \end{bmatrix}\begin{bmatrix} d_1 \\ d_2 \\ d_3 \end{bmatrix}=\begin{bmatrix} 1/60 \\ 1/140 \\ 1/105 \end{bmatrix}$$

whence $\tilde{d}_1=-12.177$, $\tilde{d}_2=48.87$, $\tilde{d}_3=-24.06$.

(*c*) The comparison, Table 13-1, brings to light one of the weaknesses of the Rayleigh–Ritz method. It is evident that $\tilde{u}_3(x)$ is a very poor approximation to $u^*(x)$. The reason is to be found in the fact that the functions $\psi_i(x)$ are linearly dependent:

$$\psi_1(x)-\psi_2(x)-\psi_3(x)=\psi_1(x)[1-x-(1-x)]=0$$

The exact equations satisfied by the d_i are

$$\begin{bmatrix} \frac{13}{630} & \frac{13}{1260} & \frac{13}{1260} \\ \frac{13}{1260} & \frac{47}{6930} & \frac{49}{13\,860} \\ \frac{13}{1260} & \frac{49}{13\,860} & \frac{47}{6930} \end{bmatrix}\begin{bmatrix} d_1 \\ d_2 \\ d_3 \end{bmatrix}=\begin{bmatrix} \frac{1}{60} \\ \frac{1}{140} \\ \frac{1}{105} \end{bmatrix}$$

Table 13-1

x	$\hat{u}_3(x)$	$\tilde{u}_3(x)$	$u^*(x)$
0	0.5384	0	0.5378
0.1	0.53071	−0.23444	0.5357
0.2	0.52302	−0.5542	0.5299
0.3	0.5153	−0.6331	0.5214
0.4	0.5076	−0.4069	0.5111
0.5	0.4999	0.0142	0.5000
0.6	0.4922	0.4332	0.4888
0.7	0.4845	0.6532	0.4785
0.8	0.4768	0.5659	0.4700
0.9	0.4691	0.2381	0.4642
1.0	0.4600	0.0	0.4600

This set of equations is inconsistent and has no solution. The system of (*b*) was obtained from this one by rounding off. Evidently, a careless application of the Rayleigh–Ritz procedure can lead to disastrous results.

The functions $\psi_1(x)$ and $\psi_2(x)$ compose an independent set and yield the approximation

$$u_2^*(x) = 7.76\,\psi_1(x) + 1.02\,\psi_2(x)$$

Now, $u_2^*(x)$ turns out to be a less accurate approximate solution than $\hat{u}_3(x)$, even though it satisfies the boundary conditions of the problem, whereas $\hat{u}_3(x)$ does not. This illustrates a second undesirable feature of the Rayleigh–Ritz approximation procedure: the success of the method is determined by the choice of trial functions. Unfortunately, there are no dependable a priori clues as to what constitutes a "good" set of trial functions.

13.4 Let Ω, S, p, q, and f be as in Problem 13.1; in addition, let $\mathbf{b} = (b_1, \ldots, b_n)$ denote a vector field defined on Ω. Consider the boundary value problem

$$-\nabla\cdot(p(x)\nabla u) + \mathbf{b}(\mathbf{x})\cdot\nabla u + q(\mathbf{x})u = f(\mathbf{x}) \qquad \text{in } \Omega \tag{1}$$

$$u = g_1(\mathbf{x}) \qquad \text{on } S_1 \tag{2}$$

$$\frac{\partial u}{\partial n} = g_2(\mathbf{x}) \qquad \text{on } S_2 \tag{3}$$

(*a*) Explicitly describe the construction of the Galerkin approximation to the weak solution of (*1*)–(*2*)–(*3*). (*b*) Show that if $\mathbf{b} = \mathbf{0}$ on Ω and $\phi_j = \psi_j$ for $j = 1, \ldots, N$, then the Galerkin approximation coincides with the Rayleigh–Ritz approximation as constructed in Problem 13.1.

(*a*) From Section 12.5, the weak formulation of (*1*)–(*2*)–(*3*) is to find u^* in $\mathscr{A}$ satisfying

$$K[u^*, v] = F[v] \qquad \text{for all } v \text{ in } \mathscr{M}$$

where

$$\mathscr{A} \equiv \{u \text{ in } H^1(\Omega) : u = g_1 \text{ on } S_1\}$$

$$\mathscr{M} \equiv \{v \text{ in } H^1(\Omega) : v = 0 \text{ on } S_1\}$$

$$K[u, v] \equiv \int_\Omega [p\nabla u\cdot\nabla v + (\mathbf{b}\cdot\nabla u + qu)v]\,d\Omega$$

$$F[v] \equiv \int_\Omega fv\,d\Omega + \int_{S_2} g_2 v\,dS$$

Let $\phi_1, \ldots, \phi_N$ denote N independent trial functions from $\mathcal{M}$ and let $\psi_1, \ldots, \psi_N$ denote N independent weight functions also from $\mathcal{M}$. Then, for an arbitrary ϕ_0 from $\mathcal{A}$, the Galerkin approximation to the weak solution of (1)–(2)–(3),

$$u_N^* = \sum_{j=0}^{N} c_j \phi_j \qquad (c_0 = 1)$$

must satisfy

$$K[u_N^*, \psi_m] = F[\psi_m] \qquad \text{for } m = 1, \ldots, N$$

i.e.

$$\sum_{j=1}^{N} A_{mj} c_j = F_m \tag{4}$$

where, for $1 \le m, j \le N$,

$$A_{mj} \equiv \int_\Omega [p \nabla \phi_j \cdot \nabla \psi_m + (\mathbf{b} \cdot \nabla \phi_j + q\phi_j)\psi_m] \, d\Omega$$

$$F_m \equiv F[\psi_m] - \int_\Omega [p \nabla \phi_0 \cdot \nabla \psi_m + (\mathbf{b} \cdot \nabla \phi_0 + q\phi_0)\psi_m] \, d\Omega$$

Note that in general $A_{mj} \neq A_{jm}$. The solution $c_1^*, \ldots, c_N^*$ of (4) yields the Galerkin approximation.

(b) Under the stated conditions, system (4) becomes identical to system (5) of Problem 13.1.

13.5 Construct the Galerkin approximation to the weak solution of the problem

$$\begin{aligned} -(u_{xx} + u_{yy}) + 2u_x - u_y &= 1 && \text{in } \Omega = \{x > 0, y > 0, x + 2y < 2\} \\ u &= 0 && \text{on } S_1 = \{x + 2y = 2\} \\ u_x(0, y) &= y && \text{on } 0 < y < 1 \\ u_y(x, 0) &= 0 && \text{on } 0 < x < 2 \end{aligned}$$

Use the single trial function $\phi(x, y) = (2 - x - 2y)(1 + x + y)$ and the single weight function $\psi(x, y) = 2 - x - 2y$.

We have $u_1^*(x, y) = c_1^* \phi(x, y)$, where c_1^* is given by

$$c_1^* K[\phi, \psi] = F[\psi]$$

Now, by computation,

$$K[\phi, \psi] = \int_\Omega [\phi_x \psi_x + \phi_y \psi_y + (2\phi_x - \phi_y)\psi] \, d\Omega = 14$$

$$F[\psi] = \int_\Omega 1\psi \, d\Omega + \int_0^1 y\psi(0, y) \, dy = -\frac{1}{3}$$

so that $c_1^* = -1/42$.

13.6 Describe explicitly the Rayleigh–Ritz procedure for approximating the eigenvalues and eigenfunctions of (12.8)–(12.9).

Evaluating the Rayleigh quotient (12.12) at

$$u_N = \sum_{j=1}^{N} c_j \phi_j$$

where the ϕ_j are linearly independent functions in $\mathcal{A}_0$, we have

$$H(c_1, \ldots, c_N) \equiv J[u_N] \equiv \frac{N(c_1, \ldots, c_N)}{D(c_1, \ldots, c_N)} \tag{1}$$

where
$$N(c_1,\ldots,c_N)=\sum_{j,k=1}^{N}\left\{\iint_\Omega [p(\nabla\phi_j\cdot\nabla\phi_k)+q\phi_j\phi_k]\,d\Omega\right\}c_jc_k\equiv\sum_{j,k=1}^{N}A_{jk}c_jc_k$$
$$D(c_1,\ldots,c_N)=\sum_{j,k=1}^{N}\left\{\iint_\Omega r\phi_j\phi_k\,d\Omega\right\}c_jc_k\equiv\sum_{j,k=1}^{N}B_{jk}c_jc_k$$

Note that both $\mathbf{A}\equiv[A_{jk}]$ and $\mathbf{B}\equiv[B_{jk}]$ are symmetric matrices. The conditions for minimizing (*1*),
$$\frac{\partial N}{\partial c_m}=H\frac{\partial D}{\partial c_m}\qquad (m=1,\ldots,N)$$
translate to the matrix eigenvalue problem $\mathbf{AX}=\mu\mathbf{BX}$, where
$$\mathbf{X}\equiv[c_1,c_2,\ldots,c_N]^T\qquad \mu\equiv H(c_1,\ldots,c_N)$$
Thus, μ_1, the smallest root (all of which are real) of the characteristic equation
$$\det(\mathbf{A}-\mu\mathbf{B})=0\tag{2}$$
is the Rayleigh–Ritz approximation to λ_1, the smallest eigenvalue of (*12.8*)–(*12.9*). And the components of $\mathbf{X}_1$, the eigenvector associated with μ_1, generate the Rayleigh–Ritz approximation to the eigenvector associated with λ_1.

Note that
$$\mu_1=\min_{\mathcal{M}_N\subset\mathcal{A}_0}J[\phi]\geq\min_{\mathcal{A}_0}J[\phi]=\lambda_1$$
The larger roots of (*2*) provide approximations for the larger eigenvalues of (*12.8*)–(*12.9*), although the accuracy of these approximations, after the first, decreases very rapidly.

13.7 Approximate the lowest fundamental frequency of a homogeneous circular membrane which is clamped at its edge.

Choose units of length and time such that $u(r,\theta,t)$, the out-of-plane displacement at position (r,θ) at time t, satisfies
$$u_{tt}=\nabla^2u(r,\theta,t)\qquad r<1,0<\theta<2\pi,t>0\tag{1}$$
$$u(1,\theta,t)=0\qquad 0<\theta<2\pi,t>0\tag{2}$$
Periodic, cylindrically symmetric solutions are of the form $u=\psi(r)\sin(\lambda^{1/2}t+\eta)$, which implies for $\psi(r)$ the Sturm–Liouville problem
$$-(r\psi')'=\lambda r\psi\qquad 0<r<1\tag{3}$$
$$\psi(0)=\text{finite},\quad \psi(1)=0\tag{4}$$
Apply to (*3*)–(*4*) the (one-dimensional) method of Problem 13.6, using the trial functions
$$\phi_1(r)=\cos\frac{\pi r}{2}\qquad \phi_2(r)=\cos\frac{3\pi r}{2}$$
Thus, for $j,k=1,2$,
$$A_{jk}=\int_0^1 r\phi_j'(r)\phi_k'(r)\,dr\qquad B_{jk}=\int_0^1 r\phi_j(r)\phi_k(r)\,dr$$
which give
$$\mathbf{A}=\begin{bmatrix}\dfrac{\pi^2+4}{16} & -\dfrac{3}{4}\\ -\dfrac{3}{4} & \dfrac{9\pi^2+4}{16}\end{bmatrix}\qquad \mathbf{B}=\begin{bmatrix}\dfrac{\pi^2-4}{4\pi^2} & -\dfrac{1}{\pi^2}\\ -\dfrac{1}{\pi^2} & \dfrac{9\pi^2-4}{36\pi^2}\end{bmatrix}$$
Solution of the characteristic equation $\det(\mathbf{A}-\mu\mathbf{B})=0$ yields the Rayleigh–Ritz approximations

$$\mu_1 = 5.790 \qquad \mu_2 = 30.578 \tag{5}$$

to λ_1 and λ_2, the two smallest eigenvalues of *(3)*–*(4)*.

For comparison, the exact solution of *(3)*–*(4)* is $\psi = J_0(\rho r)$, where ρ is a zero of the Bessel function $J_0(x)$. This implies that, to three decimals,

$$\lambda_1 = 5.784 \qquad \lambda_2 = 30.470 \tag{6}$$

which testifies to the remarkable accuracy of the Rayleigh–Ritz method *under a happy choice of trial functions.* (Bessel functions resemble sine waves.)

Supplementary Problems

13.8 Using the trial functions $\phi_1(x) = x(1-x)$ and $\phi_2(x) = x^2(1-x)$, construct one-term and two-term Rayleigh–Ritz approximate solutions to

$$-u''(x) - xu(x) = x \qquad 0 < x < 1$$
$$u(0) = u(1) = 0$$

13.9 Obtain a three-term Rayleigh–Ritz approximate solution to

$$-u''(x) + (1 + x^2)u(x) = x^2 \qquad 0 < x < 1$$
$$u'(0) = u'(1) = 0$$

using trial functions *(a)* 1, x, and x^2; *(b)* 1, $\cos \pi x$, and $\cos 2\pi x$.

13.10 Construct a three-term Rayleigh–Ritz approximate solution to

$$\begin{aligned} u_{xx} + u_{yy} &= x \qquad && x^2 + y^2 < 100,\ x > 0,\ y > 0 \\ u &= 0 && x^2 + y^2 = 100 \\ u_x(0, y) &= 0 && 0 < y < 10 \\ u_y(x, 0) &= 0 && 0 < x < 10 \end{aligned}$$

using the trial functions

$$\phi_1 = 10 - \sqrt{x^2 + y^2} \qquad \phi_2 = x\phi_1 \qquad \phi_3 = y\phi_1$$

13.11 Construct the Rayleigh–Ritz approximate solution to

$$\begin{aligned} u_{xx} + u_{yy} &= 1 \qquad && \text{in } \Omega\text{: } 0 < x, y < 1 \\ u &= 0 && \text{on } S \end{aligned}$$

using trial functions $\phi_1 = x(x-1)y(y-1)$ and $\phi_2 = x^2(x-1)y^2(y-1)$.

13.12 Construct a Rayleigh–Ritz approximation to the (singular) solution of the problem of Fig. 13-2, using $N = 2$, with $\phi_0(x, y) = 1$ and

$$\phi_1(x, y) = xy \qquad \phi_2(x, y) = xy(1 + xy)$$

13.13 Show that the Rayleigh–Ritz conditions *(13.2)* are equivalent to

$$\delta J[u_N; \phi_m] = 0 \qquad (m = 1, 2, \ldots, N)$$

i.e., the "directional derivative" of J at u_N must vanish in each of the N "directions" ϕ_m.

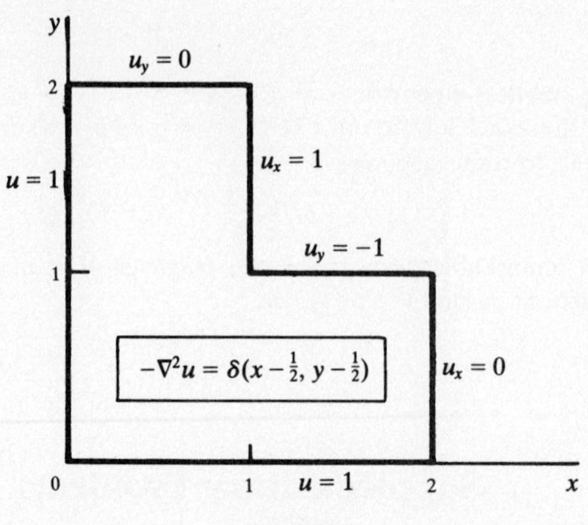

Fig. 13-2

13.14 Find the characteristic equation for the Rayleigh–Ritz approximate eigenvalues of the problem

$$-u''(x) = \lambda x u(x) \qquad 0 < x < 1$$
$$u(0) = u(1) = 0$$

if the trial functions are $\phi_1 = x(1-x)$ and $\phi_2 = x^2(1-x)$. (The true values are $\lambda_1 \approx 18.9$, $\lambda_2 \approx 81.2$.)

13.15 Show that the eigenvalues of

$$-u^{(4)}(x) = \lambda\, u(x) \qquad 0 < x < 1$$
$$u(0) = u(1) = u'(0) = u'(1) = 0$$

can be obtained by minimizing the functional

$$J[\phi] = \frac{\int_0^1 \phi''(x)^2\, dx}{\int_0^1 \phi(x)^2\, dx}$$

over an appropriate class of functions.

13.16 Construct the Galerkin approximation to the solution of

$$-u''(x) - 4u(x) = x \qquad 0 < x < 1$$
$$u(0) = u(1) = 0$$

using trial functions

$$\phi_1(x) = x(x-1) \qquad \phi_2(x) = x(x-1)(x - \tfrac{1}{2})$$

and weight functions $\psi_1(x) = 1$, $\psi_2(x) = x$.

Chapter 14

The Finite Element Method: An Introduction

The success of the approximation methods presented in Chapter 13 is largely dependent on the selection of an effective collection of trial functions ϕ_j and/or weight functions ψ_j. If these functions are chosen from certain families of piecewise polynomials, called *finite element spaces*, the following advantages are realized:

(i) It is possible to deal in a systematic fashion with regions Ω having curved boundaries of rather arbitrary shape.

(ii) One can systematically estimate the accuracy of the approximate solution in terms of the adjustable parameters associated with the finite element family.

(iii) The coefficient matrix and data vector for the system of algebraic equations defining the approximate solution can be efficiently generated by computer.

14.1 FINITE ELEMENT SPACES IN ONE DIMENSION

Suppose that the interval $[0, 1]$ is subdivided into N subintervals each of length $h = 1/N$. Let $x_j = jh \quad (j = 0, 1, \ldots, N)$ denote the nodes in the interval $[0, 1]$. Then the finite element space denoted by $S^h[k, r]$ shall consist of all functions $\phi(x)$ defined on $[0, 1]$ such that (i) on each subinterval $[x_j, x_{j+1}]$, $\phi(x)$ is a polynomial of degree at most k; (ii) $\phi(x)$ has r continuous derivatives on $[0, 1]$, which is to say, ϕ belongs to $C^r[0, 1]$.

If $r = 0$, ϕ is continuous but not necessarily differentiable at nodes. If ϕ is to be allowed to be discontinuous at nodes, we set $r = -1$. Evidently, $S^h[k, r]$ is a finite-dimensional vector space (a subspace of $L^2(0, 1)$) and so may be characterized by giving a *basis*; i.e., a linearly independent set of elements $\{\phi_j\}$ that spans the space.

Modifications for the case of nonuniform grids are easily developed.

EXAMPLE 14.1 A basis for $S^h[0, -1]$, *the piecewise constants*, is given by

$$\phi_j(x) = \begin{cases} 1 & x_{j-1} \le x \le x_j \\ 0 & \text{otherwise} \end{cases}$$

for $j = 1, 2, \ldots, N \quad (Nh = 1)$. See Fig. 14-1. The functions in $S^h[0, -1]$ are in $L^2(0, 1)$ but are not continuous.

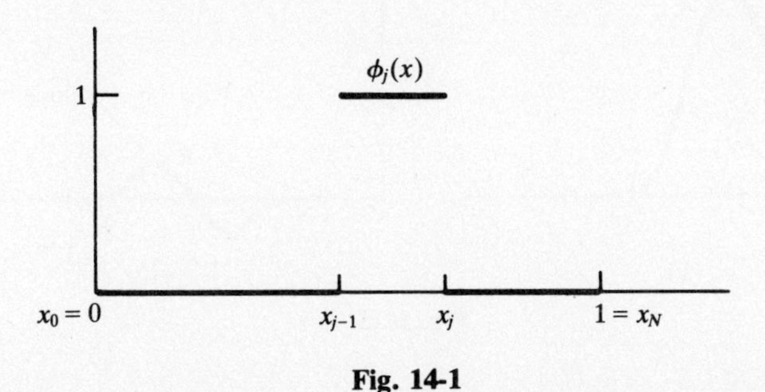

Fig. 14-1

EXAMPLE 14.2 By Problem 14.1, the "hat functions" (see Fig. 14-2)

$$\phi_0(x) = \begin{cases} (x_1 - x)/(x_1 - x_0) & x_0 \le x \le x_1 \\ 0 & \text{otherwise} \end{cases}$$

$$\phi_j(x) = \begin{cases} (x - x_{j-1})/(x_j - x_{j-1}) & x_{j-1} \le x \le x_j \\ (x_{j+1} - x)/(x_{j+1} - x_j) & x_j \le x \le x_{j+1} \\ 0 & \text{otherwise} \end{cases} \qquad (j = 1, \ldots, N-1)$$

$$\phi_N(x) = \begin{cases} (x - x_{N-1})/(x_N - x_{N-1}) & x_{N-1} \le x \le x_N \\ 0 & \text{otherwise} \end{cases}$$

where $Nh = 1$, compose a basis for $S^h[1, 0]$, *the piecewise linear functions.* Note that this space is $(N+1)$-dimensional and that the basis functions have the convenient normalization $\phi_j(x_k) = \delta_{jk}$ $(j, k = 0, \ldots, N)$. The functions in $S^h[1, 0]$ are continuous and have square-integrable first derivatives on $[0, 1]$.

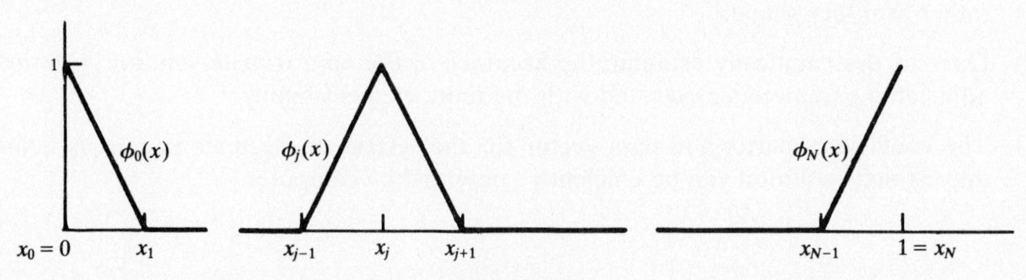

Fig. 14-2

EXAMPLE 14.3 A basis for $S^h[3, 1]$, *the piecewise cubic Hermite functions,* is jointly provided by the two families $(j = 0, \ldots, N;\ Nh = 1)$

$$\phi_j(x) = \begin{cases} (|x - x_j| - h)^2(2|x - x_j| + h)/h^3 & x_{j-1} \le x \le x_{j+1} \\ 0 & \text{otherwise} \end{cases}$$

$$\psi_j(x) = \begin{cases} (x - x_j)(|x - x_j| - h)^2/h^2 & x_{j-1} \le x \le x_{j+1} \\ 0 & \text{otherwise} \end{cases}$$

See Fig. 14-3. Note the properties

$$\phi_j(x_k) = \delta_{jk} \qquad \psi_j(x_k) = 0$$
$$\phi_j'(x_k) = 0 \qquad \psi_j'(x_k) = \delta_{jk}$$

for $j, k = 0, \ldots, N$. The functions in $S^h[3, 1]$ are continuously differentiable on $[0, 1]$ and have second derivatives which are piecewise constants (hence the second derivatives are square integrable).

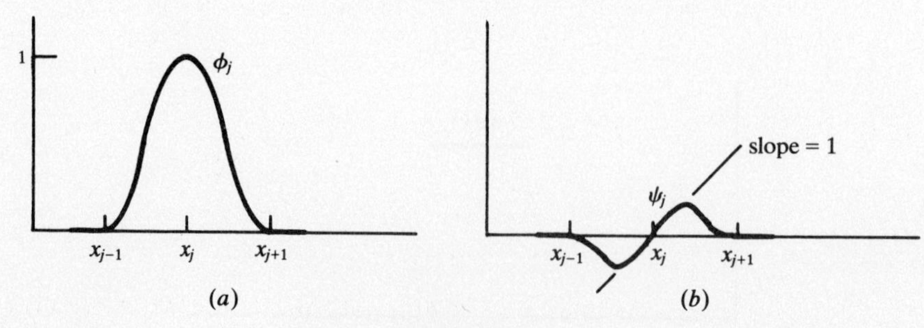

Fig. 14-3

14.2 FINITE ELEMENT SPACES IN THE PLANE

Let Ω denote a bounded region in the plane and suppose that Ω is decomposed into polygonal subregions $\Omega_1, \ldots, \Omega_N$, called *finite elements.* Let h_j denote the length of the longest side in Ω_j and let $h = \max h_j$. Finally, let Ω^h denote the polygonal region that is the union of all the Ω_j. Note that if the boundary of Ω is curved, then Ω^h may not coincide with Ω. We will denote by $S^h[k, r]$ the space of all functions $\phi(x, y)$ which are defined on Ω^h and satisfy (i) on each Ω_j, $\phi(x, y)$ is a polynomial in (x, y) of degree at most k; (ii) $\phi(x, y)$ has r continuous derivatives with respect to both x and y in Ω^h.

As in one dimension, a finite element space $S^h[k, r]$ will be specified via a basis composed of elements $\phi_j(x, y)$ associated with the nodes of the decomposition; i.e., with the vertices of the polygons.

Triangular Finite Elements

Let Ω be decomposed into triangular subregions, $\Omega_1, \ldots, \Omega_N$, where no triangle has a vertex on the side of another triangle (a *proper triangulation*; see Fig. 14-4). Euler's polyhedral formula shows that any proper triangulation into N triangles will have M nodes (vertices), where

$$\frac{N+5}{2} \leq M \leq N+2$$

Therefore, we expect $S^h(k, r)$ to be approximately N-dimensional.

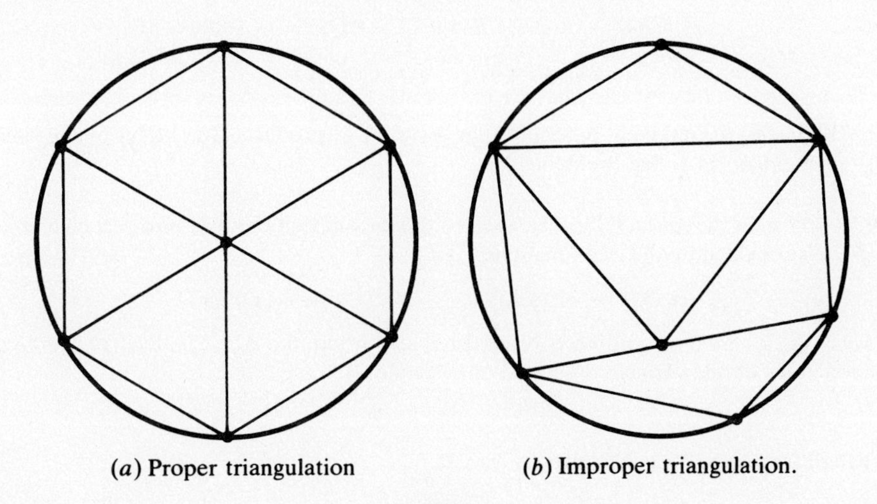

(*a*) Proper triangulation (*b*) Improper triangulation.

Fig. 14-4

EXAMPLE 14.4 Let the nodes of a proper triangulation of Ω into N triangles be labeled $z_1, \ldots, z_M$. Then a basis for $S^h[1, 0]$, *the piecewise linear functions* on Ω^h, is provided by the family $\phi_1(x, y), \ldots, \phi_M(x, y)$ that is uniquely defined by the conditions

(i) There exist constants A_{jk}, B_{jk}, C_{jk} such that

$$\phi_j(x, y) = A_{jk} + B_{jk}x + C_{jk}y \qquad \text{on } \Omega_k$$

for $1 \leq j \leq M$, $1 \leq k \leq N$; i.e., each ϕ_j is piecewise linear on Ω^h.

(ii)
$$\phi_j(z_i) = \delta_{ij} \qquad (1 \leq i,\ j \leq M)$$

The finite element space of Example 14.4 provides functions that are continuous on Ω (strictly, on Ω^h) and have first-order derivatives that are square-integrable (but are generally discontinuous). These functions would be suitable for constructing approximations to the variational solution of a boundary value problem of order two. For higher-order boundary value problems, functions having a higher degree of smoothness are required. Functions that are continuous, have continuous first derivatives, and have square-integrable second derivatives, must at the least be piecewise cubic. Such functions may be generated in more than one way.

EXAMPLE 14.5 Suppose that on each Ω_k the function $\phi_j(x, y)$ is of the form

$$\phi_j(x, y) = A_{jk} + B_{jk}x + C_{jk}y + D_{jk}x^2 + E_{jk}xy + F_{jk}y^2 + G_{jk}x^3 + H_{jk}(x^2y + xy^2) + I_{jk}y^3$$

and, together with $\partial_x\phi_j$ and $\partial_y\phi_j$, is continuous at each node z_i. Since there are 3 nodes on each triangle Ω_k, we have 3×3 conditions for determining the 9 unknowns $A_{jk}, \ldots, I_{jk}$. The functions $\phi_1, \ldots, \phi_M$ so determined constitute a basis for $S^h[3, 1]$.

Rectangular Finite Elements

Retaining the notation of the preceding subsection, we have for any proper decomposition of Ω into rectangles (where "proper" is defined as in the case of triangles)

$$N + 3 \leq M \leq 2N + 2$$

EXAMPLE 14.6 A basis for the space of piecewise linear functions on Ω^h, $S^h[1, 0]$, is provided by the family of functions $\phi_1(x, y), \ldots, \phi_M(x, y)$ defined by:

(i) On each Ω_k $(1 \leq k \leq N)$,

$$\phi_j(x, y) = A_{jk} + B_{jk}x + C_{jk}y + D_{jk}xy \qquad (1 \leq j \leq M)$$

(ii)

$$\phi_j(z_i) = \delta_{ij} \qquad (1 \leq i, j \leq M)$$

In fact, conditions (i) and (ii) uniquely determine the $\phi_k(x, y)$ as products $\phi_i(x)\phi_j(y)$ of the one-dimensional "hat functions" of Example 14.2. See Problem 14.4.

EXAMPLE 14.7 Consider the space $S^h[3, 1]$ relative to the decomposition of Ω into rectangular subregions. A basis for the space is provided by the $2M$ functions

$$\phi_k(x, y) = \phi_i(x)\phi_j(y) \qquad \psi_k(x, y) = \psi_i(x)\psi_j(y)$$

$(1 \leq k \leq M)$, where (x_i, y_j) are the coordinates of vertex z_k and where the $\phi_m(x)$ and $\psi_m(x)$ denote the piecewise cubic Hermite functions of one variable described in Example 14.3.

14.3 THE FINITE ELEMENT METHOD

When the solution of a boundary value problem is approximated by one of the techniques of Chapter 13 and when the trial functions are chosen from one of the finite element families, the approximation scheme is referred to as a *finite element method.* For problems in one dimension (i.e., for ordinary differential equations), finite element methods generally do not offer any advantage over finite difference methods. However, for certain problems in two or more dimensions, finite element methods provide distinct advantages over finite difference methods. On the other hand, the finite element approach requires complex, sophisticated computer programs for implementation, and the use of library software is to be recommended.

Solved Problems

14.1 Let $\phi_0, \phi_1, \ldots, \phi_N$ denote the hat functions of $S^h[1,0]$ defined on $[0,1]$. (*a*) Show that this is a family of $N+1$ independent functions in $S^h[1,0]$ whose span is just $S^h[1,0]$. (*b*) Describe the relation between $u(x)$ in $C[0,1]$ and $U(x)$ in $S^h[1,0]$, where

$$U(x) \equiv \sum_{j=0}^{N} u(x_j)\phi_j(x) \qquad (0 \le x \le 1)$$

(*a*) If we write $F(x) = c_0\phi_0(x) + \cdots + c_N\phi_N(x)$ and suppose that $F(x)$ vanishes for every x, then we must have $c_0 = c_1 = \cdots = c_N = 0$, since $F(x_j) = c_j \quad (0 \le j \le N)$. Thus the ϕ_j are linearly independent. To see that the ϕ_j span $S^h[1,0]$, let $v(x)$ be an arbitrary function in $S^h[1,0]$ and form the function

$$w(x) = \sum_{j=0}^{N} v(x_j)\phi_j(x) \qquad (0 \le x \le 1)$$

Now $w(x_k) = v(x_k)$ for each k, since $\phi_j(x_k) = \delta_{jk}$. In addition, $w(x)$ and $v(x)$ are linear on each subinterval $[x_{k-1}, x_k]$ and agree at the endpoints x_{k-1} and x_k. This implies that $w(x) = v(x)$ on each subinterval and hence on all of $[0,1]$; i.e., the ϕ_j span $S^h[0,1]$.

(*b*) For arbitrary $u(x)$ in $C[0,1]$, not necessarily in $S^h[1,0]$, the function $U(x)$ is the *piecewise linear interpolating approximation* to $u(x)$; see Fig. 14-5.

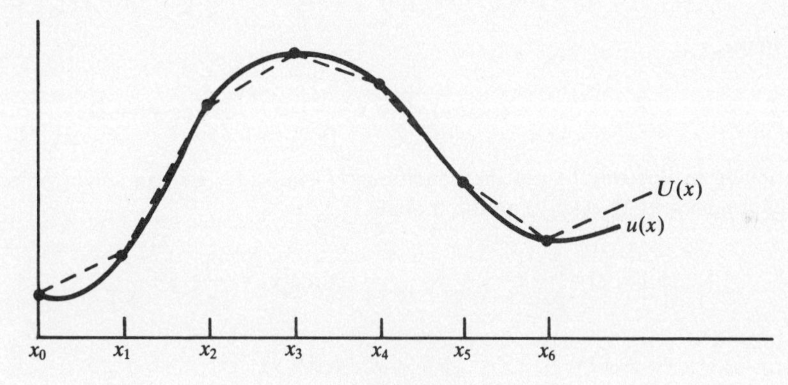

Fig. 14-5

14.2 Let a regular hexagonal region, of side 1, be properly triangulated as in Fig. 14-6. Compute a basis for $S^1[1,0]$, the piecewise linear functions on Ω^1.

The basis functions $\phi_j(x, y)$ satisfy the conditions

$$\phi_j(z_k) = \delta_{jk} \qquad (j, k = 1, \ldots, 7) \tag{1}$$

$$\phi_j(x, y) = A_{jk} + B_{jk}x + C_{jk}y \quad \text{on } \Omega_k \qquad (j = 1, \ldots, 7;\ k = 1, \ldots, 6) \tag{2}$$

The coordinates (x_k, y_k) of the node z_k are as follows:

z_k	**1**	**2**	**3**	**4**	**5**	**6**	**7**
x_k	1	2	3/2	1/2	0	1/2	3/2
y_k	$\sqrt{3}/2$	$\sqrt{3}/2$	$\sqrt{3}$	$\sqrt{3}$	$\sqrt{3}/2$	0	0

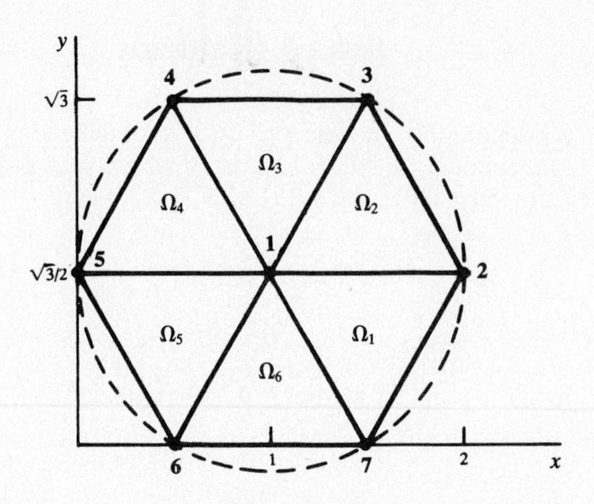

Fig. 14-6

Then, on Ω_1, $\phi_1(x, y)$ satisfies

$$\begin{aligned} \phi_1(z_1) &= A_{11} + B_{11}(1) + C_{11}(\sqrt{3}/2) = 1 \\ \phi_1(z_2) &= A_{11} + B_{11}(2) + C_{11}(\sqrt{3}/2) = 0 \\ \phi_1(z_7) &= A_{11} + B_{11}(3/2) + C_{11}(0) = 0 \end{aligned} \quad \text{or} \quad \begin{bmatrix} 1 & x_1 & y_1 \\ 1 & x_2 & y_2 \\ 1 & x_7 & y_7 \end{bmatrix} \begin{bmatrix} A_{11} \\ B_{11} \\ C_{11} \end{bmatrix} = \begin{bmatrix} 1 \\ 0 \\ 0 \end{bmatrix}$$

and hence,

$$\phi_1(x, y) = \frac{3}{2} - x + \frac{1}{\sqrt{3}} y \qquad \text{for } (x, y) \text{ in } \Omega_1$$

Continuing in this way to use the conditions (*1*) and (*2*), we can solve for each $\phi_{jk}(x, y)$, the linear function representing $\phi_j(x, y)$ on the triangle Ω_k:

$$\phi_{11}(x, y) = \frac{3}{2} - x + \frac{1}{\sqrt{3}} y \qquad \phi_{21}(x, y) = -\frac{3}{2} + x + \frac{1}{\sqrt{3}} y \qquad \cdots\cdots$$

$$\phi_{12}(x, y) = \frac{5}{2} - x - \frac{1}{\sqrt{3}} y \qquad \phi_{22}(x, y) = -\frac{1}{2} + x - \frac{1}{\sqrt{3}} y$$

$$\phi_{13}(x, y) = 2 - \frac{2}{\sqrt{3}} y \qquad \phi_{2k}(x, y) = 0 \quad (k = 3, 4, 5, 6)$$

$$\phi_{14}(x, y) = \frac{1}{2} + x - \frac{1}{\sqrt{3}} y$$

$$\phi_{15}(x, y) = -\frac{1}{2} + x + \frac{1}{\sqrt{3}} y$$

$$\phi_{16}(x, y) = \frac{2}{\sqrt{3}} y$$

Note that $\phi_2(x, y)$ vanishes identically off of Ω_1 and Ω_2. This is a result of condition (*1*). Because only Ω_1 and Ω_2 contain the vertex z_2, ϕ_2 must vanish at all three vertices of Ω_3, Ω_4, Ω_5, and Ω_6. But ϕ_2 is a linear function of x and y on each Ω_k and hence ϕ_2 must be identically zero on Ω_3 through Ω_6. It follows from this that on each triangle Ω_k there are just three of the ϕ_{jk} which are nonzero:

$$\Omega_1:\ \phi_{11}, \phi_{21}, \phi_{71} \qquad \Omega_2:\ \phi_{12}, \phi_{22}, \phi_{32} \qquad \cdots \qquad \Omega_6:\ \phi_{16}, \phi_{66}, \phi_{76}$$

Evidently, ϕ_{jk} is nonzero only if z_j is a vertex of Ω_k. This leads to the following somewhat more efficient algorithm for generating the nonzero ϕ_{jk}.

Input Data: N = number of triangular subregions
M = number of vertices

Table I. Coordinates of vertex z_j $(1 \le j \le M)$

j	x_j	y_j
1	x_1	y_1
.	....	
M	x_M	y_M

Table II. List of vertices z_i^k that belong to triangle Ω_k

k	z_1^k	z_2^k	z_3^k
1	z_1^1	z_2^1	z_3^1
.	....		
N	z_1^N	z_2^N	z_3^N

Algorithm: FOR K = 1 to N
FOR L = 1 to 3
Find J = I(L, K) such that $z_J = z_L^K$
Load Lth row of 3 × 3 coefficient matrix [M] with $(1, x_J, y_J)$
FOR L = 1 to 3

Solve:
$$[M]\begin{bmatrix} A_{LK} \\ B_{LK} \\ C_{LK} \end{bmatrix} = [e_L]$$

where $[e_L]$ = unit 3-vector with a 1 in the Lth place
Save: A_{LK}, B_{LK}, C_{LK}

Output: Three $3 \times N$ matrices, $\mathbf{A} \equiv [A_{lk}]$, $\mathbf{B} \equiv [B_{lk}]$, $\mathbf{C} \equiv [C_{lk}]$, containing the coefficients needed to form

$$\phi_{lk}(x, y) = A_{lk} + B_{lk}x + C_{lk}y \qquad (l = 1, 2, 3;\ k = 1, 2, \ldots, N)$$

Note that the index l is not necessarily the number of the vertex at which $\phi = 1$; the l-value of that special vertex may be read from Table II.

In the present problem, $N = 6$, $M = 7$. Table I has already been given; Table II is as follows:

k	z_1^k	z_2^k	z_3^k
1	1	2	7
2	1	2	3
3	1	3	4
4	1	4	5
5	1	5	6
6	1	6	7

Applying the algorithm then leads to

On Ω_1:
$$\phi_1 = \frac{3}{2} - x + \frac{1}{\sqrt{3}}y$$
$$\phi_2 = -\frac{3}{2} + x + \frac{1}{\sqrt{3}}y$$
$$\phi_7 = 1 - \frac{2}{\sqrt{3}}y$$

On Ω_2:
$$\phi_1 = \frac{5}{2} - x - \frac{1}{\sqrt{3}}y$$
$$\phi_2 = -\frac{1}{2} + x - \frac{1}{\sqrt{3}}y$$
$$\phi_3 = -1 + \frac{2}{\sqrt{3}}y$$

On Ω_3: $\phi_1 = 2 - \frac{2}{\sqrt{3}} y$

$\phi_3 = -\frac{3}{2} + x + \frac{1}{\sqrt{3}} y$

$\phi_4 = \frac{1}{2} - x + \frac{1}{\sqrt{3}} y$

On Ω_4: $\phi_1 = \frac{1}{2} + x - \frac{1}{\sqrt{3}} y$

$\phi_4 = -1 + \frac{2}{\sqrt{3}} y$

$\phi_5 = \frac{3}{2} - x - \frac{1}{\sqrt{3}} y$

On Ω_5: $\phi_1 = -\frac{1}{2} + x + \frac{1}{\sqrt{3}} y$

$\phi_5 = \frac{1}{2} - x + \frac{1}{\sqrt{3}} y$

$\phi_6 = 1 - \frac{2}{\sqrt{3}} y$

On Ω_6: $\phi_1 = \frac{2}{\sqrt{3}} y$

$\phi_6 = \frac{3}{2} - x - \frac{1}{\sqrt{3}} y$

$\phi_7 = -\frac{1}{2} + x - \frac{1}{\sqrt{3}} y$

14.3 Let Ω denote the circular region indicated in Fig. 14-6; the boundary S (dashed circle) is supposed to be partitioned into arc S_2, the minor arc between nodes 2 and 3, and the complementary arc S_1. Solve by the finite element method

$$u_{xx} + u_{yy} = 0 \qquad \text{in } \Omega$$

$$\frac{\partial u}{\partial n} = 0 \qquad \text{on } S_2$$

$$u = g \qquad \text{on } S_1$$

under the assumption that $g(z_4) = g(z_5) = g(z_6) = 100$ and $g(z_7) = -100$.

A finite element solution would consist in applying the Rayleigh–Ritz procedure (Section 13.1) to the functional

$$J[u] = \int_\Omega (u_x^2 + u_y^2)\, dx\, dy$$

with trial functions drawn from $S^1[1,0]$. In terms of the basis functions $\phi_1, \ldots, \phi_7$ ascertained in Problem 14.2, we take as $\mathcal{M}$ the subspace spanned by ϕ_1, ϕ_2, and ϕ_3 (these vanish, as is required, at nodes 4, 5, 6, and 7, where u is prescribed). As the function ϕ_0 from $\mathcal{A}$, which must assume the prescribed boundary values, choose

$$\phi_0 \equiv 100\phi_4 + 100\phi_5 + 100\phi_6 - 100\phi_7$$

Carrying out the minimization of

$$H(c_1, c_2, c_3) \equiv J[\phi_0 + c_1\phi_1 + c_2\phi_2 + c_3\phi_3]$$

we obtain $c_1^* = 42.85$, $c_2^* = 8.56$, $c_3^* = 48.56$. Our piecewise linear approximate solution is therefore

$$u_3^*(x, y) = 42.85\,\phi_1(x, y) + 8.56\,\phi_2(x, y) + 48.56\,\phi_3(x, y) + 100[\phi_4(x, y) + \phi_5(x, y) + \phi_6(x, y) - \phi_7(x, y)]$$

14.4 An L-shaped region Ω is decomposed into squares, as shown in Fig. 14-7. Construct a basis for the piecewise linear functions, $S^{1/4}[1,0]$, on $\Omega^{1/4} = \Omega$.

Let node z_k $(k = 1, 2, \ldots, 19)$ have abscissa x_i $(i = 0, \ldots, 4$; see Fig. 14-2 with $N = 4)$ and ordinate y_j $(j = 0, \ldots, 4$; replace x in Fig. 14-2 by y, and again take $N = 4)$. (It is obvious how the foregoing would read if the decomposition were into *rectangles*, with different numbers in the x- and y-directions.) Then the basis functions for $S^{1/4}[1,0]$ are given by

$$\phi_k(x, y) = \phi_i(x)\,\phi_j(y) \qquad (1)$$

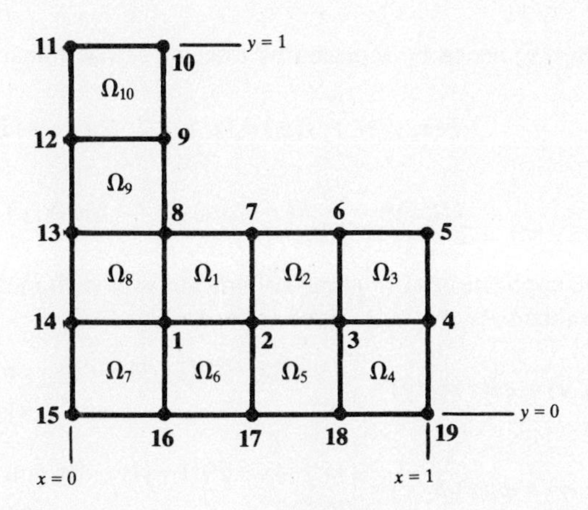

Fig. 14-7

To make (*1*) explicit, read off from Fig. 14-7 the values of i and j answering to a given k, and then appropriate $\phi_i(x)$ and $\phi_j(y)$ from Example 14.2, with $j(1/4)$ substituted for x_j or y_j. For instance,

$$\phi_8(x, y) = \phi_1(x)\phi_2(y) = \begin{cases} (1 - |4x - 1|)(1 - |4y - 2|) & 0 \le x \le 1/2,\ 1/4 \le y \le 3/4 \\ 0 & \text{otherwise} \end{cases}$$

14.5 Consider the boundary value problem

$$-u_{xx} - u_{yy} = f \qquad \text{in } \Omega \tag{1}$$

$$u = g \qquad \text{on } S_1 \tag{2}$$

$$\frac{\partial u}{\partial n} = h \qquad \text{on } S_2 \tag{3}$$

where Ω is the region of Problem 14.4, S_1 is the intersection of the boundary S with the coordinate axes, $S_2 = S - S_1$, and where f, g, and h are prescribed functions of (x, y). Set up a system of linear algebraic equations for the coefficients in a finite element approximate solution to (*1*)–(*2*)–(*3*).

The procedure parallels that of Problem 14.3; this time we appeal to Problem 13.1, wherein were developed the Rayleigh–Ritz equations for a class of boundary value problems including (*1*)–(*2*)–(*3*). Thus, in Problem 13.1, set $n = 2$, $p \equiv 1$, $q \equiv 0$, $g_1 = g$, $g_2 = h$. Rewrite (*4*) of Problem 13.1 as

$$u_{10}(x, y) = \phi_0(x, y + \sum_{k=1}^{10} c_k\phi_k(x, y) \equiv \sum_{k=11}^{19} g(z_k)\phi_k(x, y) + \sum_{k=1}^{10} c_k\phi_k(x, y)$$

where the $\phi_k(x, y)$ are the basis functions for $S^{1/4}[1, 0]$, as given by (*1*) of Problem 14.4. Then the desired linear system in $c_1, \ldots, c_{10}$ is

$$\sum_{k=1}^{10} A_{mk}c_k = F_m \qquad (m = 1, \ldots, 10)$$

with

$$\begin{aligned} A_{mk} &= \int_\Omega \nabla\phi_m(x, y)\cdot\nabla\phi_k(x, y)\,dx\,dy \\ &= \sum_{r=1}^{10}\int_{\Omega_r} \nabla\phi_m(x, y)\cdot\nabla\phi_k(x, y)\,dx\,dy \end{aligned} \tag{4}$$

and

$$F_m = \int_\Omega f(x, y)\phi_m(x, y)\,dx\,dy + \int_{S_2} h(x, y)\phi_m(x, y)\,dS - \int_\Omega \nabla\phi_m\cdot\nabla\phi_0\,dx\,dy \tag{5}$$

In (*5*), $f(x, y)$ and $h(x, y)$ are to be replaced by their piecewise linear approximants (see Problem 14.1)

$$F(x, y) = \sum_{j=1}^{19} f(z_j)\phi_j(x, y) \qquad \text{for } (x, y) \text{ in } \Omega$$

$$H(x, y) = \sum_{j=4}^{10} h(z_j)\phi_j(x, y) \qquad \text{for } (x, y) \text{ on } S_2$$

To illustrate the application of (*4*) (and to indicate why such matters are normally consigned to the computer), let us evaluate A_{12} $(= A_{21})$. Since (refer to Fig. 14-7)

$$\phi_1(x, y) = \phi_1(x)\phi_1(y) = \begin{cases} (1-|4x-1|)(1-|4y-1|) & \text{on } \Omega_1 \cup \Omega_6 \cup \Omega_7 \cup \Omega_8 \\ 0 & \text{elsewhere} \end{cases}$$

$$\phi_2(x, y) = \phi_2(x)\phi_1(y) = \begin{cases} (1-|4x-2|)(1-|4y-1|) & \text{on } \Omega_1 \cup \Omega_2 \cup \Omega_5 \cup \Omega_6 \\ 0 & \text{elsewhere} \end{cases}$$

it follows that

$$\nabla\phi_1(x, y) \cdot \nabla\phi_2(x, y) = \phi_1'(x)\phi_2'(x)[\phi_1(y)]^2 + \phi_1(x)\phi_2(x)[\phi_1'(y)]^2$$

has the representation $32(1-2x)(4x-1) - 256y^2$ in Ω_6, takes mirror-image values in Ω_1, and vanishes in all other Ω_r. Hence,

$$A_{12} = 2\int_0^{1/4} dy \int_{1/4}^{1/2} [32(1-2x)(4x-1) - 256y^2]\, dx = -\frac{1}{3}$$

In similar fashion, the constants F_m can be evaluated by one-dimensional integrations of (at worst) quadratic functions.

The analogous procedure in the case of a decomposition of Ω into triangles Ω_k is somewhat more complex, since it involves the calculation of integrals of the form

$$\iint_{\Omega_k} x^p y^q\, dx\, dy \qquad (0 \le p,\ q \le 2)$$

On the other hand, for a given degree of accuracy, the dimension M of the triangulated problem will generally be lower than that of the rectangular decomposition.

Supplementary Problems

14.6 For the triangulated hexagon ($h = 1$) shown in Fig. 14-8, use the algorithm of Problem 14.2 to generate a basis for the piecewise linear functions, $S^2[1, 0]$.

14.7 Consider the three-lobed region Ω in Fig. 14-8 bounded by the dashed curve S. Let S_1 be the circular arc centered on node 3, and let S_2 be the remainder of S. Using the approximating functions developed in Problem 14.6, solve the following problems by the finite element method:

(*a*)

$$u_{xx} + u_{yy} = 0 \quad \text{in } \Omega$$
$$u = g \quad \text{on } S$$

where $g(z_8) = g(z_9) = -g(z_{11}) = -g(z_{12}) = 50$, and $g = 0$ at all other boundary nodes.

(*b*)

$$u_{xx} + u_{yy} = 0 \quad \text{in } \Omega$$
$$\frac{\partial u}{\partial n} = 0 \quad \text{on } S_1$$
$$u = g \quad \text{on } S_2$$

where $g(z_8) = g(z_9) = -g(z_{11}) = -g(z_{12}) = 50$, $g(z_{10}) = 0$.

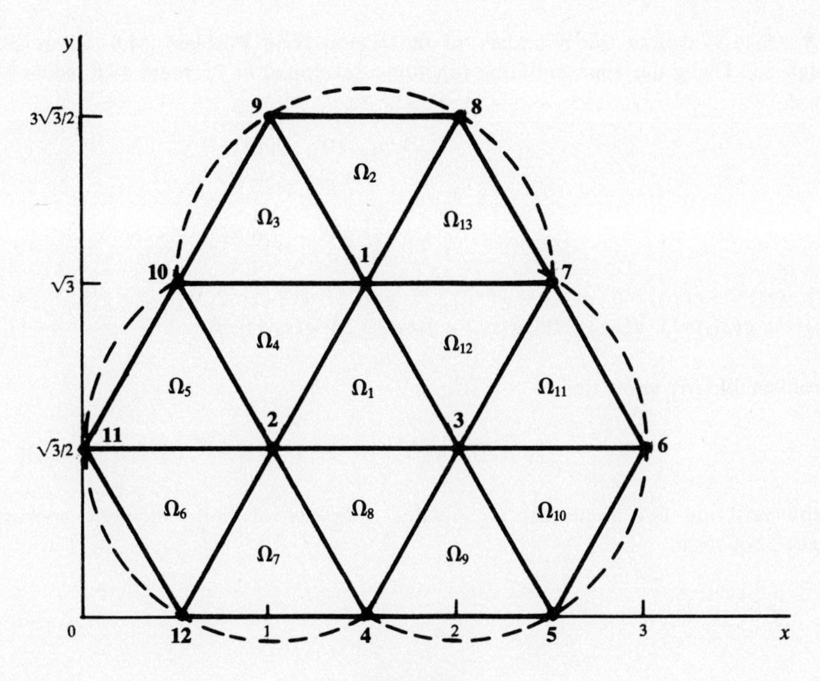

Fig. 14-8

(c)

$$u_{xx} + u_{yy} = 0 \quad \text{in } \Omega$$

$$\frac{\partial u}{\partial n} = 0 \quad \text{on } \widehat{z_4 z_5}$$

$$u = g \quad \text{on } S - \widehat{z_4 z_5}$$

where $g(z_6) = g(z_{11}) = 10$, $g(z_7) = g(z_{10}) = -g(z_{12}) = 30$, $g(z_8) = g(z_9) = 50$.

14.8 A two-dimensional region Ω, formed by cutting a groove in a rectangle, is triangulated as in Fig. 14-9 ($\Omega = \Omega^{\sqrt{5}}$). Generate by the algorithm of Problem 14.2 a basis for $S^{\sqrt{5}}[1, 0]$.

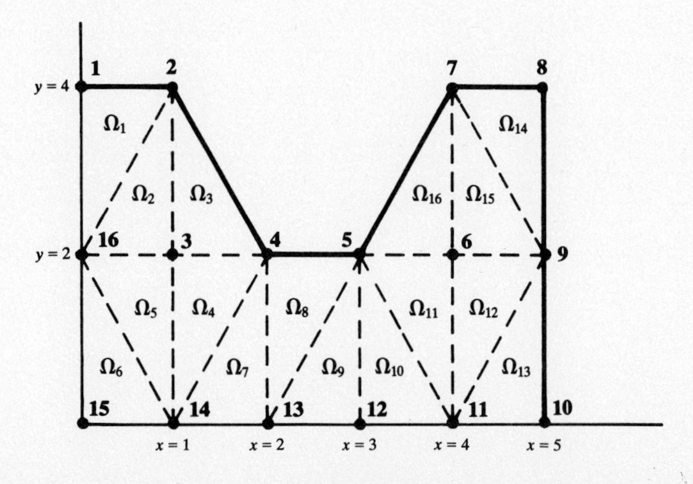

Fig. 14-9

14.9 Let $S = S_1 \cup S_2$ denote the boundary of the region Ω of Problem 14.8, where S_1 contains nodes z_9 through z_{16}. Using the approximating functions developed in Problem 14.8, solve by the finite element method

$$u_{xx} + u_{yy} = 0 \quad \text{in } \Omega$$
$$u = g \quad \text{on } S_1$$
$$\frac{\partial u}{\partial n} = 0 \quad \text{on } S_2$$

if (*a*) $g(z_9) = -g(z_{16}) = 5$, $g(z_{10}) = g(z_{15}) = 0$, $g(z_{11}) = g(z_{14}) = 2$, $g(z_{12}) = g(z_{13}) = 4$; (*b*) $g(z_9) = g(z_{10}) = g(z_{11}) = 0$, $g(z_{12}) = 5$, $g(z_{13}) = 10$, $g(z_{14}) = g(z_{16}) = 20$, $g(z_{15}) = 30$.

14.10 In Problem 14.7(*a*), show that

$$u_3^*(z_1) = \frac{1}{6}[u_3^*(z_2) + u_3^*(z_3) + u_3^*(z_7) + u_3^*(z_8) + u_3^*(z_9) + u_3^*(z_{10})]$$

thereby verifying the numerical mean-value property of finite element approximate solutions of Laplace's equation.

Answers to Supplementary Problems

CHAPTER 2

2.19 (*a*) h. everywhere; (*b*) p. everywhere; (*c*) e. everywhere; (*d*) h. $|x|>1$, p. $|x|=1$, e. $|x|<1$; (*e*) h. $y<e^{-x}$, p. $y=e^{-x}$, e. $y>e^{-x}$; (*f*) h. $x<0$, p. $x=0$, e. $x>0$; (*g*) h. $xy(xy+1)>0$, p. $xy(xy+1)=0$, e. $xy(xy+1)<0$; (*h*) h. $xy(xy-1)>0$, p. $xy(xy-1)=0$, e. $xy(xy-1)<0$

2.21 (*a*) $p(u_{xx}+u_{yy})$, elliptic; (*b*) $-p(u_{xx}+u_{yy})$, elliptic; (*c*) $u_{tt}-p(u_{xx}+u_{yy})$, hyperbolic

2.22 (*a*) $y=\text{const.}$ (*b*) $2x-y=\text{const.}$, $2x-3y=\text{const.}$ (*c*) $x\pm y=\text{const.}$ (*d*) $x\pm 2\sqrt{-y}=\text{const.}$ $(y<0)$ (*e*) $\log|y|\pm x=\text{const.}$ $(y\neq 0)$ (*f*) $x^2+y^2=\text{const.}$ (*g*) $5x\pm 2(-y)^{5/2}=\text{const.}$ $(y<0)$ (*h*) $3x\pm 2(-y)^{3/2}=\text{const.}$ $(y<0)$

2.23 $\xi=\dfrac{2}{5}x-y,\ \eta=\dfrac{4}{5}x$

2.24 $\lambda_1=1,\ \lambda_2=3,\ \lambda_3=4$ (all positive);

$$\xi_1=\frac{1}{\sqrt{6}}(x_1+2x_2+x_3)\qquad \xi_2=\frac{1}{\sqrt{2}}(x_1-x_3)\qquad \xi_3=\frac{1}{\sqrt{3}}(x_1-x_2+x_3)$$

2.25 $\eta_1=\xi_1,\ \eta_2=\xi_2/\sqrt{3},\ \eta_3=\xi_3/2$

2.26 (*a*) elliptic ($\lambda_1=2,\ \lambda_2=2,\ \lambda_3=4$ all positive)

(*b*)
$$\xi_1=\frac{1}{\sqrt{2}}(x_1-x_3)\qquad \xi_2=x_2\qquad \xi_3=\frac{1}{\sqrt{2}}(x_1+x_3)$$
$$\eta_1=\xi_1/\sqrt{2}\qquad \eta_2=\xi_2/\sqrt{2}\qquad \eta_3=\xi_3/2$$

2.27 (*a*) $\xi=4x-y,\ \eta=2x+y$
(*b*) $\xi=(x+1)^2+x\sqrt{x^2+1}+\log(x+\sqrt{x^2+1})-2y$
$\eta=(x+1)^2-x\sqrt{x^2+1}-\log(x+\sqrt{x^2+1})-2y$
(*c*) $\xi=(1+\sqrt{6}/2)x-y,\ \eta=(1-\sqrt{6}/2)x-y$
(*d*) $\xi=(1+\sqrt{2})e^x-e^y,\ \eta=(1-\sqrt{2})e^x-e^y$
(*e*) $\xi=\tan^{-1}x-\tan^{-1}y,\ \eta=\tan^{-1}x+\tan^{-1}y$

2.28 (*a*) elliptic: $\xi=y,\ \eta=x+(x^3/3)$
(*b*) elliptic: $\xi=x+2y,\ \eta=2x$
(*c*) parabolic: $\xi=x,\ \eta=y+x$
(*d*) hyperbolic: already in canonical form
(*e*) parabolic: $\xi=x,\ \eta=y/x$ $(x\neq 0)$
(*f*) hyperbolic: $\xi=xy,\ \eta=x/y$ $(xy\neq 0)$
(*g*) same as (*f*)
(*h*) parabolic: $\xi=x,\ \eta=2x+y^2$
(*i*) elliptic for $xy>0$: $\xi=3|y|^{1/2},\ \eta=|x|^{3/2}$
(*j*) elliptic for $xy<0$: $\xi=|y|^{3/2},\ \eta=|x|^{3/2}$
hyperbolic for $xy>0$: $\xi=|x|^{3/2}-|y|^{3/2},\ \eta=|x|^{3/2}+|y|^{3/2}$
(*k*) parabolic: $\xi=x,\ \eta=e^x-e^y$
(*l*) elliptic: $\xi=\log|1+y|,\ \eta=x$

(*m*) parabolic: $\xi = x$, $\eta = \sqrt{x} - \sqrt{y}$
(*n*) hyperbolic: $\xi = y + \csc x - \cot x$, $\eta = y + \csc x + \cot x$
(*o*) hyperbolic: $\xi = x^2 - 2e^y$, $\eta = x^2 + 2e^y$ $(x \neq 0)$
(*p*) parabolic: $\xi = x$, $\eta = \tan^{-1} x + \tan^{-1} y$

2.33 $u(x, t) = e^{-2x+2t} v(x, t)$

CHAPTER 3

3.38 0

3.39 4

3.42 (*b*) $$a_{mn} = \frac{16}{\pi^2} \frac{m}{m^2-1} \frac{n}{n^2-1} \frac{1}{2-m^2-n^2} \quad (m \text{ and } n \text{ even}),$$

$a_{mn} = 0$ (m or n odd and $mn \neq 1$), a_{11} arbitrary

CHAPTER 4

4.15 $u(r, t) = r^{-1}[F(r - at) + G(r + at)] \qquad (r > 0)$

4.19 (*a*) $u(x, y, t) = \exp[\lambda(\alpha x + \beta y - \mu t)]$, where $\lambda^2[\mu^2 - (\alpha a_1)^2 - (\beta a_2)^2] = b^2$; (*b*) $p^2 = b^2 - 2$

4.23 If $v_F(\mathbf{x}, t; \tau)$ solves

$$\begin{aligned} v_t(\mathbf{x}, t) &= \nabla^2 v(\mathbf{x}, t) && \mathbf{x} \text{ in } \mathbf{R}^n,\ t > \tau \\ v(\mathbf{x}, \tau) &= F(\mathbf{x}, \tau) && \mathbf{x} \text{ in } \mathbf{R}^n \end{aligned}$$

then

$$u(\mathbf{x}, t) = \int_0^t v_F(\mathbf{x}, t; \tau)\, d\tau$$

satisfies

$$\begin{aligned} u_t(\mathbf{x}, t) &= \nabla^2 u(\mathbf{x}, t) + F(\mathbf{x}, t) && \mathbf{x} \text{ in } \mathbf{R}^n,\ t > 0 \\ u(\mathbf{x}, 0) &= 0 && \mathbf{x} \text{ in } \mathbf{R}^n \end{aligned}$$

4.24

$$u(x, t) = \frac{1}{2a} \int_0^t [F(x + a(t - \tau)) - F(x - a(t - \tau))]\, d\tau$$

where $F(s)$ is an antiderivative of $f(s)$.

CHAPTER 5

5.22 Hyperbolic if $u^2 + v^2 > c^2$ (*supersonic flow*), elliptic if $u^2 + v^2 < c^2$ (*subsonic flow*).

5.23 (*d*) $\dfrac{v_1 u_1 - v_2 u_2}{u_1 - u_2}$

5.25 (*a*) $u(x, y) = f(x/y)$; (*b*) along the line $y = x/\bar{x}$

5.26 (*a*) $x^2 + y^2 = 1$, $u = 2$; (*b*) $x^2 + y^2 = 1$, $u = 2e^{(\sin^{-1} x) - \pi/2}$

5.27 (*a*) $u = b(t - a)\left(\dfrac{L - a}{L}\right)^c$; (*b*) $u = \log(x + y) + \dfrac{y}{x + y} + \left(\dfrac{x}{x + y}\right)^2$

5.28 (a) $u=\psi(x-ct)+\int_0^t \phi(cr+x-ct, r)\,dr$; (b) $u=g\left(x-\int_0^y f(z)\,dz\right)e^{cy}$

5.31 (a) $u=\frac{1}{5}[6\sin(x-3t)-6\cos(x-3t)-\sin(x+2t)+6\cos(x+2t)]$

$v=\frac{1}{5}[\sin(x-3t)-\cos(x-3t)-\sin(x+2t)+6\cos(x+2t)]$

(b) $u=\sin(x+y)$, $v=\frac{1}{6}[8\sin(x-2y)+6e^{x-2y}-8\sin(x+y)]$

5.32 See Fig. A-1.

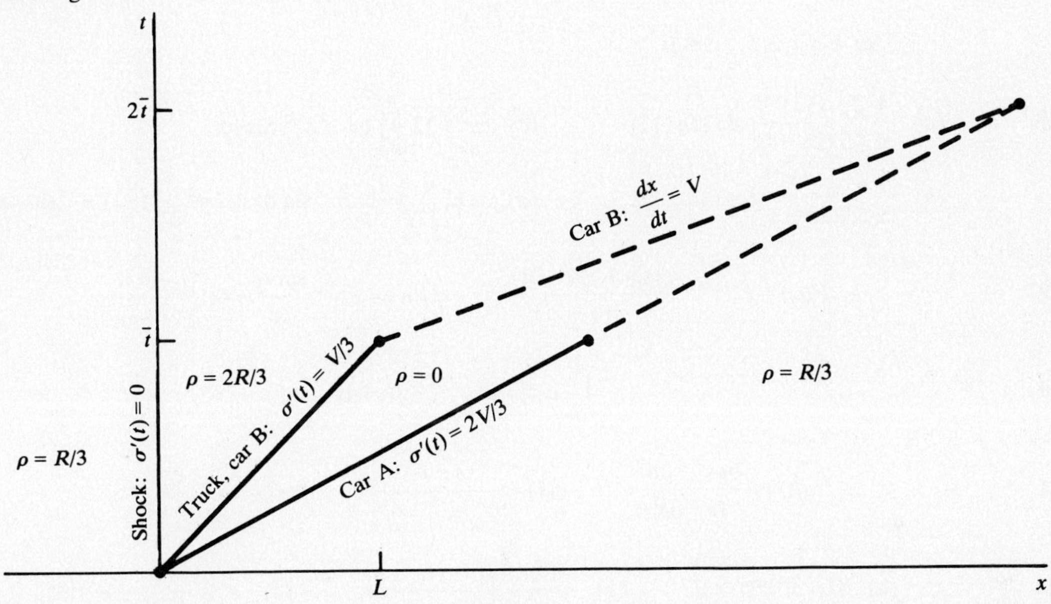

Fig. A-1

CHAPTER 6

6.8 (a) $a_0=1$, $a_n=0$ and $b_n=(1-\cos n\pi)/n\pi$ for $n=1, 2, \ldots$
(b) $a_0=\pi$, $a_n=2(\cos n\pi-1)/n^2\pi$ and $b_n=0$ for $n=1, 2, \ldots$
(c) $a_0=0$, $a_n=0$ and $b_n=-2(\cos n\pi)/n$ for $n=1, 2, \ldots$

6.9 By Theorem 6.1, the series for F and H converge pointwise, while the series for G converges uniformly.

6.10 (a) $4\pi^{-1}\sum_{n=1}^{\infty}(2n-1)^{-1}\sin(2n-1)x$; (b) 1

6.11 (a) $\lambda_n=(n\pi/\ell)^2$ and $w_n(x)=\sin\lambda_n^{1/2}x$ for $n=1, 2, \ldots$
(b) $\lambda_n=[(n-\frac{1}{2})\pi/\ell]^2$ and $w_n(x)=\sin\lambda_n^{1/2}x$ for $n=1, 2, \ldots$
(c) $\lambda_n=[(n-\frac{1}{2})\pi/\ell]^2$ and $w_n(x)=\cos\lambda_n^{1/2}x$ for $n=1, 2, \ldots$
(d) $\lambda_n=\mu_n^2$, where μ_n is the nth positive root of $\mu=\tan\mu\ell$, and

$$w_n(x)=\sin\mu_n x-\mu_n\cos\mu_n x=\frac{\sin\mu_n(x-\ell)}{\cos\mu_n\ell}$$

for $n=1, 2, \ldots$.

(e) $\lambda_{-1} = -\mu_{-1}^2$, where μ_{-1} satisfies $e^{2\ell\mu} = (\mu+1)/(\mu-1)$ and $w_{-1}(x) = \exp[-\mu_{-1}(2\ell - x)] + \exp[-\mu_{-1}x]$. For $n = 1, 2, \ldots, \lambda_n = \mu_n^2$, where μ_n is the nth positive root of $\mu \tan \mu\ell = -1$ and

$$w_n(x) = \sin \mu_n x - \mu_n \cos \mu_n x = [\cos \mu_n(x-\ell)]/\sin \mu_n \ell.$$

(f) If $\ell > \alpha - \beta$, then $\lambda_{-1} = -\mu_{-1}^2$, where μ_{-1} satisfies

$$e^{2\mu\ell} = (1+\alpha\mu)(1-\beta\mu)/(1-\alpha\mu)(1+\beta\mu)$$

and $w_{-1}(x) = (\alpha\mu - 1)(\alpha\mu + 1)e^{\mu x} - e^{-\mu x}$; if $\ell = \alpha - \beta$, then $\lambda_0 = 0$, with $w_0(x) = x - \alpha$. For $n = 1, 2, \ldots, \lambda_n = \mu_n^2$, where μ_n is the nth positive root of $(1+\alpha\beta\mu^2)\tan \mu\ell = (\alpha - \beta)\mu$ and $w_n(x) = \sin \mu_n x - \alpha\mu_n \cos \mu_n x$.

6.13 ***Theorem:*** $\tilde{F}_e(x)$ is C^k if $f^{(j)}(x)$ is continuous on $[0, \ell]$ for $j = 0, 1, \ldots, p$, and if

$$f^{(k)}(0) = f^{(k)}(\ell) = 0$$

for $k = 1, 3, 5, \ldots \le p$.

6.15 (a) $\dfrac{4}{\pi}\sum_{n=1}^{\infty} \dfrac{(-1)^{n+1}}{(2n-1)^2} \sin(2n-1)x$ (c) $\pi^2 + 12 \sum_{n=1}^{\infty} (-1)^n n^{-2} \cos nx$

(b) $\dfrac{4}{\pi}\sum_{n=1}^{\infty} \dfrac{(-1)^{n+1}}{2n-1} \cos(2n-1)x$ (d) $-12 \sum_{n=1}^{\infty} (-1)^n n^{-3} \sin nx + 2\pi^2 \sum_{n=1}^{\infty} (-1)^n n^{-1} \sin nx$

6.19
$$F(\alpha) = e^{-i2.5(\alpha - 2i)} \frac{\sin 1.5(\alpha - 2i)}{\pi(\alpha - 2i)} \qquad G(\alpha) = e^{-i2\alpha}\frac{\sin \alpha}{\pi\alpha} - 4e^{-i7\alpha}\frac{\sin\alpha}{\pi\alpha}$$

6.20
$$\frac{1}{4}\int_{-\infty}^{\infty} e^{-i4(x-y)} e^{-2|x-y|} f(y)\, dy$$

6.21
$$\hat{f}(s) = \frac{6as^2 - 2a^3}{(s+a)^3} \qquad \hat{e}(s) = \frac{s-b}{(s-b)^2 + a^2} \qquad \hat{g}(s) = \frac{e^{-s}}{s} - \frac{e^{-2s}}{s}$$

6.22 (a) $(t-3)^3 H(t-3)$ (b) $\dfrac{f(0)}{\sqrt{\pi t}} + \displaystyle\int_0^t \frac{f'(t-\tau)}{\sqrt{\pi\tau}}\, d\tau$ (c) $e^{-ht}\dfrac{k}{\sqrt{4\pi k t^3}} e^{-k^2/4t}$

6.23 Let $\alpha_n \equiv (2n+1)b - a$ and $\beta_n \equiv (2n+1)b + a$. Then,

$$F(t) = \sum_{n=0}^{\infty} [H(t-\alpha_n) + H(t-\beta_n)] \qquad G(t) = \frac{1}{\sqrt{\pi t}} \sum_{n=0}^{\infty} [\exp(-\alpha_n^2/4t) - \exp(-\beta_n^2/4t)]$$

CHAPTER 7

7.19
$$u(x,t) = u_1(x,t) + u_2(x,t) + u_3(x,t)$$
$$u_1(x,t) = g_0(t)(x - x^2/2\ell) + g_1(t)(x^2/2\ell)$$
$$u_2(x,t) = \frac{1}{2} f_0 + \sum_{n=1}^{\infty} f_n \exp(-\mu_n^2 \kappa t) \cos \mu_n x$$

where $\mu_n \equiv n\pi/\ell$ and, if $g_0(0) = g_1(0) = 0$, $f_n \equiv \dfrac{2}{\ell}\displaystyle\int_0^\ell f(x) \cos \mu_n x\, dx$

$$u_3(x,t) = \int_0^t \left[\frac{1}{2} F_0(\tau) + \sum_{n=1}^{\infty} \exp(-\mu_n^2 \kappa(t-\tau)) F_n(\tau) \cos \mu_n x\right] d\tau$$

where $F(x,t) \equiv \dfrac{\kappa}{\ell}[g_0(t) - g_1(t)] - g_0'(t)(x - x^2/2\ell) - g_1'(t)(x^2/2\ell)$ and

$$F_n(t) \equiv \frac{2}{\ell}\int_0^\ell F(x,t) \cos \mu_n x\, dx$$

7.20 $u(x, t) = u_1(x, t) + u_2(x, t) + u_3(x, t)$

$u_1(x, t) = g_0(t) + x g_1(t)$

$$u_2(x, t) = \sum_{n=1}^{\infty} f_n \exp(-\mu_n^2 \kappa t) \sin \mu_n x$$

where $\mu_n \equiv (n - \frac{1}{2})\pi/\ell$ and $f_n \equiv \frac{2}{\ell}\int_0^\ell f(x) \sin \mu_n x \, dx$

$$u_3(x, t) = \int_0^t \left[\sum_{n=1}^{\infty} \exp(-\mu_n^2 \kappa (t-\tau)) F_n(\tau) \sin \mu_n x\right] d\tau$$

where $F(x, t) \equiv -g_0'(t) - x g_1'(t)$ and $F_n(t) \equiv \frac{2}{\ell}\int_0^\ell F(x, t) \sin \mu_n x \, dx$

7.21 $u(x, t) = u_1(x, t) + u_2(x, t) + u_3(x, t)$

$u_1(x, t) = g_0(t)\Phi(x) + g_1(t)\Psi(x)$

where $\Phi(x) \equiv (e^{p(x-\ell)} - 1)/p$ and $\Psi(x) \equiv e^{p(x-\ell)}$

$$u_2(x, t) = \sum_{n=1}^{\infty} f_n \exp(-\mu_n^2 \kappa t)(\sin \mu_n x - p\mu_n \cos \mu_n x) = \sum_{n=1}^{\infty} \frac{f_n}{\cos \mu_n \ell} \exp(-\mu_n^2 \kappa t) \sin \mu_n (x - \ell)$$

where $p\mu_n = \tan \mu_n \ell$ and $f_n \equiv \frac{2}{\ell}\frac{1}{\cos^3 \mu_n \ell}\int_0^\ell f(x) \sin \mu_n (x-\ell) \, dx$

$$u_3(x, t) = \int_0^t \left[\sum_{n=1}^{\infty} \exp(-\mu_n^2 \kappa (t-\tau)) F_n(\tau) \frac{\sin \mu_n (x-\ell)}{\cos \mu_n \ell}\right] d\tau$$

where $F(x, t) \equiv \kappa[g_0(t)\Phi''(x) + g_1(t)\Psi''(x)] - g_0'(t)\Phi(x) - g_1'(t)\Psi(x)$

$$F_n(t) \equiv \frac{2}{\ell}\frac{1}{\cos^3 \mu_n \ell}\int_0^\ell F(x, t) \sin \mu_n (x-\ell) \, dx$$

7.22 $u(x, t) = \exp\left[\frac{-2bx + (4\kappa c - b^2)t}{4\kappa}\right][v_1(x, t) + v_2(x, t) + v_3(x, t)]$

$v_1(x, t) = g_0(t)(1 - x/\ell) + g_1(t)(x/\ell)$

$$v_2(x, t) = \sum_{n=1}^{\infty} f_n \exp(-\mu_n \kappa t) \sin \mu_n x$$

where $\mu_n \equiv n\pi/\ell$ and $f_n \equiv \frac{2}{\ell}\int_0^\ell e^{bx/2\kappa} f(x) \sin \mu_n x \, dx$

$$v_3(x, t) = \int_0^t \left[\sum_{n=1}^{\infty} \exp(-\mu_n \kappa (t-\tau)) F_n(\tau) \sin \mu_n x\right] d\tau$$

where $F_n(t) \equiv \frac{2}{\ell}\int_0^\ell [-g_0'(t)(1 - x/\ell) - g_1'(t)(x/\ell)] \sin \mu_n x \, dx$

7.23 $u(x, t) = \int_0^\ell \Theta(x - \xi, \kappa t) f(\xi) \, d\xi + \int_0^t \kappa \Theta(\ell - x, \kappa(t-\tau)) g_1(\tau) \, d\tau - \int_0^t \kappa \Theta(x, \kappa(t-\tau)) g_0(\tau) \, d\tau$

where $K(x, t) \equiv (4\pi t)^{-1/2} e^{-x^2/4t}$ and $\Theta(x, t) \equiv \sum_{n=-\infty}^{\infty} K(x + 2n\ell, t)$

7.24 $u(x, t) = -\kappa \int_0^t K(x, \kappa(t-\tau)) g(\tau) \, d\tau$ (see Problem 7.23)

7.25 $u(x,t) = -\int_x^\infty e^{-p(y-x)} v(y,t)\,dy$

$v(x,t) = \kappa \int_0^t h(x, \kappa(t-\tau)) g(\tau)\,d\tau$

where $h(x,t) \equiv \dfrac{x}{\sqrt{4\pi t^3}} e^{-x^2/4t}$

7.26 $u(x,t) = \kappa \int_0^t e^{c(t-\tau)} h(x + b(t-\tau), \kappa(t-\tau)) f(\tau)\,d\tau$

7.27 $u(x,t) = \dfrac{1}{2}[\tilde{F}_e(x+at) + \tilde{F}_e(x-at)] + \dfrac{1}{2a}\int_{x-at}^{x+at} \tilde{G}_e(s)\,ds$

where $\tilde{F}_e$ is the even, 2ℓ-periodic extension of f from $(0, \ell)$ to $(-\infty, \infty)$.

7.28 $u(x,t) = \dfrac{1}{2}[\tilde{F}_m(x+at) + \tilde{F}_m(x-at)] + \dfrac{1}{2a}\int_{x-at}^{x+at} \tilde{G}_m(s)\,ds$

where the mixed, 4ℓ-periodic extension of f is defined by $\tilde{F}_m(x+4\ell) = \tilde{F}_m(x)$ (all x), together with

$$\tilde{F}_m(x) = \begin{cases} f(x) & 0 < x < \ell \\ -f(2\ell - x) & \ell < x < 2\ell \\ f(-x) & -2\ell < x < 0 \end{cases}$$

7.29 $u(x,t) = \sum_{n=1}^\infty \dfrac{\ell}{n\pi a} \int_0^t \int_0^\ell \sin\dfrac{n\pi x}{\ell} \sin\dfrac{n\pi y}{\ell} \sin\dfrac{n\pi a(t-\tau)}{\ell} f(y,\tau)\,dy\,d\tau$

7.30 $u(x,t) = \sum_{n=1}^\infty g_n e^{-ct} \sin \rho_n t \sin \mu_n x$

where $\mu_n \equiv n\pi/\ell$, $\rho_n \equiv \sqrt{a^2\mu_n^2 - c^2}$, $g_n \equiv \dfrac{2}{n\pi a \rho_n}\int_0^\ell g(x) \sin \mu_n x\,dx.$

7.31 $u(x,t) = \int_0^t \left[\sum_{n=1}^\infty g_n e^{-c(t-\tau)} \sin \rho_n(t-\tau) \sin \mu_n x\right] d\tau$ (cf. Problem 7.30)

7.32 $u(x,t) = \sum_{n=0}^\infty \{g^*(t - E_n(\ell - x)) - g^*(t - O_n(x))\} + \sum_{n=0}^\infty \{f^*(t - E_n(x)) - f^*(t - O_n(\ell - x))\}$

where $E_n(x) \equiv \dfrac{2n\ell + x}{a}$, $O_n(x) \equiv \dfrac{(2n+1)\ell + x}{a}$, and

(see Table 6-3, line 6) $f^*(t-b) \equiv H(t-b) f(t-b)$.

7.33 $u(x,t) = a\sum_{n=0}^\infty \{G^*(t - E_n(\ell - x)) + G^*(t - O_n(x))\} - a\sum_{n=0}^\infty \{F^*(t - E_n(x)) + F^*(t - O_n(\ell - x))\}$

where $F(t) \equiv \int_0^t f(\tau)\,d\tau$, $G(t) \equiv \int_0^t g(\tau)\,d\tau$.

7.34 $u(x,t) = \sum_{n=0}^\infty \{(-1)^n g^*(t - E_n(\ell - x)) + g^*(t - O_n(x))\} - a\sum_{n=0}^\infty \{(-1)^n F^*(t - E_n(x)) - F(t - O_n(\ell - x))\}$

(cf. Problem 7.33)

7.35 $u(x,t) = -F^*(t - x/a)$ (cf. Problem 7.33)

7.36 $u(x,t) = -aH(t-x/a)\int_{x/a}^{t} e^{-p(t-\tau)} f(\tau - x/a)\,d\tau$

7.37 $u(x,y) = u_1(x,y) + u_2(x,y) + u_3(x,y) + u_4(x,y)$

$$u_1(x,y) = -\sum_{n=1}^{\infty} f_n \frac{\cosh n\pi(1-x)}{n\pi \sinh n\pi} \cos n\pi y$$

where $f_n \equiv 2\int_0^1 f(y)\cos n\pi y\,dy \quad (n = 0, 1, \ldots)$, $f_0 = 0$ (for compatibility)

$u_2(x,y) = -u_1(1-x, y)$ with f replaced by g

$u_3(x,y) = u_1(y,x)$ with f replaced by p

$u_4(x,y) = -u_1(y, 1-x)$ with f replaced by q

7.38 $u(r,\theta) - u(0,\theta) = -\dfrac{1}{2\pi}\displaystyle\int_{-\pi}^{\pi} \log\left[1 - 2r\cos(\theta-\phi) + r^2\right] f(\phi)\,d\phi$

7.39 $u(r,\theta) = \displaystyle\sum_{n\ \mathrm{odd}} \frac{4}{n\pi}\frac{r^n - (a^2/r)^n}{b^n - (a^2/b)^n}\sin n\theta$

7.40
$$u(x,y) = \frac{1}{2\pi}\int_0^{\infty} \log\left[(x^2 + (y+z)^2)(x^2 + (y-z)^2)\right] f(z)\,dz + \frac{1}{2\pi}\int_0^{\infty} \log\left[(y^2 + (x+z)^2)(y^2 + (x-z)^2)\right] g(z)\,dz$$

7.41 $u(x,y) = f(x)(y - \frac{1}{2}y^2) + g(x)y^2 + v(x,y)$

$$v(x,y) = \sum_{n=1}^{\infty} \frac{-1}{n\pi}\int_{-\infty}^{\infty} e^{-n\pi|x-z|}\left[\int_0^1 H(z,s)\cos n\pi s\cos n\pi y\,ds\right] dz$$

where $H(x,y) \equiv \frac{1}{2}y^2(f''(x) - g''(x)) + f(x) - g(x)$

7.42 $u(r,\theta) = \dfrac{r^2-1}{2\pi}\displaystyle\int_{-\pi}^{\pi} \frac{f(\phi)}{1 - 2r\cos(\theta-\phi) + r^2}\,d\phi$

7.43 $u(r,\theta) = +\dfrac{1}{2\pi}\displaystyle\int_{-\pi}^{\pi} \log\left[1 - 2r\cos(\theta-\phi) + r^2\right] f(\phi)\,d\phi + \text{const.}$

CHAPTER 8

8.21 (a) $2H(x) - 1$; (b) $\delta'(x)$; (c) 0

8.22 $\dfrac{2}{\ell}\displaystyle\sum_{k=1}^{\infty} \sin\frac{k\pi x}{\ell}\sin\frac{k\pi\xi}{\ell}$, $\dfrac{1}{\ell} + \dfrac{2}{\ell}\displaystyle\sum_{k=1}^{\infty}\cos\frac{k\pi x}{\ell}\cos\frac{k\pi\xi}{\ell}$

8.23 $G(\mathbf{x};\boldsymbol{\xi}) = -\displaystyle\sum_{n=1}^{\infty} \frac{u_n(\mathbf{x})u_n(\boldsymbol{\xi})}{\lambda_n^2 + c^2}$

8.24 $G(x,y;\xi,\eta) = -\dfrac{1}{ac}\displaystyle\sum_{n=1}^{\infty} e^{-c|x-\xi|}\sin\frac{n\pi y}{a}\sin\frac{n\pi\eta}{a}$

8.26 (a) $G(x,y;\xi,\eta) = G_0(x,y;\xi,\eta) - G_0(x,y;-\xi,\eta)$;
(b) $G(x,y;\xi,\eta) = G_0(x,y;\xi,\eta) + G_0(x,y;-\xi,\eta)$

8.31
(a) $L^*[u] = u_{xx} + u_{yy} - u_x + u_y + 3u$
(b) $L^*[u] = u_{xx} + u_t$
(c) $L^*[u] = u_{xx} - u_{tt}$
(d) $L^*[u] = u_{xx} + u_{yy} - (xu)_x - (yu)_y$
(e) $L^*[u] = u_{xx} + u_{yy} - yu_x - xu_y$
(f) $L^*[u] = (x^2u)_{xx} + (y^2u)_{yy}$

8.35
$$\begin{cases} aR_x + bR_y + (a_x + b_y - d)R = 0 \\ bR_x + cR_y + (b_x + c_y - e)R = 0 \end{cases}$$

must be solvable for R ($R \neq 0$).

8.36 $R = e^x$

8.37 (a) $L^*[v] = v_{xx} + v_{yy} - 2v_x - 3v_y$, $v = 0$ on $x = 0$ and $x = a$, $v_y - 3v = 0$ on $y = 0$ and $y = b$;

(b)
$$\begin{aligned} L[G] &= \delta(x-\xi)\delta(y-\eta) && 0 < x, \xi < a, 0 < y, \eta < b \\ G &= 0 && \text{on } x = 0 \text{ and } x = a \\ G_y &= 0 && \text{on } y = 0 \text{ and } y = b \end{aligned}$$

8.38 $u = e^{-3y} \sum_{n=1}^{\infty} (-1)^n \dfrac{e^{-\lambda_n |y|}}{\lambda_n} \sin n\pi x$, where $\lambda_n \equiv [9 + (n\pi)^2]^{1/2}$.

8.40 $u = \text{erf}\,(x/\sqrt{4\kappa t})$

8.41 (a) $u = U \,\text{erf}\, \dfrac{x}{\sqrt{4\kappa t}}$; (b) $u = U \,\text{erfc}\, \dfrac{x}{\sqrt{4\kappa t}}$; (c) $u = \dfrac{U}{2}\left(1 + \text{erf}\, \dfrac{x}{\sqrt{4\kappa t}}\right)$; (d) $u = \dfrac{U}{2}\left(\text{erf}\, \dfrac{\ell - x}{\sqrt{4\kappa t}} + \text{erf}\, \dfrac{\ell + x}{\sqrt{4\kappa t}}\right)$;

(e) $u = \dfrac{U}{2}\left(\text{erfc}\, \dfrac{\ell - x}{\sqrt{4\kappa t}} + \text{erfc}\, \dfrac{\ell + x}{\sqrt{4\kappa t}}\right)$; (f) $u = \dfrac{U}{2}\left(\text{erf}\, \dfrac{x-b}{\sqrt{4\kappa t}} + \text{erf}\, \dfrac{x+b}{\sqrt{4\kappa t}} - \text{erf}\, \dfrac{x-c}{\sqrt{4\kappa t}} - \text{erf}\, \dfrac{x+c}{\sqrt{4\kappa t}}\right)$

8.43 (a) $G(x, t; \xi, \tau) = K(x - \xi, t - \tau)$ and $u(x, t) = \displaystyle\int_0^t \int_{-\infty}^{\infty} G(x, t; \xi, \tau) f(\xi, \tau)\, d\xi\, d\tau + \int_{-\infty}^{\infty} G(x, t; \xi, 0) h(\xi)\, d\xi$

(b) $K(x - \xi, t - \tau) - K(x + \xi, t - \tau)$ and

$$\int_0^t \int_{-\infty}^{\infty} G(x, t; \xi, \tau) f(\xi, \tau)\, d\xi\, d\tau + \int_0^{\infty} G(x, t; \xi, 0) h(\xi)\, d\xi + \kappa \int_0^t G_\xi(x, t; 0, \tau) p(\tau)\, d\tau$$

(c) $K(x - \xi, t - \tau) + K(x + \xi, t - \tau)$ and

$$\int_0^t \int_0^{\infty} G(x, t; \xi, \tau) f(\xi, \tau)\, d\xi\, d\tau + \int_0^{\infty} G(x, t; \xi, 0) h(\xi)\, d\xi - \kappa \int_0^t G(x, t; 0, \tau) p(\tau)\, d\tau$$

(d) $\displaystyle\sum_{n=-\infty}^{\infty} [K(x - \xi + 2n\ell, t - \tau) - K(x + \xi + 2n\ell, t - \tau)]$ and

$$\int_0^t \int_0^{\ell} G(x, t; \xi, \tau) f(\xi, \tau)\, d\xi\, d\tau + \int_0^{\ell} G(x, t; \xi, 0) h(\xi)\, d\xi + \kappa \int_0^t [G_\xi(x, t; 0, \tau) p(\tau) - G_\xi(x, t; \ell, \tau) q(\tau)]\, d\tau$$

(e) $\displaystyle\sum_{n=-\infty}^{\infty} [K(x - \xi + 2n\ell, t - \tau) + K(x + \xi + 2n\ell, t - \tau)]$ and

$$\int_0^t \int_0^{\ell} G(x, t; \xi, \tau) f(\xi, \tau)\, d\xi\, d\tau + \int_0^{\ell} G(x, t; \xi, 0) h(\xi)\, d\xi + \kappa \int_0^t [G(x, t; \ell, \tau) q(\tau) - G(x, t; 0, \tau) p(\tau)]\, d\tau$$

8.44 $G(x, t; \xi, \tau) = k(x, t; \xi, \tau) - k(x, t; -\xi, \tau) = \dfrac{1}{2a}[H(a(t-\tau) - |x - \xi|) - H(a(t-\tau) - |x + \xi|)]$ for $t > \tau$.

8.45 (a) $G(x, t; \xi, \tau) = \dfrac{2}{a} \displaystyle\sum_{m \text{ odd}} \dfrac{1}{m\pi} \sin \dfrac{m\pi a(t-\tau)}{\ell} \cos \dfrac{m\pi x}{\ell} \cos \dfrac{m\pi \xi}{\ell}$ for $t > \tau$.

(b) $G(x, t; \xi, \tau) = \dfrac{1}{2a} \displaystyle\sum_{n=-\infty}^{\infty} [H(a(t-\tau) - |x - \xi - 2n\ell|) - H(a(t-\tau) - |x + \xi - (2n-1)\ell|)]$ for $t > \tau$.

CHAPTER 9

9.16 (b) It is possible for the roundoff errors to dominate the truncation errors.

9.22 (a) In the notation of Problem 9.13, with $r \equiv k/h^2$ and $s \equiv (1 + 2r) - k$,

$$\begin{bmatrix} s & -r & & & & & 0 \\ -r & s & -r & & & & \\ & -r & s & -r & & & \\ & & \cdot & \cdot & \cdot & & \\ & & & \cdot & \cdot & \cdot & \\ & & & & \cdot & \cdot & \cdot \\ & & & & -r & s & -r \\ 0 & & & & & -r & s \end{bmatrix} \begin{bmatrix} U_{1,j+1} \\ U_{2,j+1} \\ U_{3,j+1} \\ \cdot \\ \cdot \\ \cdot \\ U_{N-2,j+1} \\ U_{N-1,j+1} \end{bmatrix} = \begin{bmatrix} U_{1j} \\ U_{2j} \\ U_{3j} \\ \cdot \\ \cdot \\ \cdot \\ U_{N-2,j} \\ U_{N-1,j} \end{bmatrix}$$

(*b*) By Problem 11.11, all eigenvalues of the above matrix **C** are *greater than or equal to* 1. Hence the method is stable.

9.23 (*a*) $U_{-1,1} = 25,\ U_{01} = 50,\ U_{11} = 25$; other $U_{n1} = 0$
$U_{-2,2} = 6.25,\ U_{-1,2} = 25,\ U_{02} = 37.5,\ U_{12} = 25,\ U_{22} = 6.25$; other $U_{n2} = 0$

(*b*) $U_{-1,1} = 100,\ U_{01} = -100,\ U_{11} = 100$; other $U_{n1} = 0$
$U_{-2,2} = 100,\ U_{-1,2} = -200,\ U_{02} = 300,\ U_{12} = -200,\ U_{22} = 100$; other $U_{n2} = 0$

9.27 With $h = \ell/N$, $(x_n, t_j) = (nh, jk)$, $r \equiv a^2k/h^2$:

$$U_{n,j+1} = rU_{n-1,j} + (1-2r)U_{nj} + rU_{n+1,j} \qquad (n = 0, 1, \ldots, N-1;\ j = 0, 1, 2, \ldots) \tag{1}$$

$$U_{n0} = f(x_n) \qquad \text{[starting values for (1)]} \tag{2}$$

$$\alpha U_{0j} + \beta \frac{U_{1j} - U_{-1,j}}{2h} = p(t_j) \qquad \text{[used to eliminate } U_{-1,j} \text{ from (1)]} \tag{3}$$

$$U_{Nj} = q(t_j) \qquad \text{[used to eliminate } U_{Nj} \text{ from (1)]} \tag{4}$$

9.28 With $h = 1/N$, $(x_n, t_j) = (nh, jk)$, $r \equiv a^2k/h^2$, $s \equiv ck/2h$:

$$(-r-s)U_{n-1,j+1} + (1+2r)U_{n,j+1} + (-r+s)U_{n+1,j+1} = U_{nj} \qquad (n = 1, 2, \ldots, N-1;\ j = 0, 1, 2, \ldots)$$
$$U_{n0} = 0 \qquad (n = 1, 2, \ldots, N-1)$$
$$U_{0j} = 1,\ U_{Nj} = 0 \qquad (j = 0, 1, 2, \ldots)$$

CHAPTER 10

10.14 $|\xi| = \left| \dfrac{\cos(\beta/2) - isa \sin(\beta/2)}{\cos(\beta/2) + isa \sin(\beta/2)} \right| = 1$

10.15 (*a*) $\delta_t^2 U_{nmj} - s^2c^2(\delta_x^2 U_{nmj} + \delta_y^2 U_{nmj}) = 0 \quad (s \equiv k/h)$; (*b*) $c^2s^2 \le \frac{1}{2}$

10.16 (*a*) $u(x,t) = \frac{1}{2}[\cos(x-2t) + \cos(x+2t)]$, $u(0, 0.04) = \cos 0.08 \approx 0.99680$
(*b*) $U_{02} \approx 0.99680$; (*c*) $U_{02} \approx 0.99840$

10.17 (*a*) $\delta_t^2 U_{nj} = (c^2s^2)\delta_x^2 U_{nj} - kb(U_{n,j+1} - U_{n,j-1}) \qquad (s \equiv k/h)$
(*b*) Stable if $c^2s^2 \le 1$.

10.18 (*a*) $u = \log x + 2\log y$; (*b*) $\dfrac{y}{x} = \beta = \text{const.}$, $xy = \alpha = \text{const.}$

10.19 (*a*)

$$y_\alpha = \frac{y}{x} x_\alpha \qquad xy(u_x)_\alpha - y^2(u_y)_\alpha = y_\alpha$$

$$y_\beta = -\frac{y}{x} x_\beta \qquad -xy(u_x)_\beta - y^2(u_y)_\beta = y_\beta$$

and $u_\alpha = u_x x_\alpha + u_y y_\alpha$ or $u_\beta = u_x x_\beta + u_y y_\beta$

(*b*) $P = (\sqrt{4.2}, \frac{1}{2}\sqrt{4.2})$

(*c*)

	Exact	Numerical
$u_x(P)$	0.48795004	0.48795181
$u_y(P)$	1.95180015	1.95181085
$u(P)$	0.33281394	0.37422268 $(\int_O^P)$ 0.34661688 $(\int_R^P)$

10.21 (*b*) $ye^{-x^2/2} = \beta = \text{const.}$, $ye^{x^2/2} = \alpha = \text{const.}$, $P = (\sqrt{5/2}, 2e^{3/4})$

10.22

	$x(P)$	$y(P)$	$u_x(P)$	$u_y(P)$	$u(P)$
Exact	$\sqrt{5/2} \approx 1.581$	$2e^{3/4} \approx 4.234$	$2e^{3/4} \approx 4.234$	$\sqrt{5/2} \approx 1.581$	$xy \approx 6.694$
Numerical*	$5/3 \approx 1.666$	$10/3 \approx 3.333$	$10/3 \approx 3.333$	$5/3 \approx 1.666$	$50/9 \approx 5.555$

* The iterative method of Problem 10.5 could be used to improve these values.

10.23 (*a*) $U_{21} = 0.2727$, $u(1, 0.2) = 0.2076$; (*b*) $U_{10,1} = 0.4925$, $u(1, 0.04) = 0.4906$

10.24 (*a*) $U_{21} = 0.7518$, $u(1, 0.2) = 0.7646$; (*b*) $U_{10,1} = 0.8350$, $u(1, 0.04) = 0.8356$

10.25 $U_{11} = 0.4991 \times 10^{-3}$, $u(0.1, 0.1) = 0.1666 \times 10^{-3}$

10.26 $U_{91} = 0.12333$, $u(0.9, 0.2) = 0.18$

10.27 $U_{11} = 0.7633$, $u(0.5, 0.2) = 0.7592$

10.28 $U_{21} = 0.1680$, $u(0.2, 0.2) = 0.1666$

10.29

	Exact	Numerical
$x(P)$	1.044	1.0
$u(P)$	0.3132	0.3

10.30 In both solutions, $x(P) = 13/6$, $u(P) = -5/6$.

CHAPTER 11

11.17
$$\frac{a_{m+1/2,n}U_{m+1,n}-(a_{m+1/2,n}+a_{m-1/2,n})U_{m,n}+a_{m-1/2,n}U_{m-1,n}}{h^2}$$
$$+\frac{b_{m,n+1/2}U_{m,n+1}-(b_{m,n+1/2}+b_{m,n-1/2})U_{mn}+b_{m,n-1/2}U_{m,n-1}}{k^2}+c_{mn}U_{mn}=f_{mn}$$

where, e.g., $a_{m-1/2,n}$ represents a mean value (possible choices include the arithmetic, geometric, or harmonic mean) between $a_{m-1,n}\equiv a(x_{m-1}, y_n, U_{m-1,n})$ and $a_{mn}\equiv a(x_m, y_n, U_{mn})$.

11.18 Set $c=f\equiv 0$ in Problem 11.17.

11.19 (*a*)
$$\begin{aligned}
-4U_{11}+U_{21}+U_{12}&=0\\
-3U_{21}+U_{11}+U_{31}&=0\\
-4U_{31}+U_{21}+U_{32}&=-0.25\\
-3U_{12}+U_{11}+U_{13}&=0\\
-3U_{32}+U_{31}+U_{33}&=-0.5\\
-4U_{13}+U_{12}+U_{23}&=-0.25\\
-3U_{23}+U_{13}+U_{33}&=-0.5\\
-4U_{33}+U_{23}+U_{32}&=-1.5
\end{aligned}$$

(*b*)
$$\begin{aligned}
U_{11}^{k+1}&=(U_{21}^{k}+U_{12}^{k})/4\\
U_{21}^{k+1}&=(U_{11}^{k+1}+U_{31}^{k})/3\\
U_{31}^{k+1}&=(U_{21}^{k+1}+U_{32}^{k}+0.25)/4\\
U_{12}^{k+1}&=(U_{11}^{k+1}+U_{13}^{k})/3\\
U_{32}^{k+1}&=(U_{31}^{k+1}+U_{33}^{k}+0.5)/3\\
U_{13}^{k+1}&=(U_{12}^{k+1}+U_{23}^{k}+0.25)/4\\
U_{23}^{k+1}&=(U_{13}^{k+1}+U_{33}^{k}+0.5)/3\\
U_{33}^{k+1}&=(U_{23}^{k+1}+U_{32}^{k+1}+1.5)/4
\end{aligned}$$

(*c*)

	U_{mn}		
3	0.1874	0.4249	0.5874
2	0.0749	–	0.4249
1	0.0374	0.0749	0.1874
n / m	1	2	3

	$u=mn/16$		
3	0.075	0.375	0.5625
2	0.05	0.25	0.375
1	0.01	0.05	0.075
n / m	1	2	3

11.20 (*a*) $-4U_{22}+U_{32}+U_{23}+U_{12}+U_{21}=0$; (*b*) $-4U_{35}+U_{34}+2U_{25}=0$;
(*c*) $-4U_{51}+U_{41}+U_{52}+2=0$
(*d*)

	U_{mn}				
5	0.0822	0.1543	0.1943		
4	0.1746	0.3406	0.4689		
3	0.2755	0.5648	1.0		
2	0.3629	0.6429	0.8748	0.9552	0.9854
1	0.5330	0.7693	0.9012	0.9607	0.9865
n / m	1	2	3	4	5

11.22 $U_i^{k+1}=U_i^k+R_{ii}^{k+1}a_{ii}^{-1}$

11.23 $U_i^{k+1}=U_i^k+\omega R_{ii}^{k+1}a_{ii}^{-1}$

11.24 In (*11.8*) and (*11.9*), replace $m+\mu$, $n+\nu$ by $m+\nu$, $n+\mu$.

11.25
$$\mathbf{L} = \begin{bmatrix} 1 & 0 & 0 \\ 2 & 1 & 0 \\ 1 & 3 & 1 \end{bmatrix} \qquad \mathbf{U} = \begin{bmatrix} 1 & 0 & 1 \\ 0 & 3 & 2 \\ 0 & 0 & 2 \end{bmatrix}$$

11.26 (a) $(c-4)U_{mn} + U_{m,n+1} + U_{m-1,n} + U_{m,n-1} + U_{m+1,n} = F_{mn}$

(b)
$$\mathbf{T}_J = \frac{1}{c-4} \begin{bmatrix} 0 & -1 & -1 & 0 \\ -1 & 0 & 0 & -1 \\ -1 & 0 & 0 & -1 \\ 0 & -1 & -1 & 0 \end{bmatrix}$$

(c) $0, 0, \dfrac{2}{c-4}, \dfrac{-2}{c-4}$

(d) $c > 6$ or $c < 2$

CHAPTER 12

12.12 $\mathcal{M} \equiv \{v \text{ in } H^1(\Omega) : v = 0 \text{ on } x^2 + y^2 = 1\}$

$$\delta J[u; v] = 2\int_\Omega (y^2 u_x v_x + x^2 u_y v_y)\, d\Omega \qquad (u \text{ in } \mathcal{A},\ v \text{ in } \mathcal{M})$$

12.13 $\mathcal{D} = \{u \text{ in } H^2(\Omega) : u = x^2 \text{ on } x^2 + y^2 = 1\}$, $\nabla J[u] = -y^2 u_{xx} - x^2 u_{yy}$ for u in $\mathcal{D}$

12.14
$$\begin{aligned} -y^2 u_{xx} - x^2 u_{yy} - F \quad & \text{in } \Omega \\ y^2 x u_x + x^2 y u_y = 0 \quad & \text{on } S \end{aligned}$$

12.15 $\mathcal{M} = \mathcal{A}_1$, $\delta J[u; v] = 2\int_\Omega [(\nabla^2 u)(\nabla^2 v) - Fv]\, d\Omega \qquad (u, v \text{ in } \mathcal{M})$

12.16 $\mathcal{D} = \left\{u \text{ in } H^4(\Omega) : u = \dfrac{\partial u}{\partial n} = 0 \text{ on } S\right\}$, $\nabla J[u] = 2(\nabla^4 u - F)$ for u in $\mathcal{D}$

12.17 $\mathcal{D} = \left\{u \text{ in } H^4(\Omega) : \nabla^2 u = \dfrac{\partial(\nabla^2 u)}{\partial n} = 0 \text{ on } S\right\}$, $\nabla J[u] = 2(\nabla^4 u - F)$ for u in $\mathcal{D}$

12.19 Minimize

$$J[u] = \int_\Omega (\nabla u \cdot \nabla u - 2Fu)\, d\Omega - 2\int_{S_2} g_2 u\, dS + \int_{S_3} (pu^2 - 2g_3 u)\, dS$$

over $\mathcal{A} = \{u \text{ in } H^1(\Omega) : u = g_1 \text{ on } S_1\}$.

12.23 (a) Find u in $\mathcal{A} = \{u \text{ in } H^1(\Omega) : u = 0 \text{ on } S\}$ such that

$$\int_\Omega (u_x v_x + u_y v_y + 2u_x v)\, dx\, dy = \int_\Omega v\, dx\, dy$$

for all v in $\mathcal{M} = \mathcal{A}$. (b) Find u in $\mathcal{A} = \{u \text{ in } H^1(\Omega) : u(x, 2-x) = 0 \text{ for } 0 < x < 2\}$ such that

$$\int_\Omega (u_x v_x + u_y v_y + 2u_x v)\, dx\, dy = \int_\Omega v\, dx\, dy + \int_0^1 (y-1)v(0, y)\, dy + \int_0^1 x(x-1)v(x, 0)\, dx$$

for all v in $\mathcal{M} = \mathcal{A}$.

CHAPTER 13

13.8 $u_1^*(x) = 0.263\,x(1-x)$, $u_2^*(x) = 0.177\,x(1-x) + 0.173\,x^2(1-x)$

13.9 (*a*) $0.217 + 0.064\,x - 0.007\,x^2$; (*b*) $0.245 - 0.032\cos\pi x - 0.00156\cos 2\pi x$

13.10 $(7.794 + 0.1702x + 1.0666y)\phi_1$

13.11 $-0.0142\,\phi_1 - 0.0075\,\phi_2$

13.12 $u_2^*(x, y) = 1 + 0.029\,xy + 0.011\,xy(1+xy)$

13.14

$$\det\begin{bmatrix} \frac{1}{3}-\frac{\mu}{60} & \frac{1}{6}-\frac{\mu}{105} \\ \frac{1}{6}-\frac{\mu}{105} & \frac{2}{15}-\frac{\mu}{168} \end{bmatrix} = 0$$

13.16 $u_2^* = \frac{3}{8}\phi_1 - \frac{5}{14}\phi_2$

CHAPTER 14

14.6

$$\mathbf{A} = \begin{bmatrix} -1 & 3 & 0.5 & -1.5 & 0.5 & -0.5 & 0 & 1.5 & 0 & 2.5 & 3.5 & 1.5 & 3.5 \\ 2.5 & -2.5 & -2 & 2 & -1 & 0.5 & -0.5 & -1.5 & 2.5 & 1 & -1.5 & 2 & -0.5 \\ -0.5 & 0.5 & 2.5 & 0.5 & 1.5 & 1 & 1.5 & 1 & -1.5 & -2.5 & -1 & -2.5 & -2 \end{bmatrix}$$

$$\mathbf{B} = \begin{bmatrix} 0 & 0 & 1 & 1 & 1 & 1 & 0 & -1 & 0 & -1 & -1 & -1 & -1 \\ -1 & 1 & 0 & 0 & 0 & -1 & 1 & 1 & -1 & 0 & 1 & 0 & 1 \\ 1 & -1 & -1 & -1 & -1 & 0 & -1 & 0 & 1 & 1 & 0 & 1 & 0 \end{bmatrix}$$

$$\mathbf{C} = \frac{\sqrt{3}}{3}\begin{bmatrix} 2 & -2 & -1 & 1 & -1 & 1 & 2 & 1 & 2 & 1 & -1 & 1 & -1 \\ -1 & 1 & 2 & -2 & 2 & 1 & -1 & 1 & -1 & -2 & -1 & -2 & -1 \\ -1 & 1 & -1 & 1 & -1 & -2 & -1 & -2 & -1 & 1 & 2 & 1 & 2 \end{bmatrix}$$

14.7 (*a*) $u_3^*(x, y) = 14.3\,\phi_1(x, y) - 14.3\,\phi_2(x, y) + 50\,\phi_8(x, y) + 50\,\phi_9(x, y) - 50\,\phi_{11}(x, y) - 50\,\phi_{12}(x, y)$

(*b*) $u_7^*(x, y) = 16.32\,\phi_1(x, y) - 16.32\,\phi_2(x, y) - 14.25\,\phi_4(x, y) - 2.85\,\phi_5(x, y) + 2.85\,\phi_6(x, y) + 14.25\,\phi_7(x, y) + 50\,\phi_8(x, y) + 50\,\phi_9(x, y) - 50\,\phi_{11}(x, y) - 50\,\phi_{12}(x, y)$

(*c*) $u_5^*(x, y) = 31.25\,\phi_1(x, y) + 10.7\,\phi_2(x, y) + 16.77\,\phi_3(x, y) + 6.23\,\phi_4(x, y) + 12.44\,\phi_5(x, y)$

14.8

$$\mathbf{A} = \begin{bmatrix} -1 & -1 & -1 & 1 & -1 & 0 & 0 & 2 & 0 & 0 & 4 & 4 & 0 & 5 & 6 & 4 \\ 0 & 1 & 3 & -1 & 1 & 1 & -1 & -2 & -2 & -3 & -4 & -4 & -4 & -6 & -1 & -2 \\ 2 & 1 & -1 & 1 & 1 & 0 & 2 & 1 & 1 & 4 & 1 & 1 & 5 & 2 & -4 & -1 \end{bmatrix}$$

$$\mathbf{B} = \begin{bmatrix} -1 & 0 & 0 & -1 & 1 & 1 & 0 & -1 & 0 & 0 & -1 & -1 & 0 & -1 & -1 & -1 \\ 1 & 1 & -1 & 1 & 0 & -1 & 1 & 1 & 1 & 1 & 1 & 1 & 1 & 1 & 0 & 1 \\ 0 & -1 & 1 & 0 & -1 & 0 & -1 & 0 & 0 & -1 & 0 & 0 & -1 & 0 & 1 & 0 \end{bmatrix}$$

$$\mathbf{C} = \frac{1}{2}\begin{bmatrix} 1 & 1 & 1 & 1 & 1 & 0 & 1 & 1 & 1 & 1 & 0 & 1 & 1 & 0 & -1 & 0 \\ 0 & -1 & -1 & 0 & -1 & -1 & -1 & 0 & -1 & 0 & 1 & 0 & -1 & 1 & 1 & -1 \\ -1 & 0 & 0 & -1 & 0 & 1 & 0 & -1 & -1 & -1 & -1 & -1 & 0 & -1 & 0 & 1 \end{bmatrix}$$

14.9 (*a*) $u_8^* = -2.625\,\phi_1 + 1.25\,\phi_2 - 1.5\,\phi_3 + 0.438\,\phi_4 + 2.532\,\phi_5 + 3.213\,\phi_6 + 3.258\,\phi_7 - 1.606\,\phi_8$

(*b*) $u_8^* = 14.389\,\phi_1 - 5.00\,\phi_2 + 13.779\,\phi_3 + 10.7\,\phi_4 + 4.888\,\phi_5 + 1.955\,\phi_6 + 1.222\,\phi_7 - 0.977\,\phi_8$

Index